INDUSTRIAL ELECTRONICS AND CONTROL

J. A. SAM WILSON

Training Director
International Society of Certified Electronics Technicians
National Electronic Service Dealers' Association

 SCIENCE RESEARCH ASSOCIATES, INC.
Chicago, Palo Alto, Toronto, Henley-on-Thames, Sydney, Paris

A Subsidiary of IBM

To Sharon

Compositor	Graphic Typesetting Service
Acquisition Editor	Alan W. Lowe
Project Editor	Jay Schauer
Technical Artist	John Foster
Cover Designer	Barbara Ravizza

Acknowledgments

Figures 12-10 and 12-13 courtesy of Continental Specialties Corporation.

Library of Congress Cataloging in Publication Data

Wilson, J. A.
  Industrial electronics and control.

  Includes index.
  1. Industrial electronics.   2. Electronic
control.   I. Title.
TK7881.W56   621.381   78-7424
ISBN 0-574-21515-8

10 9 8 7 6 5 4 3 2

PREFACE

The lack of instructional material with up-to-date coverage of solid state devices motivated the development of this book: a fundamentals treatment for effectively preparing service technicians. This survey of industrial electronics follows a components-circuits-systems approach. Basic control theory has been melded into the overall presentation, to give students an appreciation of the basic control concepts prevalent in industrial electronic equipment.

The only prerequisite for this book is basic electrical theory (dc/ac). The book reviews some basic electrical concepts in Chapter 2. The instructor may wish to skip this review material, as well as the basic electron device theory in Chapter 3.

Industrial electronics today is a mixture: electrical items, many in the power class; and electronic items, with digital logic making a heavy impact. This diversity was recognized in planning this book. What some instructors call industrial electronics others might call industrial electricity, electromechanics, or simply industrial control, depending on the focus of the program. While the main concern in this book is electronics, the book includes industrial electricity items that interface with the electronics. Thus, this book is also ideal for use in training industrial maintenance electricians or electrical technicians, since most electrical training programs try to expose students to related electronic theory.

The material in this book reflects the results of a survey conducted by the author for NESDA (the National Electronic Service Dealers' Association). As the chairman of the industrial electronics test committee of ISCET (the International Society of Certified Electronics Technicians, an affiliate of NESDA), the author queried technicians around the country about their jobs, the components and systems they work on, and the test equipment they use. This information provided an insight into the common factors affecting most electronics technicians. It became the basis for ISCET's Certified Electronic Technician Examination, and for this book.

Although binary arithmetic and Boolean algebra are traditionally covered in industrial electronics courses, the survey showed that the practical usefulness of these subjects varied from job to job. For this reason, the binary number system is included in an appendix, but the book doesn't require an understanding of the subject.

Some mention should be made about the reviews at the end of each chapter. A summary of the chapter details the major points. A programmed review and self-test allow students to test their understanding on their own.

Since the programmed review also develops new theories and concepts, students should be encouraged to use these learning aids.

A lab manual that will provide practical hands-on experience with industrial electronic circuits and system applications is being prepared.

The author and publisher thank all those whose careful reviewing and editing helped to shape this text, especially:

Tom Bingham, Saint Louis Community College at Florissant Valley
Wayne Dunbar, Saint Paul Area Technical Vocational Institute
Kenneth R. Edwards, International Brotherhood of Electrical Workers
Stanley J. Nawrocki, Milwaukee Area Technical College
Rex W. Pershing, University of Northern Iowa
Charles Shoemaker, State University of New York College at Oswego

The patience and understanding of Alan Lowe, SRA's Technical Education Editor, is gratefully acknowledged. Finally, this book could not have been completed without the help, suggestions, typing, and dedication of Sharon Wilson.

CONTENTS

CHAPTER 1

INTRODUCTION TO INDUSTRIAL ELECTRONICS

Industrial electronics is the application of electronic systems to the control of machinery and processes in industry. Industrial electronics technicians, however, would consider this definition too limiting. For they may work on a closed circuit television system or an audio intercommunication system, or they may troubleshoot test equipment, none of which is strictly in the domain of industrial electronics.

If industrial electronics is primarily concerned with control circuitry, then what about garage-door openers? Certainly this is an electronic system that controls the motor that opens the door. And what about automotive electronic systems that control the rate of fuel injection and the spark advance for the engine? Aren't these also control circuits that control machinery? In the broader definition, these subjects come under the heading of industrial electronics.

To cope with the wide variety of systems that the technician will encounter, an industrial electronics technician must have a background in basic electricity (including circuits), must know how dc and ac machines work, and must have a good understanding of electronic components, basic electronic circuitry, and troubleshooting techniques.

Control circuits can be classified into two types: *open loop* and *closed loop*. Both types will be discussed in this chapter. Then the block diagram of a numerical control system will be discussed to give an overall view of the units used in industrial electronic systems.

OPEN-LOOP CONTROL

An open-loop control system is one that is not automatically varied or controlled by a signal taken from the output. An example of a simple open-loop control system is the gas stove illustrated in Figure 1-1. To boil the water in the pan the gas flame is set to the desired height by the manual control. The water in the pan becomes hotter and hotter until eventually it starts to boil. Theoretically, once the boiling point is reached, the gas flame could be reduced to a much lower value to maintain the water in a boiling condition. However, since a gas stove has no automatic system to reduce the flame, any changes in the setting must be brought about by a manual adjustment in the control.

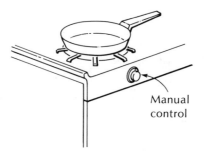

Figure 1-1. A simple open-loop control

Open-loop controls are used in many industrial electronics applications. Electronic controls are used to control machinery for a desired operation. With an open-loop control it is presumed that the machine can carry out its function without the need for constant monitoring of the output.

CLOSED-LOOP CONTROL

A closed-loop system is one in which the output is continually measured, and the input is adjusted on the basis of these measurements. An example of a simple closed-loop system is a thermostatically controlled gas furnace in a home. This system is illustrated in Figure 1-2.

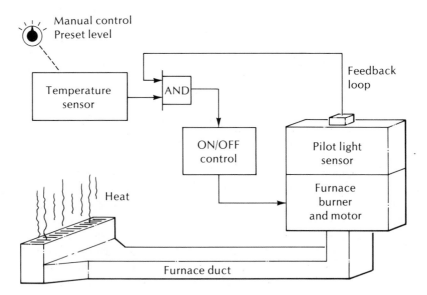

Figure 1-2. A closed-loop system

The operator adjusts the manual temperature control to the desired value of room temperature. A unit in the temperature sensor measures the temperature of the room. If it is below the desired temperature, the furnace is turned on to bring the temperature up to the correct value. Once the correct value is reached, the control circuit turns the furnace off. The furnace is turned on and off periodically to maintain the room at the desired temperature.

A closed-loop system has two basic features that are apparent in Figure 1-2.

1. It has a method of measuring the conditions that are present at the output. In the closed-loop furnace system the temperature sensor is located in the wall unit where the temperature adjustment is made.
2. It has a provision for using this measurement to control the operation of the system.

The sensor is often called a *transducer*. A sensor, or transducer, is a component in which the energy of one system controls the energy in another system.

In the imaginary system of Figure 1-2 the furnace burner is ignited with a pilot light. Because it would be undesirable for the gas to come on in the furnace burner if the pilot light were out, a pilot light sensor is provided. When the pilot light is on, the sensor delivers an ON signal to the system. Two ON signals are delivered to a circuit referred to in the block diagram as an *AND*. AND circuits are forms of basic logic that are discussed in another chapter of this book. An AND circuit will only produce an output signal when all input signals are present. In other words, there can be no output from this AND circuit unless signals from both sensors are present at the input. If only one signal is present, there is no output and the furnace cannot come on.

Assuming that both signals are present, an AND circuit produces an output signal that turns the furnace burner on and starts the fan motor. The fan circulates hot air through the hot air ducts, out the register, and eventually back to the transducer. When the temperature raises the sensor heat to a point where the ON signal goes off, the furnace goes off. This is true even though there is still an ON signal from the pilot light. Remember, the AND circuit, which is also called an *AND gate,* must have both signals present to produce an output.

EXAMPLES OF INDUSTRIAL CLOSED-LOOP SYSTEMS

Motor Speed Control

An example of a closed-loop feedback system for controlling the speed of an electric motor is shown in Figure 1-3. The speed of a motor may change when its load changes. The *load* is the opposing force that the motor is turning against (not shown in the illustration).

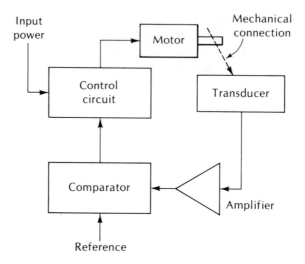

Figure 1-3. Closed-loop motor speed control

Input power for operating the motor is delivered to a *control circuit*. The purpose of this circuit is to control the amount of power sent to the motor as required for maintaining a desired motor speed.

The motor speed is monitored by a transducer connected to the motor shaft. The purpose of the transducer in this case is to sense the motor speed. The electrical output of the transducer is related to the speed in some way. For example, for some transducers the amplitude of the output voltage is related to the motor speed, and in others the output frequency is related to the motor speed.

The transducer output is amplified, and the amplified output is delivered to a circuit called the *comparator*. This circuit compares the output signal from the transducer amplifier with a reference signal. The *reference signal* is a voltage or frequency equal to the input from the transducer amplifier when the motor speed is correct. When the amplified transducer signal is different from the reference—either in voltage or frequency—then a correction signal will be sent from the comparator to the control circuit. The correction signal causes a change in power to the motor that corrects its speed. If the motor speed is too high, the comparator output will cause the control circuit to reduce power to the motor, and thereby reduce its speed to the correct value. If the motor speed is too low, the comparator will cause the control circuit to increase power to the motor, and thus increase its speed to the correct value.

The overall result is that the closed-loop system will maintain the motor at a constant speed regardless of the amount of load delivered to the motor.

NUMERICAL CONTROL OF DRILL PRESS

The drill press shown in Figure 1-4 represents one example of how electronics can be used to control manufacturing. This particular application is

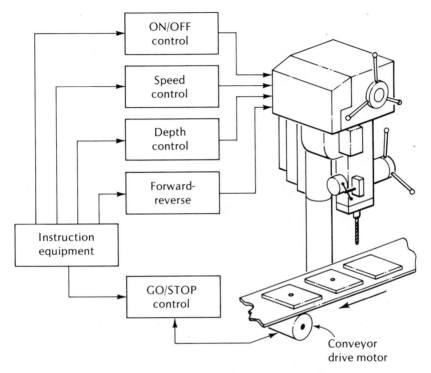

Figure 1-4. Drill press with open-loop control

called *numerical control* because the instructions for the machine are in the form of *binary numbers*. Other applications are: mixing chemicals *(process control)*, automatic drafting machines, control of assembly lines, and control of many other manufacturing systems.

Numerical control will be used as an example to introduce some of the important terms used in control systems.

Instruction Equipment

The first step in the block diagram of Figure 1-4 is called *instruction*. This part of the system determines what the drill press is doing at all times. For example, it instructs the machine to be ON or OFF, run *fast* or *slow, lower* or *raise* the drill bit, and run forward or reverse.

MACHINE LANGUAGE

In order for an electronic system to control a machine, there must be some way for man to communicate his ideas to the machine. In other words, there must be a *machine language*. In the case of a numerical control system, the machine language is in the form of digits or numbers, each representing some command in the system.

In our everyday life we consider the decimal system to be most convenient for counting money, making weights and measurements, etc. However, in the electronics world the decimal system is too complex because there are no simple devices which are able to indicate 10 individual discrete steps. Most electronic devices have only two steps: *off* and *on, conducting* and *nonconducting, current* and *no current*. For this reason, the most convenient numbering system used in digital control is *binary*. In this numbering system there are only 2 digits, just as there are only 10 digits in the decimal system. Since there are only two digits, 0 and 1, a wide variety of components can be used to represent them. Figure 1-5 shows a few methods of representing binary digits. In each case the binary digit 0 is represented by one condition of the device and the binary digit 1 is represented by another condition.

Binary numbers greater than 1 are made by combining the binary digits. Here are a few binary numbers and the corresponding decimal numbers.

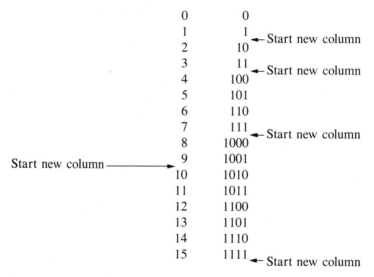

From the numbering system shown it is clear that four components are needed to represent decimal number 8 (that is, binary number 1000). Figure 1-6 shows how decimal numbers greater than 1 can be represented with four lamps.

As in the decimal column, a new column must be started when all of the binary digits have been used. A complete system of binary numbers and binary arithmetic has been developed.

Using the binary system of numbering, it is possible to perform any of the mathematical computations that are performed by the decimal system. A more extensive treatise on the binary system of numbering is covered in Appendix I, but in this application the important thing is to understand that the binary numbers represent on and off conditions. These numbers translate into machine action.

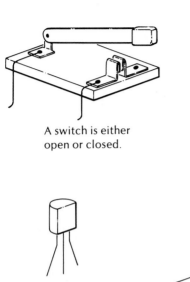

A switch is either open or closed.

A lamp is either glowing or not glowing.

A transistor may be used as a switch.

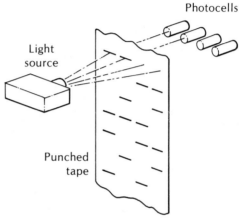

Photocells

Light source

Punched tape

When light passes through the slot the photocell conducts; when there is no slot, the photocell does not conduct.

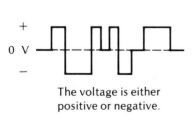

An SCR may be used as a switch.

The voltage is either positive or negative.

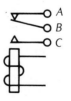

A B C

The relay is either deenergized (A and B closed) or energized (B and C closed).

Figure 1-5.   Examples of components used for binary numbers

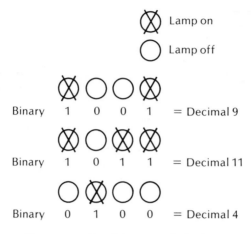

Figure 1-6. Use of lamps to display binary numbers

Suppose, for example, that binary number 3 is delivered to the machine. This number may be translated into an increase in speed or a change in position, or some other function performed by the machine.

PROGRAM STORAGE

Once you realize that the machine can be controlled by combinations of binary numbers, the next step is to find a way to store the binary numbers so that the machine operations can be repeated whenever necessary.

Figure 1-7 shows two methods used for storing machine control information. The punched tape and punched cards both represent binary numbers. There is either a hole in the tape or card, or there is not a hole. The hole represents a 1 and the place where there is no hole represents a 0. Therefore, the tape can be thought of as a method of storing the information for operating the machine.

The process of putting the numbers on the tape is called *programming*. So, a program for numerical control is simply a set of digits that tell the machine what to do.

TAPE READERS

The punched tape or punched card has all the information needed for operating the machine. This may be thought of as being a *memory* for the system because it remembers the instructions for performing a certain manufacturing process. For the example of Figure 1-4 it will be assumed that the program is on a punched tape.

The next step is to be able to get the information off the tape and into the system. A *tape reader* is needed. Figure 1-8 shows an example of how a tape reader can operate. Small metal feelers rest on the tape as it is pulled

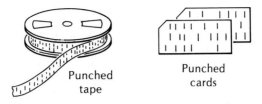

Punched
tape

Punched
cards

Figure 1-7. Two methods of storing programs

through the tape machine. The paper insulates feeler *a* from the drum and produces a zero output corresponding to binary 0. The hole permits feeler *b* to touch the drum and completes the circuit. This circuit produces an output of +3 volts (V) to represent the binary 1. The equivalent switching circuit is also shown in Figure 1-8.

The output of the tape reader, then, is a set of digital numbers that correspond to the different control instructions. These digital numbers are fed to a control circuit.

Control Circuit

The earliest equipment for controlling machinery used relays and electromagnetic switches to turn the machines on and off and control their speed. Systems using these components are still in use, and they are sometimes referred to as *magnetic control* or *electromagnetic control*. In modern electronic systems these electromagnetic components have been replaced

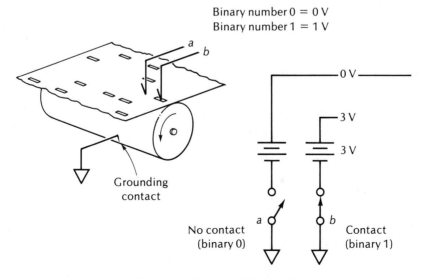

Binary number 0 = 0 V
Binary number 1 = 1 V

Figure 1-8. One type of tape reader

with electronic circuits that have no moving parts. These circuits are sometimes referred to as *static control*. In the newer systems static controls are preferred because they have no moving parts to wear out, and have no switch contacts to become coated or worn. In other words, static controls are more reliable. However, they have the disadvantage that they are more difficult to service and require the skills of trained technicians for their maintenance.

The static (or electromagnetic) control circuitry takes the output from the tape reader and converts it into electrical signals which operate the motor of the drill press, control the motor speed, determine the depth of drilling, and control the motion of the object being drilled on the conveyor system.

The control circuit is normally a feedback system. Signals are returned to the control circuit from the drill press and from the conveyor belt. These returned signals permit the control circuit to change the command signals as necessary for proper operation. A good example is the speed control that was discussed earlier. In this system a *tachometer* would feed a signal representing motor speed back to the control circuit where it would be compared in the comparator to determine if the speed was proper. If not, the speed would be changed by the control circuitry.

X-Y-Z Coordinates

In the simple system of Figure 1-4 only one hole is being drilled in the block, and then it is moved on down the conveyor belt. In a typical numerical control system, it is more likely that a large number of holes would be drilled in the block. Furthermore, different sizes of holes may be required and they may be drilled at different depths. One means of accomplishing this would be to set up one drill press for each hole being drilled, using a separate numerical control system for each hole. However, such a method would be highly uneconomical and certainly would slow the system.

A better method is to use one drill press and one numerical control system. The block being drilled is positioned for each hole. A system of coordinates is used that is much like the coordinates used in a basic algebra graphing problem. Figure 1-9 shows the coordinates under the drill bit. In order to drill a hole at some position the control circuit positions the block first in the X plane, then in the Y plane, and finally in the Z plane. As the block is moved, the feedback signal is returned to the control to tell the control circuit exactly what its position is. The drill cannot start until the block is in exactly the right position for being drilled. Feedback signals to the control system are used for positioning the block.

In the example of Figure 1-9 the zero or reference point is at the corner of the block. Assuming that the block is in the correct position (which is accomplished by positioning controls), the drill bit is moved along the X axis to a point that is 1.3 cm (centimeters) from zero. Then, it is moved along the Y axis a distance. Finally, the drill makes a hole 2 cm deep which is along the Z axis.

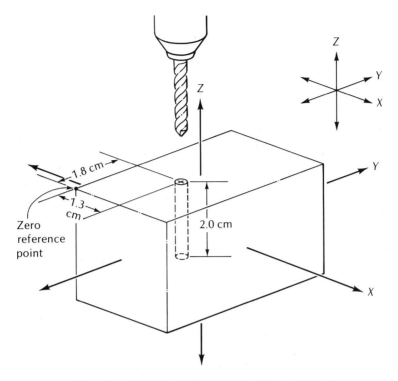

Figure 1-9. Location of drilled hole by *XYZ* coordinates

The punched tape delivers signals to the control circuits that position the block and position the drill bit. Feedback signals establish the reference points so that the drill bit is moved to the exact position.

In the imaginary system just discussed, it was presumed that the drill bit is moved along the *X, Y,* and *Z* axes. Actually, it is simpler to hold the drill in a single position and move the block. The overall result is the same. The drill bit still ends up at the same position and drills to the same depth.

To be able to drill different size holes, the drill press might have a number of drill bits mounted on a spindle. The taped program calls up the right size drill for each hole being drilled, and sends the proper instruction to another control system (not shown in Figure 1-4). After the process is completed the signal goes to the control circuitry, telling it to move the block out and position a new block in place ready for drilling.

SYSTEMS, UNITS, CIRCUITS, AND COMPONENTS

The complete assembly of equipment in Figure 1-4 is an example of a system. In other words, the term *system* refers to everything related to making a complete product or process.

In this book, the word *unit* is used to mean one complete section within a system. Thus, in Figure 1-4 the *speed control* and the *depth control* are units of the system.

Each unit contains individual *circuits*. As an example, the speed control unit shown in Figure 1-10 has logic circuits, switching circuits, amplifier circuits, and *interfacing* circuits. (An interfacing circuit allows two different kinds of circuits to be connected. For example, if one circuit has a 5 V output, and the input stage of the next circuit can only handle 3 V, then the two circuits may be connected through a third circuit (or component) called an interface.)

Components are the building blocks of circuits. An amplifier circuit may have such components as resistors, capacitors, inductors, and transistors. *Discrete components* are individual parts that can be replaced when they are faulty. Components in integrated circuits are built into the circuit in such a way that they cannot be replaced, so they are not discrete components.

A good place to start the study of industrial electronics is to review the types of components available, and to learn how they work and what they do. The next three chapters cover this subject. Then, units and systems will be taken up in that order.

MAJOR POINTS

1. Industrial electronics is the application of electronics to industrial processes. Much of the industrial electronic circuitry is involved with control of machinery.
2. Feedback is used in closed-loop systems, but not in open-loop systems.
3. A transducer, which is also called a *sensor,* is used to sense the output in a closed-loop system. The output of a transducer is a voltage or frequency that is related to the system output.
4. In a closed-loop system a comparator compares the outputs of a transducer and a reference. The output of a comparator is usually called an

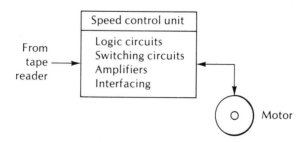

Figure 1-10. Makeup of control unit

error signal because it is related to the error between the actual output and the desired output of a system.

5. Two examples of closed-loop control circuits are the motor speed control and the numerical control systems discussed in this chapter.
6. In a numerical control system the program for operating a machine is in the form of binary digits.
7. A binary numbering system has only two digits: 0 and 1. Any number can be made by combining binary digits.
8. The reason binary numbers are popular in industrial electronics is that many components have two states—on and off, low and high, etc.—which can be represented by binary digits.
9. Programs for numerical control can be stored on punched tapes or punched cards. The holes represent the digit 1, and the places where there are no holes represent zeros.
10. When numerical control is used with a drill press (and with other machines), the hole is identified by *X, Y,* and *Z* coordinates.

PROGRAMMED REVIEW

This section reviews the material covered in this chapter. In addition, the section introduces new material. The ease with which you complete this section is a measure of your understanding of the material you have studied.

The review is programmed to assist you. Every time you choose an answer to a question, you will be directed to a different block. If you answer the question incorrectly, you will be told to re-read the question and choose a different answer. When you choose the correct answer, you will find an extended analysis of the answer, and a new question.

Start with block 1. In this case, if you feel the correct answer is A, then proceed to block 6. If you feel that choice B is correct, then proceed to block 13. Continue until you reach the end of the review.

---

1. Since it requires more digits to represent binary numbers (greater than number 1) than is required for the decimal equivalent, why does so much industrial electronic equipment and computer equipment use binary numbering?

   A. It is easier to find components with 2 states than it is to find components with 10 states. (Go to block 6.)
   B. The equipment does not take up as much space when designed for use with binary numbers. (Go to block 13.)

---

2. Your answer to the question in block 16 is wrong. Read the question again, then go to block 22.

---

3. The correct answer to the question in block 19 is B. Today you can buy comparators in integrated circuits like the integrated circuit used in the experiment for this chapter.
Here is your next question.
What are the two digits in a binary numbering system?

A. 1 and 2. (Go to block 10.)
B. 0 and 1. (Go to block 16.)

4. Your answer to the question in block 24 is wrong. Read the question again, then go to block 19.

5. Your answer to the question in block 18 is wrong. Read the question again, then go to block 12.

6. The correct answer to the question in block 1 is A. Binary numbers are used in industrial electronics, but when numbers are presented to humans, they are better in decimal form. Special circuits are designed for converting binary numbers to decimal numbers and also to convert decimal numbers back to binary numbers.
Here is your next question.
A row of lamps is used to represent binary numbers.

When the lamp is ON, the binary 1 is represented and when the lamp is OFF, binary 0 is represented. What is the decimal equivalent of the binary number illustrated?

A. 10. (Go to block 17.)
B. 13. (Go to block 25.)

7. The correct answer to the question in block 25 is A. A transducer is a device which permits the energy of one system to control the energy of another system. Transducers are a very important part of industrial electronic systems because they perform approximately the same function as the sense organs in a human. In other words, they make it possible to tell how things are going in the manufacturing process.

Another field of electronics that uses sensors or transducers extensively is instrumentation. The transducers in that field are used to sense the thing being measured.

Here is your next question.

The temperature of the tap water in a kitchen sink is varied by adjusting the hot and cold controls. This is an example of:

A. open-loop control. (Go to block 18.)
B. closed-loop control. (Go to block 26.)

---

8. The correct answer to the question in block 21 is A. It is not necessary to use 0 volts to represent binary 0. In fact, any two voltages could be used to represent 0 and 1.

Here is your next question.

What is the name of the circuit that can be used to connect two circuits which have different voltage levels? _____ (Go to block 28.)

---

9. Your answer to the question in block 22 is wrong. Read the question again, then go to block 27.

---

10. Your answer to the question in block 3 is wrong. Read the question again, then go to block 16.

---

11. Your answer to the question in block 24 is wrong. Read the question again, then go to block 19.

---

12. The correct answer to the question in block 18 is B. The system for an electric oven is very similar to the thermostatic control system for the home.

Here is your next question.

Static controls:

A. operate with sources of static electricity. (Go to block 20.)
B. have no moving parts. (Go to block 24.)

---

13. Your answer to the question in block 1 is wrong. Read the question again, then go to block 6.

14. Your answer to the question in block 25 is wrong. Read the question again, then go to block 7.

15. Your answer to the question in block 21 is wrong. Read the question again, then go to block 8.

16. The correct answer to the question in block 3 is B. There is no digit 2 in the binary system. The digit 2 is represented by binaries as 10. Here is your next question.
The stored program in a numerical control system is sometimes called *software*. Which of the following is a method of storing programs?

A. Colored tape. (Go to block 2.)
B. Punched tape. (Go to block 22.)

17. Your answer to the question in block 6 is wrong. Read the question again, then go to block 35.

18. The correct answer to the question in block 7 is A. At first you might think that the water control illustrated is a closed-loop system because a human being can readjust the faucets as necessary to maintain a certain temperature control. That, of course, would not be an electronic system. The term *closed-loop control* refers to systems in which control is automatic without the need for intervention by humans.
Here is your next question.
The temperature of an electric oven is set at 300°F. The oven circuitry maintains this temperature by turning the heating element on and off as needed. This is an example of

A. open-loop control. (Go to block 5.)
B. closed-loop control. (Go to block 12.)

19. The correct answer to the question in block 24 is C. The Z axis is the vertical axis, and the depth of the hole is measured along the Z axis.
Here is your next question.
In a closed-loop speed control the comparator compares the output signal from the transducer (or transducer amplifier) with the signal from

A. the control circuit. (Go to block 23.)

B. a reference circuit. (Go to block 3.)

---

20.   Your answer to the question in block 12 is wrong. Read the question again, then go to block 24.

---

21.   The correct answer to the question in block 27 is that static controls have greater reliability. This is an important advantage because the repair time for electronic equipment in industrial electronics can be very expensive. The disadvantage of static controls is that they are somewhat more difficult to service; an electronics technician is required to perform the necessary troubleshooting procedures.

Here is your next question.

A certain component has only two output states: 6 volts and 12 volts. Could this component be used for an industrial electronic system that uses the binary system?

A. Yes. (Go to block 8.)

B. No. (Go to block 15.)

---

22.   The correct answer to the question in block 16 is B. Other methods of storing programs are magnetic tapes and disks. Regardless of the method used, the objective is still the same—that is, to deliver binary numbers to the system in order to control the system operations.

Here is your next question.

In this chapter two numbering systems were discussed. The binary system has 2 digits and the decimal system has 10 digits. Another popular system in industrial electronics is the octonary system which has 8 digits. Which of the following would *not* be a digit in the octonary system?

A. 8. (Go to block 27.)

B. 0. (Go to block 9.)

---

23.   Your answer to the question in block 19 is wrong. Read the question again, then go to block 3.

---

24.   The correct answer to the question in block 12 is B. The term *static control* is not as popular now as it used to be. At one time, a great majority of control systems operated with relays and electromagnetic switches, and static controls represented a small part of the overall field control. Today, however, the majority of the systems being designed are electronic.

Here is your next question.

In a numerically controlled drill press the depth of the hole being drilled is measured along the

A. *X* axis. (Go to block 4.)
B. *Y* axis. (Go to block 11.)
C. *Z* axis. (Go to block 19.)

---

25.   The correct answer to the question in block 6 is B. The binary number displayed is 1101. This is the same as decimal number 13. Here is your next question.
Another name for sensor is

A. transducer. (Go to block 7.)
B. relator. (Go to block 14.)

---

26.   Your answer to the question in block 7 is wrong. Read the question again, then go to block 18.

---

27.   The correct answer to the question in block 22 is B. Just as there is no digit 2 in the binary system and no digit 10 in the decimal system, there is no digit 8 in the octonary system. The digits in the octonary system are 0 through 7 inclusive.
Here is your next question.
What is an advantage of static control over electromagnetic control? (Go to block 21.)

---

28.   The correct answer to the question in block 21 is interface. Interfaces can be purchased in integrated circuit packages.

You have now completed the programmed section.

SELF-TEST WITH ANSWERS

(Answers to this test are given at the end of the chapter.)

1.   The name for a numbering system with only two digits is
_____ .

2.   The device used to sense the output of a system is called a
_____ or a _____ .

3.   The type of electronic control system that automatically maintains the output is called _____ .

4. The shaft speed of an electric motor can be sensed with a tachometer. In a closed-loop motor speed control, the output of the tachometer goes to _____ .

5. The largest decimal number that can be represented with four binary digits is _____ .

6. Two different signals can be compared in a circuit called a _____ .

7. Two circuits having different voltage levels can be connected through an _____ .

8. In a closed-loop motor speed control, the amount of electric power delivered to the motor is determined by the input signal to the _____ .

9. The three axes used to locate a hole of given depth are _____ , _____ , and _____ .

10. Another name for a numerical control program stored on tape is _____ .

Answers to Self-Test

1. Binary system.
2. Sensor or a transducer.
3. Closed-loop control.
4. An amplifier (or else directly to the comparator).
5. 15.
6. Comparator.
7. Interface.
8. Control circuit.
9. X, Y, and Z.
10. Software. (This term is also used to refer to computer programs.)

CHAPTER 2

COMPONENTS USED IN INDUSTRIAL ELECTRONICS

Electronics may be thought of as being a field of science that puts the electron to work. The first practical applications were in radio transmitters and receivers. Tubes were used as the amplifying components.

The types of circuits used in early radio systems are sometimes called *analog,* or linear. This simply means that the output signal is directly dependent upon the input signal. If the input signal increases in amplitude, the output signal also increases in amplitude.

The word *linear* is also used in electronics to mean that the output signal is an exact replica of the input signal. That is not the meaning intended here. To avoid confusion, the term *analog* will be used. It means that the output signal is a direct function of the input signal at all times.

The use of electronics in industry was very limited at first. The reliability of tube circuits was poor, so maintenance costs were high. Even though most of the repairs involved a simple replacement of tubes, the "downtime" could be very expensive.

Another problem was in the fact that analog circuits are not easily adapted to industrial machinery. *Digital circuits* are more useful in manufacturing and process control. A digital circuit is one in which the outputs are at fixed levels, such as ON and OFF. Also, the output *may* remain at a fixed level even though the input changes.

Once the use of digital circuitry was well understood, and once more reliable transistors and integrated circuits became available, the field of industrial electronics expanded rapidly.

Readers with a good background in basic electronics can skip this chapter, which is a broad review of basics. It does not include switching components.

THEORY

Electronic Components

Resistors, capacitors, and inductors are sometimes called *two-terminal passive components.* As their name implies, they are connected into the circuitry by only two leads or terminals. The word *passive* means that they do not generate a voltage and that they have no gain characteristics.

RESISTORS

A resistor is a component that opposes the flow of current. Two effects always occur when current flows through a resistor:

- Heat is produced.
- A voltage drop occurs.

It is useful to keep these effects in mind when analyzing a circuit. Resistors are used in circuits for the following reasons:

- To generate heat
- To introduce a voltage drop
- To limit current

The heat produced by current flowing through a resistor is often undesirable. It represents wasted power that cannot be used for anything practical. However, in a few applications resistors are used for producing heat. Examples are an electric stove and an electric oven.

Figure 2-1 shows some examples of how resistors are used to limit and control current. A *rheostat* is a variable resistor connected in such a way that

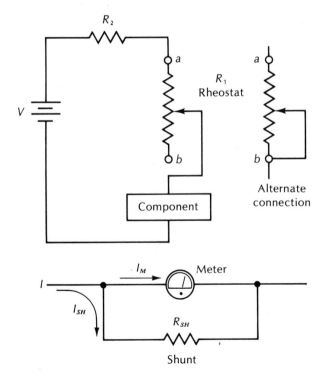

Figure 2-1. Circuits in which resistors control current

it controls the current in the circuit. Although there are three terminals on rheostat R_1, only two of the terminals are used. With the alternate connection there are still only two connections to the rheostat. However, if the *arm* (arrow on symbol) of the rheostat should stop making contact, the alternate connection would still maintain a current path through the resistance. This can be an advantage in some circuits.

The arm of the rheostat is moved toward point *a* to reduce the resistance of the circuit and therefore increase the current. The arm of the rheostat is moved toward point *b* to increase the circuit resistance and decrease the current. By adjusting rheostat R_1 to the proper value, the desired current through the component can be obtained.

Resistor R_2 is used to limit the current through the component to a maximum value. This means that when the rheostat arm is adjusted to point *a,* the amount of current flowing through the circuit is limited only by series-limiting resistor R_2.

In the meter circuit of Figure 2-1 the shunt resistor (R_{SH}) is used to limit current through the meter movement. The input current I divides into two currents, I_M and I_{SH}. Resistor R_{SH} shunts the undesired current (I_{SH}) around the meter movement, and therefore protects the meter movement from excessive current flow. Shunt resistors are used in ammeter circuits.

Figure 2-2 shows two examples of how resistors are used to introduce a voltage drop. In the lamp circuit, the applied voltage (V) is greater than the voltage rating of the lamp. Resistor R_1 drops the applied voltage to the desired lamp voltage. As an example, if the lamp voltage rating is 6.3 V and

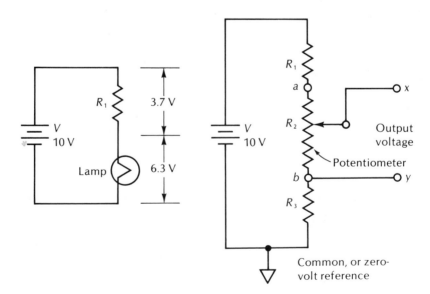

Figure 2-2.　Resistors used to drop voltage

the battery voltage is 10 V, then the amount of voltage that must be dropped across R_1 is $10 - 6.3 = 3.7$ V.

It might be argued that resistor R_1 is used to limit the current since it is *in series* with the lamp—that is, since the same current flows through both components. However, lamps are usually rated by allowable filament voltage, not current. So, the resistor is used to protect the lamp from an overvoltage.

In the potentiometer circuit of Figure 2-2, variable resistor R_2 is used to control the amount of output voltage. When the arm of the resistor is moved toward point *a,* all of the voltage drop across R_2 is delivered to the output terminals. When the arm of the resistor is moved to point *b,* the output voltage is zero volts.

Resistor R_3 is used to assure that the output voltage at terminal *x* can never be lower than some predetermined value. For example, if the voltage drop across R_1 is 2 V, then point *x* can never be adjusted to a point less than 2 V with respect to common.

Resistor R_1 assures that the arm can never be adjusted above a predetermined value. All the resistors in the potentiometer circuit are used for introducing voltage drops.

There are two uses of the word "potentiometer" in electronics. It is a voltage-adjusting circuit as shown in Figure 2-2; it is also a very sensitive voltage-measuring instrument.

BOLOMETERS

The *temperature coefficient* of a resistor or conductor is a measure of how much its resistance changes for a given change in temperature. A *positive* temperature coefficient means that the resistance increases when the temperature increases, and a *negative* temperature coefficient means that the resistance decreases when the temperature increases. A *bolometer* is a component that has a large temperature coefficient. There are two forms of bolometers: *barretters* and *thermistors*. Thermistors are made with beads of semiconductor materials, and barretters are made with a fine wire or metal film. Both are components with a wide temperature coefficient. This means that a small change in temperature produces a large change in resistance.

Figure 2-3 shows the symbol for a thermistor and a typical characteristic curve. The curve shows that the temperature coefficient is negative, which is usually the case for thermistors. However, thermistors with positive temperature coefficients are also made.

Since the resistance of the thermistor depends on its temperature, it can be used as a temperature measurement probe. This application is discussed in Chapter 4.

Thermistors are also used in protective circuits. For example, they can be mounted in a motor frame to sense the temperature of an electric motor. If the motor temperature increases, because of excessive load or for some other

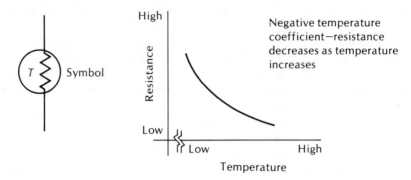

Figure 2-3. Symbol and curve for a thermistor

reason, the thermistor resistance change controls a circuit that turns off the power to the motor.

An ordinary tungsten filament lamp can be used as a barretter. It has a positive temperature coefficient. The symbol and typical characteristic curve are shown in Figure 2-4.

It is not unusual to use tungsten lamps in electronic circuits. In addition to tungsten lamps, barretters are also made with very fine platinum wires suspended between electrodes in a glass enclosure. They look like small electric fuses. As a matter of fact, some small wire fuses can be used as barretters.

Barretters, like thermistors, are used to sense temperature and temperature changes.

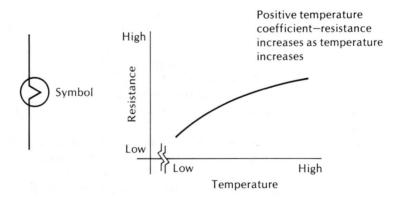

Figure 2-4. Symbol and curve for tungsten filament lamp

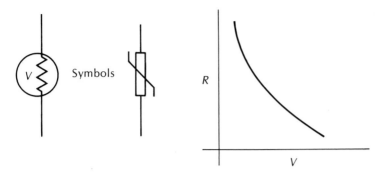

Figure 2-5. Symbol and curve for a voltage-dependent resistor (VDR)

VOLTAGE-DEPENDENT RESISTORS

Figure 2-5 shows the symbols used for a voltage-dependent resistor (VDR). This component is also known by such names as *varistor, globar, stabistor,* and *thyrite* resistor.

As shown on the characteristic curve of Figure 2-5, the resistance across the device decreases as the voltage increases. This makes it useful for preventing high voltages from destroying semiconductor components. A good example of the high voltages that are troublesome are the noise signals riding on ac power waveforms. Such noise signals are called *transients* or *spikes*. (See Figure 2-6.) Noise spikes are usually caused by machinery or other electrical equipment operating in the vicinity. A voltage-dependent resistor connected across the power line permits the sine wave voltage to be passed but eliminates the voltage spikes.

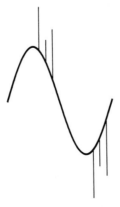

Figure 2-6. Transients on the power line voltage

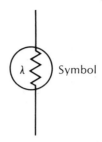 Symbol

Figure 2-7.   Symbol for a light dependent resistor

Voltage-dependent resistors are also used for *voltage regulators*. In this application they hold the voltage of a system constant even though the load current is changing.

LIGHT-DEPENDENT RESISTORS

It is possible to make a resistor which changes in resistance value in accordance with changes in almost any type of energy. Thus, the thermistor changes its resistance with heat energy, and the VDR changes its resistance with electrical energy. The LDR, or light-dependent resistor, changes its resistance in accordance with the amount of light striking it.

The LDR is an example of an optoelectronic component. Figure 2-7 shows its symbol. It can be used as a light sensor. There are many applications in industrial electronics where this feature is useful. Other optoelectronic components are discussed later in this chapter, and in Chapter 3.

Capacitors

A capacitor is a component used to store electric energy. A definition of capacitor is: *a component that opposes any change in voltage across its terminals*. This definition makes it easier to understand the use of capacitors in electronic circuits.

Basically, a capacitor consists of two or more metal plates separated by an insulating material called the *dielectric*. The energy is stored in the dielectric.

An easy way to understand how capacitors are used to store energy is to study the basic oscillator shown in Figure 2-8. Electron current flow will be used in this discussion. When the switch is closed, the capacitor begins to *charge*. The positive terminal of the battery attracts electrons away from one plate of the capacitor, and the negative terminal of the battery forces electrons onto the other plate. As a result, there is a positive capacitor plate with a deficiency of electrons and a negative capacitor plate with an excess of electrons. So, there is a voltage across the capacitor.

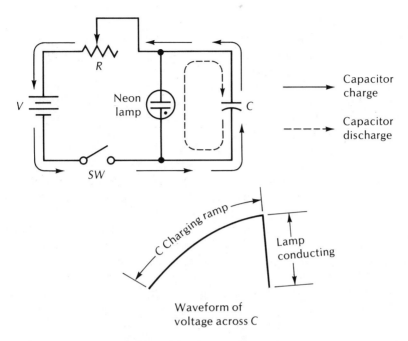

Figure 2-8. Neon lamp oscillator

The voltage across the capacitor increases as more electron carriers move into and out of the capacitor plates. The solid arrows show the path of the charging current. Variable resistor *R* (connected as a rheostat) is used to limit the amount of charging current. The larger the resistance, the longer it takes the voltage across the capacitor to reach a certain predetermined value.

A neon lamp is a component that conducts only when the voltage across it reaches a certain minimum value called the *firing potential*. A neon lamp is one of many components used in electronics that can be referred to as *breakover devices*. They do not conduct until the voltage across them reaches the breakover or firing potential.

As a capacitor continues to charge, the voltage across it increases. Eventually, it reaches the firing potential of the neon lamp. Once the neon lamp fires, as indicated by a glow, it offers a low resistance path for electron flow. During this period of time, the excess electrons that were stored on the negative plate of the capacitor flow through the neon lamp to equalize the positive charge. This motion of charge carriers is shown by the dashed arrow in Figure 2-8.

The voltage across the capacitor is shown to be a sawtooth waveform. The long part of the sawtooth, called a *ramp,* represents the charging of the capacitor. The short part of the sawtooth, called the *flyback,* represents the period of time when the neon lamp is conducting. The neon lamp does not discharge the capacitor to zero volts. Once the voltage across the capacitor is below the *extinguishing voltage* of the neon lamp, it stops conducting and the capacitor begins charging again.

The purpose of the capacitor in Figure 2-8 is to store energy for a certain period of time and then release the energy when the neon lamp fires. In most applications in electronics the capacitor is used for storing energy. This important characteristic is sometimes overlooked because of the extremely short period of time that the energy is stored. For example, in a circuit operating at a frequency of 1 megahertz (1,000,000 cycles per second) the capacitor may be called upon to store energy for one-half cycle out of each cycle. Since one cycle only takes a millionth of a second, and a capacitor stores energy for only half of this time, the capacitor is actually storing energy for only 1/2,000,000 of a second during each cycle. This extremely short period of time may be difficult to imagine, but in reality it is a relatively long time for many of the electronic circuits in use.

Capacitors can also be used as voltage dividers. A circuit is shown in Figure 2-9. When two capacitors are placed across any voltage source, as shown in this illustration, the lower voltage value always occurs across the larger capacitor. This is because the *reactance* of the capacitor—which is the opposition the capacitor offers to the flow of alternating current—is inversely proportional to the capacitance value. Mathematically,

$$X_C = \frac{1}{2\pi f C}$$

This equation indicates that larger capacitance values *(C)* give lower reactance *(X_C)* values. Lower reactance means less opposition to the flow of alternating current.

In the circuit of Figure 2-9, if C_1 were a very large capacitor, the voltage drop across it would be so small that it could be neglected. If the two

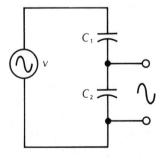

Figure 2-9. A capacitive voltage
divider

capacitors are equal in value, then one-half of the ac voltage will appear across each.

An advantage of using capacitors as voltage dividers in comparison to using resistors is in the fact that the capacitors dissipate practically no heat. Remember that heat is a loss of energy in electronic circuits and is usually undesired.

If the capacitor was theoretically perfect, no heat would be dissipated. In practice, however, capacitors do become warm when connected across an ac voltage source. One reason for this is the heating of the dielectric. A simplified illustration showing the cause of dielectric heating is shown in Figure 2-10. Only one of the atoms in the dielectric is shown, but of course there are many, many millions of atoms in this material.

When the capacitor is charged with a negative voltage (excess electrons) on one plate and positive voltage on the other (deficiency of electrons), the orbit of the electrons in the atom is pulled into an elliptical shape. This is illustrated by the dashed line in Figure 2-10. On the next half cycle when the voltage across the capacitor reverses, the orbit of the electrons around the atoms in the nucleus is pulled into an elliptical orbit in the reverse direction. This is also shown with a dashed line.

If the voltage changes very rapidly, as it will with a high frequency, the electrons' orbits will change very rapidly. This motion produces heat. Dielectric heaters used for industrial applications work on this principle. The material to be heated is inserted between two plates of a capacitor and a high-voltage, high-frequency ac is applied across the capacitor. This heats the material very rapidly.

Capacitors can be used to introduce a phase difference or time lag. This important application is illustrated in Figure 2-11. Here an *RC* circuit is connected across an ac line that has a sinusoidal voltage. The waveform of the voltage across the capacitor is shown with dashed lines in the illustration. Note that it lags behind the applied voltage. The amount of lag depends upon the value of resistance and the value of capacitative reactance. If the resistance was zero ohms, the amount of phase lag would be zero. By increasing

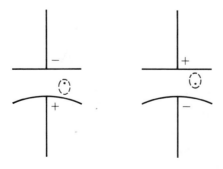

Figure 2-10.   Principle of dielectric heating

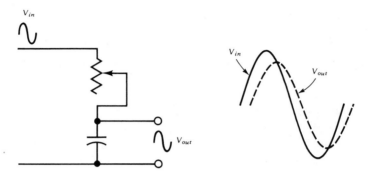

Figure 2-11.   An *RC* phase shift network

the resistance the phase difference increases. The phase shift network of Figure 2-11 is used in motor control and power control systems.

Since a capacitor opposes any change in voltage across its terminals, it can be used as a *filter*. A simple capacitive filter is shown in Figure 2-12. The input voltage has a sawtooth waveform that can be considered to be an ac sitting upon a dc voltage. The result of filtering is a smoother dc at the output. One way to interpret this operation is that the capacitor will readily pass higher frequencies and oppose the passage of lower frequencies. For the input signal the higher frequencies represent the ripple and the lowest frequency (zero hertz) is the dc voltage. The dc voltage cannot pass through the capacitor, but the high frequency ripple can. Therefore, at the output of the circuit, only the dc voltage is left.

All of the discussions in this chapter that relate to capacitors are based on a capacitor model. In reality, the capacitor's operation is much more complex than indicated here. This is an important point because the technician must realize that the objective is to teach him how to interpret the use of capacitors in circuits. These models enable him to make a fast interpretation of what the capacitor is used for, and in troubleshooting, to interpret the operation of the circuit if the capacitor is not working properly. If the capacitor is studied more in depth, the model that he is working from must be expanded and improved upon.

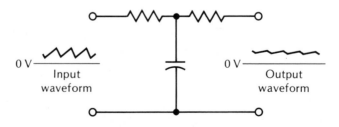

Figure 2-12.   A capacitive filter

Inductors

Inductors are usually made by coiling wire. They are components that store energy in the form of an electromagnetic field surrounding the inductor. An inductor may also be thought of as being a component that opposes any change in current through its coil.

The most important application of inductors in electronics is as components that store energy. Inductors can be used to store electromagnetic energy in the same way that capacitors are used to store electrostatic energy. Because of problems inherent with inductors, their use is limited in modern circuits. This is especially true in integrated circuit systems. The problems are:

- Large size and weight (compared to other components)
- Chance of interference due to stray electromagnetic fields
- Chance of damage to components due to inductive kickback

Although inductors as components are not used extensively in some of the newer integrated circuits, many components have built-in inductance. Motors, relays, some types of heating elements, and wire-wound resistors all have inductance. This inductance must be taken into consideration when analyzing circuits.

The concept of inductive kickback is illustrated in Figure 2-13. When a dc current is flowing through a coil, the current produces a magnetic field that surrounds the coil. This magnetic field represents energy stored.

The resistor in the circuit is simply for the purpose of limiting current through the coil. Since a coil consists of nothing more than a large number of turns of wire, it could be practically a short circuit across a battery without the limiting resistor.

When the switch for the circuit is opened, the magnetic field collapses around the coil. According to Faraday's law, *any time a magnetic field cuts across a conductor, a voltage is induced*. The collapsing magnetic field cuts

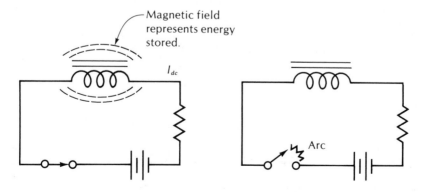

Figure 2-13. Inductive circuits produce switching arcs

across all of the turns of wire in the coil and produces a very large voltage, called a *countervoltage*.

An arc is developed between the switch contacts because of this countervoltage. At one time this was known as a *counter-electromotive force*, or *EMF*. The term *EMF* is no longer in popular use because it is misleading. Voltage is not a unit of *force*, but rather, it is a unit of work.

Inductive kickback can also be produced when the switch is closed to produce a current through the coil. Assume for a moment in the circuit of Figure 2-13 that there is no current flowing in the circuit. When the switch is closed, the current starts to flow. A magnetic field starts to expand around the coil. This moving magnetic field cuts across the turns of wire and produces a large countervoltage.

To summarize, an inductor is a component that opposes any change in current through it. When the current through it is maximum as shown in Figure 2-13, and the switch is opened, the coil tries to keep the current flowing. If no current is flowing in the circuit when the switch is closed, the coil tends to try to keep the current at zero.

In electric motors, for example, the field windings and armature windings have a large number of turns of wire wrapped around a magnetic material. The use of the iron core material increases the number of wire-flux linkages and produces a larger countervoltage than would occur for an air core. Thus, starting a motor or stopping a motor can produce inductive kickbacks that are large enough to destroy other components in the circuit.

To avoid this possibility, several steps can be taken. Special starting circuits can be used to gradually increase or decrease the current through the motor. Also, a component can be connected across the switch as shown in Figure 2-14. The capacitor stores the extra energy produced by the inductive kickback, and then releases it to the circuit slowly at a later time. The voltage across the VDR becomes very high when the switch is opened, and thus offers a low resistance path for the energy produced by the inductive kickback.

Current through a coil lags 90° behind the voltage across it, so inductors can be used to produce a phase difference. A circuit similar to the one shown in Figure 2-11 can be used. If an inductor replaces the capacitor in that circuit, the voltage across the inductor will lead the line voltage.

Figure 2-14. Two methods of protecting switch contacts in inductive circuits

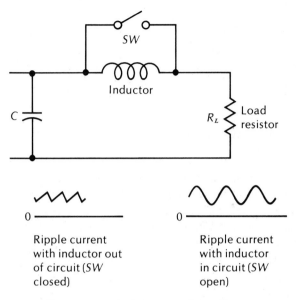

Figure 2-15.   An inductor used as a filter

Inductors can also be used as filters (Figure 2-15). The input current waveform is a pulsating dc. The coil opposes the rapid changes in current. This filters the current variations.

A comparison of the circuit of Figure 2-15 with the circuit of Figure 2-12 shows that the coil removes ripple *current* variations, whereas the capacitor removes ripple *voltage* variations. The two circuits can be combined into a single filter to produce very effective filtering. An example, shown in Figure 2-16, is called a *pi-section filter* because it looks like the Greek letter π (pi). The result of filtering is that the voltage across the load resistor (R_L) is almost a pure dc value. This pure dc is necessary for operating electronic amplifying components.

Inductors can also be used as heaters for conducting materials. In the application illustrated in Figure 2-17, the material to be heated is inserted into

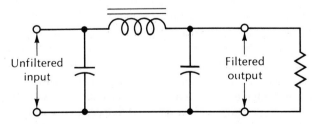

Figure 2-16.   An LC filter circuit

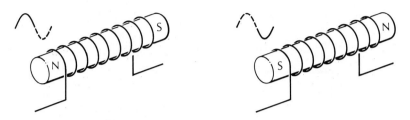

Figure 2-17. With ac current the magnetism reverses each half cycle

the center of the inductor. On one half cycle the current through the coil causes the material to be magnetized north and south in one direction. On the next half cycle the current through the coil reverses and this causes the material to be magnetized in the opposite direction. The rapid switching of an inductance causes its core to be heated very rapidly and to a very high temperature. Joints in iron or steel pipes can be welded efficiently with this type of heating.

TRANSFORMERS

Transformers are a special application of inductors. Current flowing in the primary winding of the transformer produces an expanding and contracting magnetic field around the primary. This magnetic field cuts across the turns of wire in the secondary and induces a voltage across the secondary winding.

If there are more turns of wire in the secondary winding than there are in the primary, then the voltage can be stepped up, or increased. This principle is easy to understand because the greater number of secondary turns means that the lines of flux are cutting across more turns of wire, and the voltage for each turn adds to a larger value. Figure 2-18 shows the relationship between primary and secondary turns and voltages.

At first it might seem that the transformer makes it possible to get something for nothing. However, when the voltage across the secondary is increased, the amount of current that the secondary can carry is decreased. The total power $(P = V \times I)$ is theoretically the same in the primary and in the secondary. This assumption would be true if the transformer was perfect, but in practice transformers do dissipate some power in the form of heat. So, the output power is always something less than the input power.

The relationships between primary and secondary currents, and between primary and secondary impedances are also shown in Figure 2-18. Impedance is the opposition that a component offers to the flow of alternating current. It is not often that a technician must calculate turns ratios for transformers and voltage ratios. However, it is important to understand these basic relationships because it shows the importance of replacing transfor-

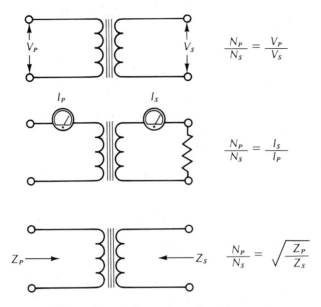

$$\frac{N_P}{N_S} = \frac{V_P}{V_S}$$

$$\frac{N_P}{N_S} = \frac{I_S}{I_P}$$

$$\frac{N_P}{N_S} = \sqrt{\frac{Z_P}{Z_S}}$$

Figure 2-18. Basic transformer equations

mers with identical replacement parts. It is *not* possible to determine the turns ratio by measuring the resistance of the primary and the resistance of the secondary, then taking the ratio of these resistances. It *would* be possible if the same wire size was used for the primary and secondary. However, in practice the windings have different wire sizes and a simple resistance ratio does not yield the simple turns ratio. To get the turns ratio (N_P/N_S) it is best to measure the primary and secondary voltages, then calculate from the equation in Figure 2-18.

Figure 2-19 shows a transformer with the same number of turns on the primary and secondary. Thus, the primary and secondary voltages are equal. This is an *isolation transformer* that is used to electrically isolate the

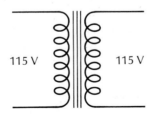

Figure 2-19. Isolation transformer

circuitry at the secondary from the ac power line. This safety precaution is used extensively by technicians when working with equipment that may be using the metal chassis as a common circuit point.

Figure 2-20 shows two examples of *autotransformers*. They use a single winding for both primary and secondary. As shown in the illustration, no isolation is provided by the transformer between primary and secondary circuits. Thus, if a technician accidentally touches one of the secondary connections, he is being connected directly into the ac power line. One-half of the ac power line is earth grounded and if a polarized plug is used, it is possible to connect the secondary circuit in such a way that its ground point is the same as earth ground. A plug is polarized if it is designed in such a way that it can only be inserted into a socket one way. However, if the plug is not polarized, the transformer can be connected into the power line in such a way that point *a,* which would normally be the chassis ground point, is connected to the high tension side of the power line. A technician who touches the chassis and some ground point could be seriously hurt if this happened. That is why polarized plugs are very important in electronic systems.

The transformers discussed so far are for single-phase circuits. In three-phase circuits the transformer primaries and secondaries can be connected in two different ways: delta (Δ) or wye (Y). Since the primaries can be connected in two different ways, and the secondaries can be connected in two different ways, there are actually four possible ways to connect a transformer in a three-phase circuit. These four methods are shown in Figure 2-21 along with the equations for the secondary voltage.

In electronic circuits the connection most often used is delta to wye. One reason for this is that the wye circuit has a common line and this is important in electronic circuitry. Another reason is the more beneficial turns ratio relationship.

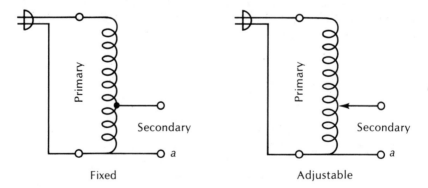

Figure 2-20. Examples of autotransformers

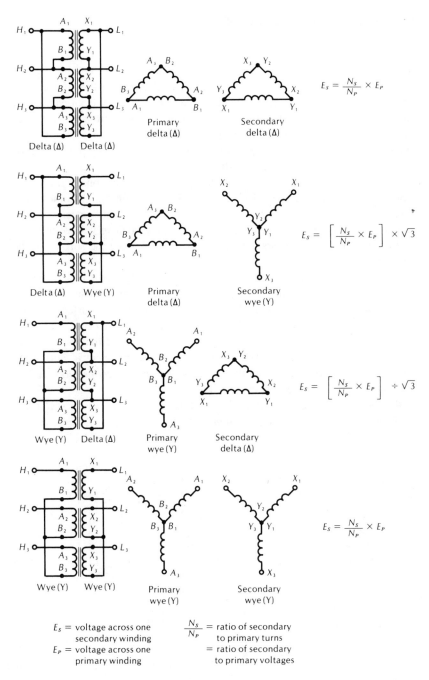

$$E_S = \frac{N_S}{N_P} \times E_P$$

Primary delta (Δ)   Secondary delta (Δ)

Delta (Δ)   Delta (Δ)

$$E_S = \left[\frac{N_S}{N_P} \times E_P\right] \times \sqrt{3}$$

Delta (Δ)   Wye (Y)   Primary delta (Δ)   Secondary wye (Y)

$$E_S = \left[\frac{N_S}{N_P} \times E_P\right] \div \sqrt{3}$$

Wye (Y)   Delta (Δ)   Primary wye (Y)   Secondary delta (Δ)

$$E_S = \frac{N_S}{N_P} \times E_P$$

Wye (Y)   Wye (Y)   Primary wye (Y)   Secondary wye (Y)

E_S = voltage across one secondary winding

E_P = voltage across one primary winding

$\frac{N_S}{N_P}$ = ratio of secondary to primary turns = ratio of secondary to primary voltages

Figure 2-21. Three-phase transformer connections

ELECTRONIC COMPONENTS

Tube Diodes

All tubes operate on the same basic theory. Electrons are emitted from a *cathode* material and attracted to a positive electrode called the *anode*. These electrons may or may not be controlled as they travel from cathode to anode. The starting point is to obtain electron emission from the cathode. There are several ways of doing this.

a. *Thermionic emission* occurs when a metal is heated to a point where it emits electrons from its surface. This is the most common method.

b. *Secondary emission* occurs when high velocity electrons collide with molecules or metal surfaces. The impact of the electron knocks other electrons, called *secondary electrons,* loose from the point of impact.

c. *Field emission* is obtained by placing a highly positive voltage near the surface of a metal. The voltage tears electrons away from the surface of the material.

Certain materials emit electrons when they are exposed to light. These *photoemissive materials* are used in certain types of phototubes.

Radioactive materials continually emit electrons from their surface. Radioactive emission is not used extensively in electron tube devices.

At one time almost all diodes used in electronics were the vacuum-tube type shown in Figure 2-22. Their purpose is to permit electric current to flow in one direction, but not in the other. This is called *rectification*. In this application they act as switches or valves.

Vacuum-tube diodes are of two kinds: those with *directly heated* cathodes, and those with *indirectly heated* cathodes. In the directly heated type

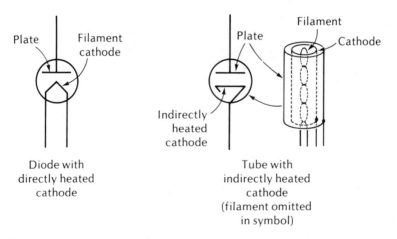

Figure 2-22. Vacuum-tube diodes

the filament current causes the filament to heat. The heated filament evaporates electrons off its surface, forming a cloud of electrons (called a *space charge*) around the filament. When the plate of the tube is made positive, electrons in the space charge are attracted to the plate. When the plate is made negative, the electrons are repelled by the negative plate and no electron current flows through the tube. Therefore, the diode will permit current flow only when the plate is positive with respect to the cathode.

With the indirectly heated type the filament heats a surrounding sleeve called the *cathode*. Emission of electrons is directly from the cathode surface.

The symbol for a diode is in no way related to its construction, which is shown by the inset. Normally the plate and cathode are concentric with the filament in the center. The same is true with the directly heated type in which the plate surrounds the filament. This construction gives a larger plate area and a larger current-carrying capacity for the tube.

Sometimes an inert gas is placed inside the envelope. This type of tube is called a *phanatron*. The symbol is shown in Figure 2-23. An inert gas is one that does not readily combine with other materials, so it does not cause metals in the tube to corrode. The presence of the inert gas produces *avalanching* which is also shown in Figure 2-23. In this process a fast-moving electron from the cathode collides with a gas molecule. The impact knocks an electron from the molecule, so two electrons, both of them moving at high velocities, are now moving toward the anode. When they strike molecules, each liberates additional electrons which also strike molecules. As seen in the illustration, one electron very rapidly produces many electrons. This simple process is repeated many times in the phanotron tube. The result is that a high amount of current reaches the plate.

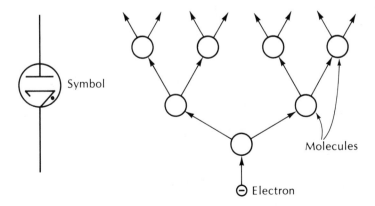

Figure 2-23.   Phanatron operation showing avalanche effect

Not all tubes have heated cathodes. Figure 2-24 shows an example of a cold cathode neon tube. This is the type used in the oscillator circuit of Figure 2-8.

To start the tube action, assume that neon gas atoms are moving around between the electrodes. At normal room temperature some free electrons will also exist in the area. When a positive voltage is placed on one of the electrodes and a negative voltage on the other, as shown in the illustration, the negative electrons begin to drift. They move away from the negative electrode toward the positive electrode. When the voltage is low, the electron drift is of a low velocity. As the voltage is increased, the *firing potential* is reached. When this occurs, the electron motion is sufficiently high to produce avalanching, and the tube glows.

Each electron-atom collision produces a positive molecule (called an *ion*) that drifts toward the negative electrode. As the ions approach the negative electrode, their positive potential pulls the electrons off the surface by field emission.

The electrons that leave the negative cathode combine with the positive ions and give up energy, and light is produced. This is a very important point: *When electrons change from a high energy level to a low energy level, they give off light or electromagnetic energy at some frequency.* Because current avalanching occurs in the neon lamp, its resistance is low during conduction. It can readily conduct electricity, even though there is no heated cathode. Conduction through the neon lamp discharged the capacitor in the circuit of Figure 2-8.

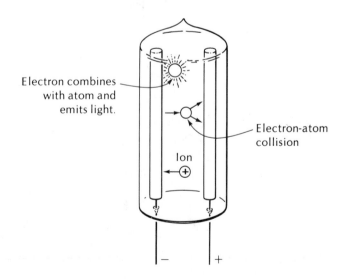

Electron combines with atom and emits light.

Electron-atom collision

Ion

−   +

Figure 2-24. A neon lamp

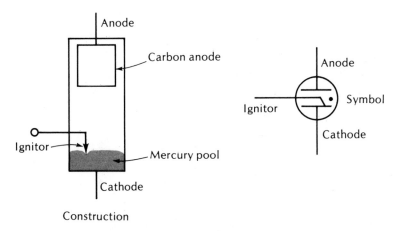

Figure 2-25. High-current ignitron

Another example of cold cathode tubes is shown in Figure 2-25. This is a high current *ignitron* that can readily produce current to 10,000 amperes. It does not have a heated cathode. In its place there is a cone-shaped electrode called the *ignitor*. This electrode penetrates the surface of the mercury pool but the mercury does not wet it.

When a voltage is placed on the ignitor there is an arc discharge between it and the mercury pool. This discharge releases electrons that are attracted to the positive carbon anode. As the electrons move toward the anode they collide with mercury vapor molecules near the surface of the ignitor. The result is avalanching and a very high current value. The carbon anode can dissipate a large amount of heat, so it can carry the large amount of current.

The symbol for the ignitron shows that it is a gas-type diode. This is indicated by the dot within the circle.

Tubes are being rapidly replaced by semiconductor devices in industrial electronic systems. However, much tube equipment is still in existence, so a technician should be familiar with tube theory.

Semiconductor Diodes

The principle of operation for a semiconductor diode is shown in Figure 2-26. It consists of the junction of two types of materials: *P* and *N*. These materials may be either germanium, silicon, or gallium arsenide. The *N*-type material should be thought of as having free electrons. The electrons are not floating around in the material, but they are available for electron current when a voltage is applied across the material.

The *P*-type material contains positive charge carriers called *holes*. The holes, like the electrons, are not floating around in the material, but they are available for current when a voltage is placed across the material. Holes move away from a positive voltage and move toward a negative voltage.

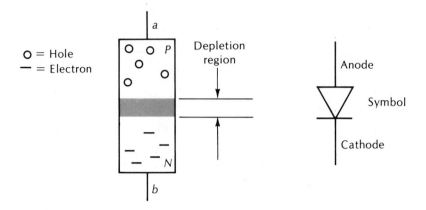

Figure 2-26. Model and symbol for semiconductor diode

When a diode is first constructed, an area is formed between the P and N regions that is called a *depletion region*. It does not contain any charge carriers (electrons or holes), so it can be thought of as being an insulating region between the two semiconductor materials.

When point a in Figure 2-26 is made negative and point b positive, the holes and electrons are pulled away from the center of the diode. The positive holes are attracted to the negative voltage and the negative electrons are attracted to the positive voltage. This condition is known as *reverse biasing*. The effect is to increase the depletion region because the charge carriers are pulled away from the junction.

When point a is made positive and point b is made negative, the charge carriers are pushed toward the depletion region. The holes are pushed away from the positive voltage and the electrons are pushed away from the negative voltage. If the voltage is high enough, the electrons and holes will be forced to cross through the barrier. When this occurs, current flows in the circuit. The minimum voltage necessary to produce current is about two-tenths of a volt for a diode made of P and N germanium, and about seven-tenths of a volt for a diode made of P and N silicon. When charge carriers are forced across the depletion region, the diode is said to be *forward biased*.

The symbol for a semiconductor diode looks like an arrow. The arrow points in the direction of $+$ to $-$ in conventional current flow. Electron flow is aiways against the direction of the arrow.

Figure 2-27 shows the semiconductor action with forward bias and reverse bias across the diode. The resistor in this circuit is used for limiting the current when the diode is forward biased. When an ac voltage is placed across a diode, current flows only on the half cycle when the anode is positive with respect to the cathode. In other words, the semiconductor diode does the same job as the tube diode.

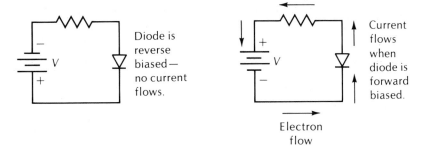

Figure 2-27. Reverse bias and forward bias for a semiconductor diode

At one time diodes were used almost exclusively as *rectifiers* or *detectors*—that is, they converted alternating current to a form of pulsating direct current. Because of rapid advances in semiconductor technology, many new types of diodes are available.

Figure 2-28 shows the symbol and characteristic curve for a specialized diode called a *zener*. The curve shows that when the anode voltage is positive with respect to the cathode, current will flow through the diode. This operation is in the *first quadrant,* and in this operation the zener diode is no different than a rectifier diode.

When the anode is slightly negative, the diode does not conduct. As the amount of negative voltage is increased, a certain point (called the *zener potential* or *zener voltage*) is reached, and the diode begins to conduct very heavily in the reverse direction. This is operation in the *third quadrant.*

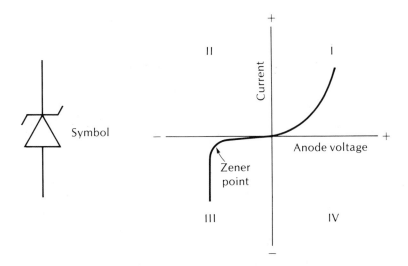

Figure 2-28. Symbol and curve for a zener diode

All semiconductor diodes have a zener point. For many types, once the zener potential is reached, the diode is destroyed due to the high current flow. The zener diode is made specifically for the operation in the third quadrant, so it does not burn out.

An important feature of this diode is the fact that within specified current limits the voltage across the diode is nearly constant. Therefore, the zener diode is used as a *voltage regulator*—that is, a circuit that maintains the voltage at a constant value provided the amount of current flow in the circuit is within specified limits.

Figure 2-29 shows the zener diode in a simple voltage regulator circuit. The output voltage is taken directly across the zener. If the unregulated input voltage starts to increase, the increased voltage occurs across R and not across the zener diode.

Figure 2-30 shows two *optoelectronic diodes*. The term *optoelectronic* is used for components that depend upon light or give off light in their operation. The photodiode is constructed so that light can be directed to the junction of the P and N regions. When light falls on this region the resistance of the depletion region is lowered, and current through the forward-biased diode increases. Thus, the conduction of the diode is directly dependent upon the amount of light falling on it.

A light-emitting diode (LED) also has a *PN* junction. As the electrons travel across the depletion region, they give up energy. It was noted in the discussion of neon lamps that when electrons give off energy it is in the form

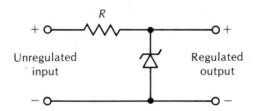

Figure 2-29. Zener voltage regulator circuit

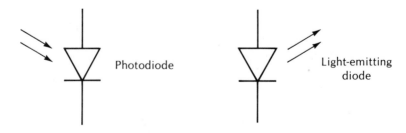

Figure 2-30. Optoelectronic diodes

of electromagnetic radiation, often in the form of light. So, light is produced when current flows through the forward-biased diode. This light is visible externally.

AMPLIFYING COMPONENTS

Amplifying components are three-terminal devices that are designed to control the amount of charge carrier flow from one point to another. Charge carriers are holes and electrons. They are considered to be carriers of electricity.

With amplifying components a small amount of voltage or current can be used to control a large number of charge carriers. Hence these components are called *amplifiers*. The three most important types of amplifying components in use today are *tubes, bipolar transistors,* and *field-effect transistors*. The basic difference between the three amplifying components is shown in Figure 2-31. The current through the tube, marked *I,* is normally considered to be an electron current flow from a cathode to a positive plate. The amount of current flowing through the tube is controlled by an input control voltage *(V)*. A small change in this control voltage produces a large change in the output current.

In the bipolar transistor the current may consist of either holes or electrons within the device. External to the device the circuit may be analyzed in terms of either electron flow or conventional current (hole) flow. The output

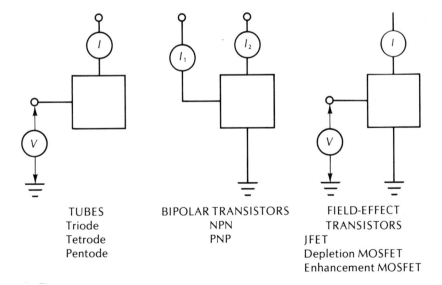

TUBES	BIPOLAR TRANSISTORS	FIELD-EFFECT
Triode	NPN	TRANSISTORS
Tetrode	PNP	JFET
Pentode		Depletion MOSFET
		Enhancement MOSFET

Figure 2-31.   Comparison of basic amplifier components

current in Figure 2-31 is marked as current I_2. It is controlled by a second current (I_1). A very small change in I_1 will produce a large change in I_2, giving it the amplifying characteristics. Two types of bipolar transistors are in use: the *NPN* and the *PNP*. These letters refer to the construction of the transistor. This will be discussed in reference to Figure 2-35.

Field-effect transistors have characteristics similar to those of the tube. The output current through the device is marked I. This current is controlled by a voltage V. A very small change in the amount of V at the input electrode can produce a relatively large change in I at the output. There are three kinds of field-effect transistors in use: the *JFET,* the *depletion MOS-FET,* and the *enhancement MOSFET.* These terms refer to their construction.

The operation of the amplifying components can best be understood from a discussion using models. The models represent the operation, but do not represent the actual construction of the device.

Tubes

Figure 2-32 shows the model for a *triode* tube. The term *triode* implies that there are three electrodes: plate, grid, and cathode. Electron flow through the tube is from the cathode to the plate in the same way that electron current flows through a diode. The grid is a construction of wires placed between the cathode and the plate. The grid voltage is normally negative with respect to the cathode. In fact, for all discussions of tubes the cathode can be considered to be zero volts, the plate positive, and the grid negative.

Since the grid is negative, it has an electric field that opposes the flow of electrons. The more negative the grid voltage the fewer the number of electrons that can pass from the cathode to the plate. If the grid is made sufficiently negative, a point called *cutoff* is reached and no plate current can flow.

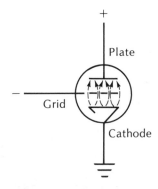

Figure 2-32. Electron flow in a triode tube

A small change in negative grid voltage causes a large change in the number of electrons that can actually reach the plate. The plate current is actually electron current flow in the plate circuit.

Figure 2-33 shows the schematic symbol and characteristic curve of a tetrode tube. The screen grid of the tetrode acts as a shield between the control grid and plate. Since the control grid and plate are made of metal and are separated by a vacuum dielectric, they form a capacitor. In a triode tube this capacitance can provide a path for high-frequency signals. The result is a loss of gain. The screen grid in the tetrode eliminates this undesired capacitance coupling.

The disadvantage of a tetrode is that the screen grid can attract more electrons than the plate under certain conditions; high-velocity electrons knock secondary electrons off the plate surface and these secondary electrons are attracted to the positive screen grid. When this happens, making the plate more positive causes a decrease in plate current for a portion of its operating range. The result is the negative resistance on the curve. With normal resistance, increasing the plate voltage would always increase the plate current, so the term *negative resistance* is used to describe the decrease in current when the plate is made more positive.

A negative grid between the screen and the plate will repel the secondary electrons back to the plate. This is the theory of the pentode tube of Figure 2-34. Since the secondary electrons return to the plate, they cannot produce an increase in screen current. In this manner the negative resistance of the tetrode is eliminated.

Tetrodes and pentodes have a higher gain than triodes. Along with their higher gain, they also introduce more noise into the circuit. In electronics the term *noise* means an undesired electric signal.

Bipolar Transistors

Figure 2-35 shows the model for a bipolar transistor. The name *bipolar* means that two kinds of current carriers, or *charge carriers,* are used in its

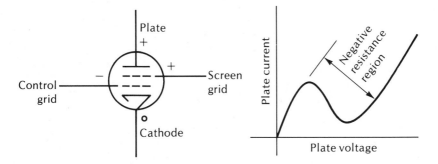

Figure 2-33. Symbol and characteristic curve for a tetrode tube

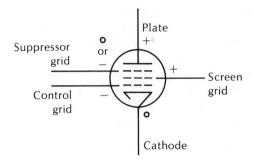

Figure 2-34.   Symbol for a pentode tube

operation. Electrons carry the current through the N material and holes carry the current through the P material. Both N and P materials are used in making bipolar transistors, so both types of charge carriers are used. An *NPN* transistor is shown. For a *PNP* transistor the emitter and collector would be P material and the base material would be N material.

Note that the P base region is small compared to the N collector and N emitter regions. In normal operation the collector of the *NPN* transistor is made positive with respect to the emitter. The emitter is considered to be at zero volts. The number of charge carriers that can actually cross the emitter-base junction depends upon how much base current flows. Increasing the base current increases the number of electrons to cross into the base.

The positive connection to the collector is marked $++$, which means that it is more positive than the $+$ for the base connection. Once the electrons get

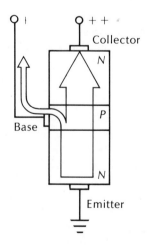

Figure 2-35.   Distribution of current in a bipolar transistor

into the base they are attracted by the two positive voltages, but over 95% go to the more positive collector. The large arrow represents the collector current and the small arrow represents the base current. A small change in base current causes a large change in the number of electrons that cross into the collector region.

If the transistor was a *PNP* type, current flow through the device would consist of holes. The arrows would be in the same direction as shown in Figure 2-35, but they would be showing direction of hole flow.

Optoelectronic Transistors

Figure 2-36 shows two examples of phototransistors. Both are optoelectronic components. When light strikes the base region of the phototransistor, charge carriers are released. The result is an increase in collector current corresponding with an increase in the amount of light.

The photodarlington is comprised of a phototransistor and a conventional transistor. The word *darlington* means that the two collectors are connected, and the emitter of one transistor goes to the base of the other. This type of connection gives a very high gain.

Current through the phototransistor sets the base current of the second transistor. A small change in the amount of light causes a larger change in phototransistor conduction, and an even greater change in conduction for the second transistor.

Figure 2-37 shows an example of an optical coupler. It has an LED and a phototransistor in a package that does not permit outside light to enter. Although the external circuitry is not shown, it is assumed that the LED is forward biased and the phototransistor is in a circuit that makes its collector positive with respect to its emitter. When current flows in the LED circuit, it emits light. This light causes the phototransistor to conduct. Varying the LED current causes the collector current to vary also.

The advantage of the optical coupler is that there is no electrical connection between the input (LED) and output (phototransistor circuit). The two

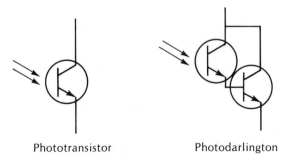

Phototransistor   Photodarlington

Figure 2-36. Two examples of optoelectronic transistors

Figure 2-37.   An optical coupler

circuits are isolated from each other, and the device is sometimes called an *optoisolator*. Instead of the phototransistor, some versions of the device use a photodiode and other versions use a photodarlington.

Field-Effect Transistors

Figure 2-38 shows a model for a junction field-effect transistor (JFET). It has a P-type material embedded into an N-type material. The N material in this device is called the *channel*, and the device is called an *N-channel JFET*. The three connections to an FET are called *source, gate,* and *drain*. The names of the three electrodes for the amplifying components are summarized in Table 2-1.

The source is considered to be zero volts. A positive voltage on the drain attracts the electrons from the source. A small negative voltage on the gate reverse biases the *PN* junction and causes a depletion region as indicated by the shaded area.

The size of the depletion region depends upon the amount of negative voltage on the gate. Making the gate more negative increases the size of the de-

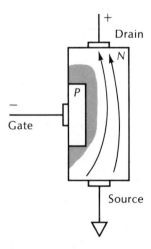

Figure 2-38.   Electron flow in an *N*-channel junction field effect transistor (JFET)

Table 2-1. Electrodes of Amplifying Components

	Electrode for Entering Current	Electrode for Controlling Current	Electrode for Leaving Current
Tube	Cathode	Grid	Plate
Bipolar transistor	Emitter	Base	Collector
FET	Source	Gate	Drain

pletion region and decreases the size of the conduction paths for the electrons. This operation is shown in Figure 2-39. The gate voltage directly controls the number of electrons that can reach the gate.

There is an important limit to the amount of input gate signal the JFET can handle. If the gate is accidentally forward biased by making the gate positive with respect to the channel, a large amount of current flow will flow across the gate-to-channel junction. When this happens, the JFET is usually destroyed. The input signal must not be allowed to make the gate of the N-channel JFET positive at any time.

To get around the disadvantage of the JFET, an insulating material can be added at the gate-to-channel junction. This insulating material is shown as a heavy dark line in Figure 2-40. The term *IGFET*, which stands for insulated gate field-effect transistor, was originally used for this type of device. Since the insulating material is made of a metal oxide, the term *metal oxide semiconductor field-effect transistor*, or MOSFET, also describes the component. The term *MOSFET* is now generally accepted.

The gate is negative with respect to the source in the N-channel MOSFET of Figure 2-40. The drain is positive with respect to the source. The opera-

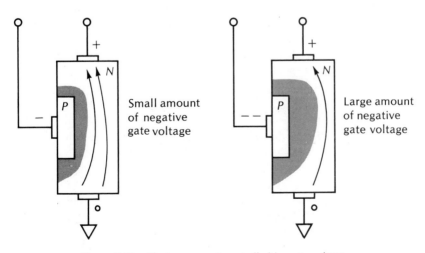

Small amount of negative gate voltage

Large amount of negative gate voltage

Figure 2-39. Electron current controlled by gate voltage

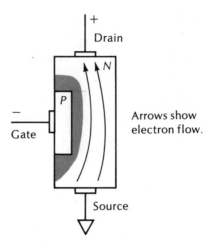

Figure 2-40. Electron flow in an *N*-channel MOSFET

tion is the same as for the *N*-channel JFET. Making the gate more negative increases the size of the field, indicated by the shaded area, and limits the number of electrons that can flow to the drain. A small change in gate voltage produces a large change in drain current. Even though the gate is made positive with respect to the drain, no current can flow across the gate-to-source junction. The insulating material prevents the current flow, but at the same time allows the electric field around the gate to control electron flow from the source to the drain.

The disadvantage of the MOSFET is in the fact that static electricity can destroy the very thin insulating material. Making measurements with a voltmeter that has a static charge on its probe can quickly destroy a MOSFET. Some newer MOSFET's have zener diodes connected internally between the gate and the source and also between the gate and the drain. When the voltage goes above a predetermined value, the zener diode conducts and prevents a high voltage from destroying the insulating gate.

The *N*-channel JFET and *N*-channel MOSFET are similar in operation to triode tubes. If *P*-channels are used, the charge carriers are holes, and the voltages on the electrodes are reversed.

A type of MOSFET that has no similarity to tubes is shown in Figure 2-41. This is called an *enhancement MOSFET*. When no voltage is applied to the gate, no current can flow through this device since the depletion region extends completely through the channel. Making the gate slightly positive decreases the depletion region and permits charge carriers to flow from the source to the drain. Note that this is an *N*-channel device, but a positive voltage must be applied to the gate in order to reduce its depletion region and permit current to flow.

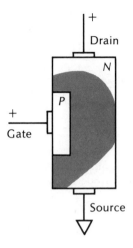

Figure 2-41.   An enhancement MOSFET

The name *enhancement MOSFET* means that a voltage is needed on the gate to enhance, or increase, channel current flow. The component in Figure 2-40 is called a *depletion MOSFET* because a voltage is used on the gate to reduce the channel current flow.

All of the amplifying components have dc operating voltages to get them into operation. Signal voltages are usually added to the input control voltage or current. This causes the signal through the component to vary with the signal on the control electrode. A small change in signal input causes a large change in signal current output, so these components can all be used as amplifiers.

It is important for a technician to memorize the polarities of dc voltages required on amplifying components. These voltages and symbols are summarized in Figure 2-42. The dc operating voltages on the input electrodes (grid, base, and gate) are called *bias voltages*. When analyzing a circuit that uses amplifying components, troubleshooting normally occurs in two stages: First, the dc voltages are measured. If the correct operating voltages are not present on the various electrodes of the amplifying component, it is not possible for the component to work properly in the circuit. However, caution must be used here. The voltages for operating *may* be derived from the signals. For example, in a diode a positive anode voltage is needed for forward biasing. This positive voltage may come from the input ac voltage.

The second step in analyzing and troubleshooting a circuit that has amplifying components is to trace the signal paths into and out of the component. For most applications the input signal will have a lower amplitude than the output signal. It is not always possible to discern this, however, if an oscilloscope is used because the output *current* may be greater than the *input* current, but an oscilloscope shows only *voltage* waveforms.

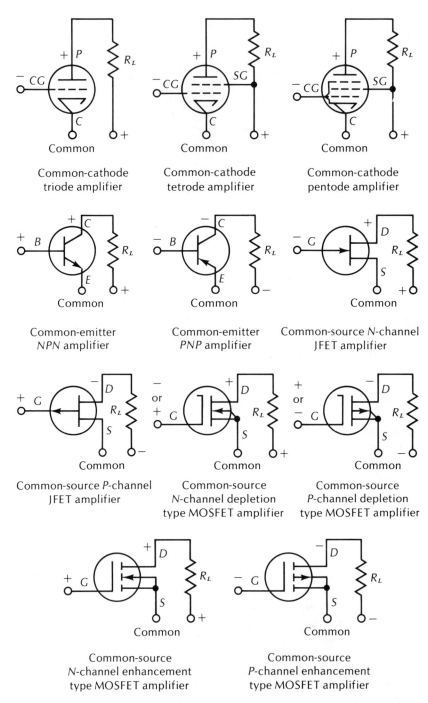

Figure 2-42. Dc operating voltages and symbols for amplifying components

BASIC AMPLIFIER CIRCUITS

Figure 2-43 shows basic amplifiers using a tube, a bipolar transistor, and an FET. The signal paths for the three circuits are the same. The input signal is delivered to the control electrode through capacitor C_1, and the output signal is taken through C_2. The output signal is developed across R_L in each amplifier. The signal voltages are developed across resistors connected to the grid, base, or gate.

This is an important thing to remember about the three types of amplifiers: *The signal paths are identical!* The difference between the circuits is the dc voltages and the type of amplifying component used.

In the tube circuit, a dc cathode current flows through R_2 in the direction shown by the arrow. The voltage drop across R_2 makes the cathode positive with respect to ground. The grid is at ground potential because there is almost no grid current flow. (A few microamperes of current flow because electrons strike the grid in their path to the plate. This small current must return to the cathode through R_1, but it is usually disregarded.)

Since the cathode is positive with respect to ground, and the grid is at ground potential, it follows that the grid is negative with respect to the cathode. Therefore, the tube is properly biased for operation as an amplifier.

The positive bias needed for operating the *NPN* transistor is obtained from two resistors (R_1 and R_2) connected as a voltage divider. The circuit is designed so that the base voltage is more positive than the emitter but less positive than the collector. The emitter resistor (R_3) is used to make the amplifier independent of changes in the positive power supply voltage, and also independent of temperature changes.

Operation of the FET circuit is similar to the tube amplifier. Current through source resistor R_2 makes the source positive with ground. Since the gate is at ground potential, it is negative with respect to the source.

When troubleshooting electronic circuits, it is always a good idea to start by measuring the dc operating voltages. Values of voltage, and also of circuit components, vary from circuit to circuit. The literature supplied by manufacturers of industrial electronic equipment usually gives the values to be expected.

LASERS

A *laser* is a component that can produce a very high amount of light energy. Its operation is based on principles that have already been discussed.

It has been noted that electrons can give off energy in the form of light when they pass from a high energy state to a low energy state. In the case of neon lamps this light is *random* or *incoherent*. This simply means that at any one instant the number of electrons giving off light can be different than for

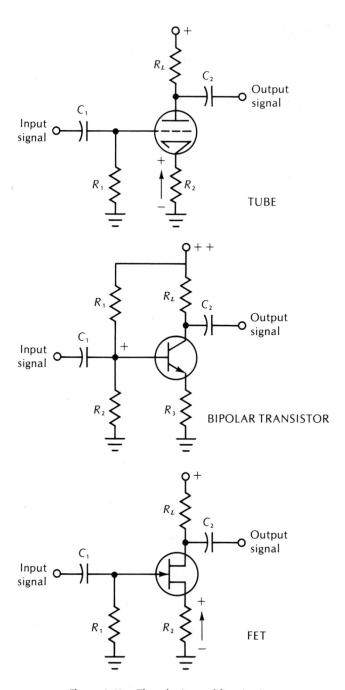

Figure 2-43. Three basic amplifier circuits

any other instant. The light waves emitted from the neon lamp are not *in phase*—that is, light produced by waveforms in phase would contain very high energy.

The basic principle of a ruby laser that produces coherent light is shown in Figure 2-44. Other materials may be used, but the principle of operation is the same. The first step in getting the laser action is to raise the electrons in the ruby to a higher energy level. There are a number of ways to do this. For example, they could be heated as in the case of thermionic emitters, or they could be exposed to light as in the case of photodiodes. Electrons have the peculiar characteristic that they can only exist at certain energy levels. So, a certain minimum amount of energy is required to raise the energy level of the electron one notch, or *quantum.*

The ruby laser raises the energy of the electrons by exposing them to a high-intensity light. This light is produced by a *light pump* which is usually a xenon light source surrounding the laser material.

Once the electrons are raised to the higher energy state, the light pump is turned off and the electrons are permitted to return to their lower state, at which point they all give off light. All of the electrons return to the lower state at the same instant of time, and the light waves are in phase. The mirror at one end of the ruby prevents light from leaving in that direction, thus all of the light is emitted from the other end. This intense light is used for a number of applications in industrial electronics such as cutting, drilling, and heating.

MAJOR POINTS

1. Resistors are used for limiting current, dropping voltage, and generating heat.
2. Capacitors are used most often for storing energy. They are also used for dropping voltage and, with resistors, to introduce phase or time delay.

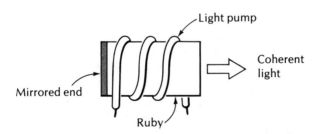

Figure 2-44. The ruby laser

3. Inductors are used most often for storing energy. They are also used for introducing phase or time delay.
4. Light-dependent resistors, photodiodes, phototransistors, photodarlington transistors, and optical couplers are all examples of optoelectronic components.
5. Zener diodes are designed to operate with reverse bias. The voltage across a zener is constant regardless of the amount of reverse current flow, so they are used as voltage regulators.
6. Avalanching in gas-filled diodes and certain semiconductor diodes (such as avalanche rectifier diodes) results in a very large current value.
7. Vacuum tubes and field-effect transistors are voltage-controlled amplifying components. A small change in voltage on their control electrodes (grid or gate) results in a large change in current through them.
8. Bipolar transistors are current-controlled amplifying components. A small change in base current causes a large change in current through the transistor.
9. Depletion MOSFET's will conduct a current from source to drain when there is no gate voltage. Enhancement MOSFET's cannot conduct a current from source to drain unless a voltage is present at the gate.
10. In comparing tube, bipolar transistor, and field-effect transistor amplifiers, the ac signal paths are the same. The dc voltages used for operating the components may be different.

PROGRAMMED REVIEW

(Instructions for using this programmed section are given in Chapter 1.)

The important concepts of this chapter are reviewed here. If you have understood the material, you will progress easily through this section. Do not skip this material, because some additional theory is presented.

---

1. A certain resistor has a decrease in resistance when its temperature decreases. This resistor has

    A. a positive temperature coefficient. (Go to block 10.)
    B. a negative temperature coefficient. (Go to block 25.)

---

2. The correct answer to the question in block 23 is B. For most applications the input signal is delivered to the base of a transistor. For tube amplifiers the signal usually goes to the grid.
    Here is your next question.

Another name for a linear circuit, in which the output is a direct function of the input, is

A. digital circuit. (Go to block 17.)
B. analog circuit. (Go to block 27.)

---

3. The correct answer to the question in block 31 is C. Transformer turns ratios are seldom accurately determined by measuring primary and secondary resistance values because separate wire sizes are likely to be used.
   Here is your next question.
   Which of the following can be thought of as being a component that opposes any change in current through it?

   A. Inductor. (Go to block 22.)
   B. Capacitor. (Go to block 28.)

---

4. The correct answer to the question in block 22 is A. A varistor, or voltage-dependent resistor, puts a very low resistance across the switch contacts when there is a high countervoltage. The low resistance path around the switch contacts prevents the arcing that destroys switch contacts.
   Here is your next question.
   To forward bias a diode, make its cathode

   A. negative with respect to its anode. (Go to block 19.)
   B. positive with respect to its anode. (Go to block 26.)

---

5. The correct answer to the question in block 19 is A. The larger voltage is always across the smaller capacitor. This is true regardless of whether the applied voltage is dc or ac.
   Here is your next question.
   A variable resistor that is connected into a circuit to control current is called a

   A. potentiometer. (Go to block 20.)
   B. rheostat. (Go to block 23.)

---

6. The correct answer to the question in block 29 is A. The ac current through a capacitor leads the ac voltage across it. Another way of saying this is that the voltage lags behind the current.
   With an inductor the current lags behind the voltage. Another way of saying this is that the voltage leads the current.
   Here is your next question.

In the illustration shown here, to increase the current through the meter, the value of shunt resistance must be

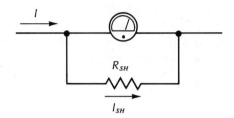

A. increased. (Go to block 16.)
B. decreased. (Go to block 24.)

---

7. Your answer to the question in block 16 is wrong. Read the question again, then go to block 31.

---

8. Your answer to the question in block 10 is wrong. Read the question again, then go to block 29.

---

9. Your answer to the question in block 19 is wrong. Read the question again, then go to block 5.

---

10. The correct answer to the question in block 1 is A. The temperature coefficient is positive when the resistance increases with an increase in temperature. Therefore, it must also be positive when the resistance decreases with a decrease in temperature.
Here is your next question.
The component shown here is properly biased when

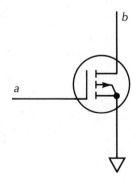

A. *a* and *b* are both positive. (Go to block 8.)
B. *a* is positive and *b* is negative. (Go to block 15.)
C. *a* is negative and *b* is positive. (Go to block 21.)
D. *a* and *b* are both negative. (Go to block 29.)

---

11.   Your answer to the question in block 23 is wrong. Read the question again, then go to block 2.

---

12.   Your answer to the question in block 29 is wrong. Read the question again, then go to block 6.

---

13.   Your answer to the question in block 22 is wrong. Read the question again, then go to block 4.

---

14.   Your answer to the question in block 31 is wrong. Read the question again, then go to block 3.

---

15.   Your answer to the question in block 10 is wrong. Read the question again, then go to block 29.

---

16.   The correct answer to the question in block 6 is A. The lower the value of shunt resistance, the lower the value of current through the meter movement. If the shunt resistance is zero ohms (a short circuit), no current flows through the meter. If the shunt resistance is infinitely large (an open circuit), all of the current flows through the meter.
Here is your next question.
In their normal operation, LED's are

A. forward biased to emit light. (Go to block 31.)
B. reverse biased. (Go to block 7.)

---

17.   Your answer to the question in block 2 is wrong. Read the question again, then go to block 27.

---

18.   Your answer to the question in block 31 is wrong. Read the question again, then go to block 3.

---

19.   The correct answer to the question in block 4 is A. The anode must be more positive than the cathode in order for a diode to conduct. Another way of saying this is that the cathode must be more negative than the anode.
Here is your next question.
Which capacitor in this circuit has the larger voltage across it?

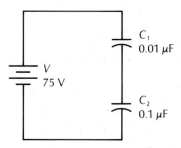

A. C_1. (Go to block 5.)
B. C_2. (Go to block 9.)

---

20. Your answer to the question in block 5 is wrong. Read the question again, then go to block 23.

---

21. Your answer to the question in block 10 is wrong. Read the question again, then go to block 29.

---

22. The correct answer to the question in block 3 is A. Inductors oppose a change in current, and capacitors oppose a change in voltage.
Here is your next question.
Which of the following components might be used to protect switch contacts from inductive kickback?

A. A varistor. (Go to block 4.)
B. An isolation transformer. (Go to block 13.)

---

23. The correct answer to the question in block 5 is B. A potentiometer controls voltage and a rheostat controls current.
Here is your next question.
The control electrode for a *PNP* transistor is called the

A. grid. (Go to block 30.)
B. base. (Go to block 2.)
C. gate. (Go to block 11.)

---

24. Your answer to the question in block 6 is wrong. Read the question again, then go to block 16.

---

25. Your answer to the question in block 1 is wrong. Read the question again, then go to block 10.

---

26. Your answer to the question in block 4 is wrong. Read the question again, then go to block 19.

---

27. The correct answer to the question in block 2 is B. Both analog and digital circuits are used in industrial electronics. However, the rapid advances in digital circuits were responsible for the wide variety of applications in modern systems.
Here is your next question.
In the ruby laser the electrons are raised to a higher energy level by using ———————— . (Go to block 32.)

---

28. Your answer to the question in block 3 is wrong. Read the question again, then go to block 22.

---

29. The correct answer to the question in block 10 is D. Symbols and voltage polarities are shown in Figure 2-42. It is important for technicians to memorize these symbols and voltages.
Here is your next question.
Which type of component has a voltage across it that lags behind the current through it by 90°?

A. Capacitor. (Go to block 6.)
B. Inductor. (Go to block 12.)

---

30. Your answer to the question in block 23 is wrong. Read the question again, then go to block 2.

---

31. The correct answer to the question in block 16 is A. A forward current is needed to produce light with an LED. Charge carriers moving across the junction give up energy in the form of light.
Here is your next question.
An ohmmeter is used to measure the primary resistance (25 ohms) and secondary resistance (100 ohms) of a transformer. The turns ratio (N_P/N_S) of this transformer

A. is 1:4. (Go to block 14.)
B. is 4:1. (Go to block 18.)
C. cannot be determined from this information. (Go to block 3.)

---

32. A light pump. This is a powerful light obtained from an intense lamp.

You have now completed the programmed section.

SELF-TEST

(Answers to this test are given at the end of the chapter.)

1. Is this statement true or false: Light is given off when an electron collides with an atom.

2. Name four methods used to obtain electron emission.

3. The primary voltage of a certain transformer is 117 volts and the secondary voltage is 936 volts. What is the turns ratio of this transformer?

4. A type of resistor used for removing spikes from powerline voltages and other sine-wave voltages is the _____ .

5. A barretter is an example of a _____ .

6. A diode used for regulating a dc voltage is the _____ .

7. Name an advantage and a disadvantage of a pentode as compared to a triode.

8. Undesired transient voltages in electronic circuits are referred to as _____ .

9. Which type of bipolar transistor requires a negative voltage on its base and collector with respect to the voltage on the cathode?

10. What is the name for the high intensity in-phase light from a laser?

11. Draw a basic amplifier circuit using a *PNP* transistor.

12. Draw a basic amplifier circuit using an *N*-channel MOSFET.

Answers To Self-Test

1. False. Light is given off when an electron combines with an atom that has lost an electron. (An atom that has lost an electron is called a *positive ion*.)

2. • By heating a metallic material (thermionic emission)
   • By focusing light upon certain materials (photoemission)
   • By directing a stream of high-speed electrons against a material (secondary emission)
   • By placing a strong positive voltage near the surface of a material (field emission)

3. Turns ratio $= \dfrac{N_P}{N_S} = \dfrac{0.125}{1} = \dfrac{1}{8}$

4. Voltage-dependent resistor.

5. Bolometer.

6. Zener diode.

7. Advantage: High amplification (gain)
   Disadvantage: High noise

8. Noise.

9. A *PNP* transistor.

10. Coherent light. The intensity of this light is so great that it can be used to drill holes in metals.

11. The circuit should be the same as the *NPN* circuit in Figure 2-43 with the following changes:
    • Change the symbol to *PNP* type.
    • Reverse the dc supply polarity.

12. The circuit should be the same as the JFET circuit in Figure 2-43 except that the *N*-channel MOSFET symbol is used.

CHAPTER 3

THYRISTORS AND OTHER SWITCHING COMPONENTS

Many applications in industrial electronics require the use of components which are capable of switching very rapidly and, in some cases, of switching high power equipment on and off. *Thyristors* are semiconductor components used specifically for switching. They have only two stable states of operation: *conducting* and *nonconducting*, or *cutoff*.

A mechanical switch that turns lights on and off is directly analogous to a thyristor. The switch, like the thyristor, has no in-between level of operation. Two types of thyristors are discussed in this chapter: those with two terminals, and those with three or more terminals.

Relays are remotely operated electromechanical switches. Despite the progress in semiconductor technology, relays still find extensive use in industrial electronic circuitry. One reason for their popularity is that a single relay can control a large number of circuits. Some of the more popular relays are discussed in this chapter.

Readers with an extensive background in electronics should go through the programmed section and review questions. If the material is familiar, the theory section can be skipped.

THEORY

Tube Forerunners

As with most active semiconductor components, there were tube devices that did the same job as the newer thyristors. Many systems are still in operation that use tube circuits. Although the basic theory of operation of the tube switching components is different from thyristors, and the dc operating voltages are different, the circuitry is very similar. Therefore, the thyristor circuits described in this chapter also apply for the tube equivalents.

NEON LAMPS

A good example of a two-terminal tube device that is still used in many applications is the neon lamp. Figure 3-1 shows the symbol for this lamp and a typical characteristic curve. This curve is important because of its similarity to the ones for thyristors shown later in this chapter.

The voltage across the neon lamp is plotted in a horizontal direction, or *x* axis, and the current through the lamp is plotted in a vertical direction, or *y*

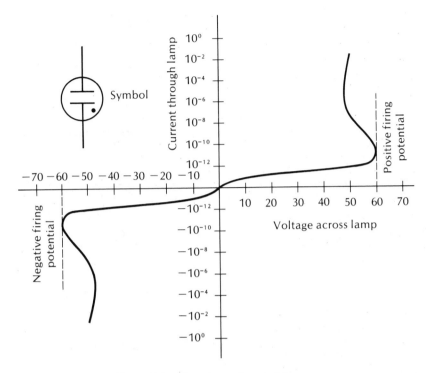

Figure 3-1. Neon lamp characteristic curve

axis. Moving along the voltage axis from the 0 point, it is seen that the current through the lamp is extremely low as long as the voltage is below 60 V. The current is in the order of 1 picoampere (pA), that is,

$$\frac{1}{1,000,000,000,000} \text{ ampere (A)}$$

However, when the voltage across the lamp reaches 60 V, which is called the *firing potential*, the lamp glows and begins to conduct current. Once the firing potential has been reached, the voltage across the lamp decreases to about 50 V.

The voltage across the lamp remains at about 50 V—even though the current through the lamp changes from 10^{-6} to 10^{-2} A. This is a wide range of current values: 0.000001 to 0.01 A, yet the voltage hardly changes. It is this feature that makes it possible to use a neon lamp as a *voltage regulator*.

When the voltage across the lamp is reversed, the same thing occurs. The lamp fires at –60 V, and after firing, the voltage across the lamp decreases to about 50 V. To turn the lamp off it is necessary to reduce the voltage across it to some value below 50 V or above –50 V. In this particular lamp the 50

V point is called the *normal glow voltage*. The current flow in the lamp during normal glow is called the *holding current*. The value of voltage that the lamp voltage must be reduced in order to turn the lamp off is called the *extinguishing voltage*. The extinguishing voltage is always some value below the normal glow voltage.

The particular lamp shown in Figure 3-1 is bilateral; that is, it will conduct current readily in either direction once it is fired. Neon lamps are also made which are unilateral. This is accomplished by shaping the plates so the current will flow more readily from a smaller plate to a larger one.

A very important point to note about the operation just described is that the neon lamp has two stable states of operation. Below 60 V it is considered to be *nonconducting* even though there is a very small current through it. After it reaches the firing potential it goes into a *conducting* state.

Neon Lamp Circuit

Keep in mind the fact that the circuits studied for neon lamps can be duplicated by using the equivalent thyristors.

Figure 3-2 shows a neon oscillator circuit. This type of circuit is sometimes called a *relaxation oscillator*. By definition, a relaxation oscillator produces a square wave, pulse, or sawtooth waveform. The frequency of the waveform is dependent primarily upon the time constant of its *RC* or *RL* circuit, but this frequency may be modified with certain limits by the use of a

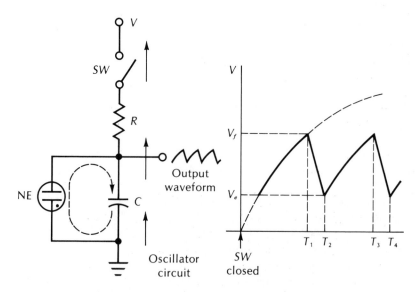

Figure 3-2.   Neon lamp oscillator and related curve

synchronizing signal. In this application, the lamp acts as a switch across the capacitor. When switch *SW* is closed, capacitor *C* charges toward voltage *V*. The *RC* time constant curve is shown with a dashed line on the graphs. The neon lamp is not conducting at this time, so it becomes like an open switch. Electron flow is shown with solid arrows.

When the firing potential (V_f) is reached at time T_1, the neon lamp conducts. It behaves like a closed switch to discharge the capacitor. The discharge electron current path is shown with the dashed arrow on the schematic. On the curve the rapidly discharging capacitor voltage is represented between T_1 and T_2.

At time T_2 the extinguishing potential (V_e) of the lamp is reached. The lamp stops conducting again and acts as an open switch. This allows the capacitor to charge again. The cycle of charge and discharge repeats, and a sawtooth waveform appears at the output.

Figure 3-3 shows two sensing circuits—one for heat and one for light. In the heat sensing circuit a thermistor, which was discussed in Chapter 2, has been added to the neon oscillator. The time constant of the *RC* network now depends upon the thermistor resistance as well as upon *R* and *C*. Assuming that the thermistor has a negative temperature coefficient, the higher the temperature the lower the resistance and the shorter the time required to charge *C* to the firing potential.

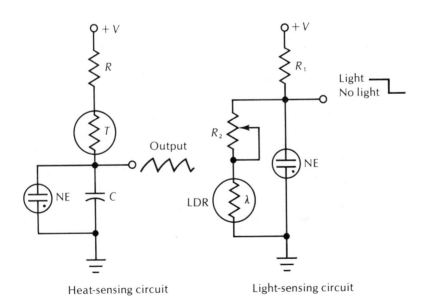

Heat-sensing circuit   Light-sensing circuit

Figure 3-3. Heat and light indicators

Decreasing the time for a cycle is the same as increasing the frequency. Frequency and time are related by the equation:

$$T = \frac{1}{f}$$

where T is the period, or time for one full cycle in seconds, and f is the frequency in hertz.

The overall result is that the frequency of the thermistor circuit is increased when the temperature is increased, and decreased when the temperature is decreased. A measuring system that relates frequency to temperature can be connected to the output. The thermistor can be located at some remote position where it is inconvenient to make a direct temperature measurement.

Since the accuracy of the measurement is questionable, especially over a wide range of temperatures, the circuit is more useful in applications where it is desired to determine if the temperature has risen above a certain value.

In the light-dependent resistor (LDR) circuit of Figure 3-3 the neon lamp is used as an indicator to show whether there is light or darkness at some remote point where the LDR is located. When there is no light the resistance of the LDR is high. The voltage drop across the LDR and R_2 in series is large enough to cause the neon lamp to fire.

When a light strikes the LDR its resistance is lowered. The voltage across the LDR and R_2 in series is no longer high enough to cause the neon lamp to glow. Thus, the lamp is ON if the (remote) light is off, and the lamp is OFF if the (remote) light is on.

Neon lamps are used as triggering devices in three-terminal thyristor circuits and this will be illustrated later in the chapter.

Disadvantages of Neon Lamps

One disadvantage of the neon lamp is that it is susceptible to radiant energy such as heat and light. The firing potential can be reduced by heating the lamp a small amount—such as with body temperature—or by shining a light on it. The added light energy releases free electrons in the neon gas and these electrons contribute to the avalanching that takes place within the lamp.

Another disadvantage of the neon lamp is that it requires a relatively large voltage to fire it. This voltage is too high for many applications in semiconductor technology. The equivalent two-terminal thyristors can be triggered into conduction at a much lower potential.

THYRATRONS

Figure 3-4 shows the characteristic curve and symbol for a *thyratron*, a gas-filled triode tube. Unlike the neon device, it has a heater (not shown in the symbol) that raises the cathode temperature to a point where it can read-

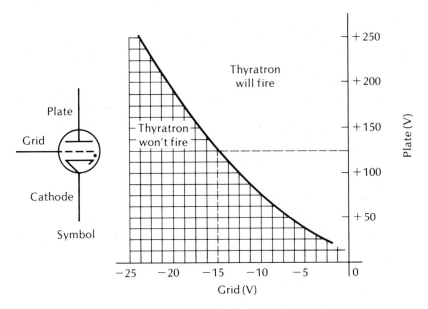

Figure 3-4. Symbol and characteristic curve of a thyratron

ily emit electrons. If the plate is made positive with respect to the cathode, it can attract these electrons. However, no current flows through the thyratron when it is first energized because the grid exercises complete control in the off condition.

To start the plate current flowing through the thyratron, it is necessary to raise the grid potential to a voltage value called the *firing potential*. A very important feature of the thyratron is that once the grid has started plate current flowing, it is no longer able to exercise control over the current flow through the tube.

The graph shows that the firing potential on the grid is directly related to the anode-to-cathode voltage in such a way that a larger anode voltage requires less grid voltage than a low anode voltage for producing anode current. As an example, when the anode voltage is 125 V positive with respect to the cathode, the thyratron will fire at any time that the grid voltage is raised above a negative −15 V. Again, once a thyratron is in conduction, the grid can no longer shut the conduction down.

In order to stop the thyratron from conducting it is necessary to take one of the following actions:

1. Open the plate circuit or cathode circuit.
2. Decrease the plate voltage to zero volts or negative with respect to the cathode.
3. Increase the cathode voltage so that it is equal to or more positive than the plate voltage.

Example of Thyratron Circuit

Figure 3-5 shows a thyratron circuit in which a dc voltage is used to control an ac load voltage. Since the thyratron is unilateral, it will only conduct when the plate is positive with respect to the cathode. Also, the value of plate voltage required for conducting is controlled by the grid voltage.

Figure 3-6 shows waveforms for two different settings of grid voltage (V_g). When the grid voltage is set at -7.5 V, the thyratron fires each time the plate voltage reaches about $+50$ V. When the grid voltage is set at -12.5 V, the thyratron fires each time the plate voltage reaches about 100 V. (These values can be verified on the typical curve of Figure 3-4.)

The overall result is that current is delivered to the lamp load only during the unshaded portions of the waveforms in Figure 3-6. The greater the amount of unshaded waveform, the brighter the lamp. Therefore, the brightness of the lamp can be controlled by varying the dc grid voltage. The value of the circuit lies primarily in the fact that a very low current dc circuit can be used to control a high ac load current.

Two-Terminal Thyristors

DIACS

The solid-state equivalent of the neon lamp is the *diac*, or *three-layer diode*. Its symbols, model, and characteristic curve are shown in Figure 3-7. Both of the symbols shown are in common use. Note that diacs are two-terminal devices. The model shows that this is a three-layer device.

When a voltage is placed across the diac, it will not conduct readily because one of the PN junctions is reverse biased. However, if the voltage is made sufficiently large, the zener point is reached and the diac begins to conduct. In this case the conduction point is called V_{BO} which stands for the

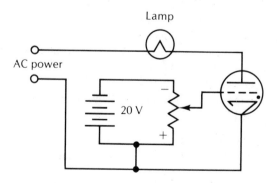

Figure 3-5. Thyratron with dc control

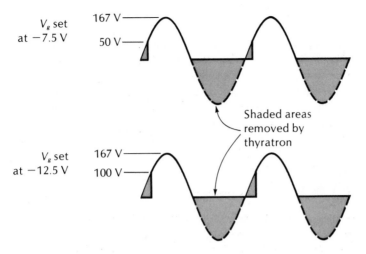

V_g set at -7.5 V

167 V

50 V

Shaded areas removed by thyratron

V_g set at -12.5 V

167 V

100 V

Figure 3-6. Waveforms for the circuit of Figure 3-5

breakover voltage. The breakover voltage has the same meaning for a diac as the firing potential of a neon lamp.

Once the breakover voltage is reached, the voltage across the diac is reduced to the normal operating value, and it is nearly constant for different values of current through it. It is necessary to limit the current through the diac by some means. This constraint is usually accomplished by placing a resistor in series with it.

Once a diac has begun to conduct, it cannot be reduced to the nonconducting state unless the voltage across it is reduced to zero or nearly zero volts.

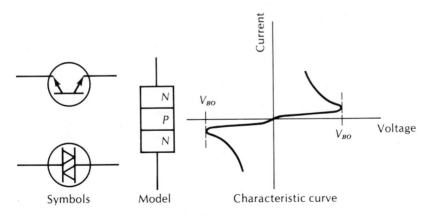

Current

V_{BO}

V_{BO}

Voltage

N

P

N

Symbols

Model

Characteristic curve

Figure 3-7. The diac, or three-layer, diode

SHOCKLEY DIODE

A *four-layer diode* or *Shockley diode* is illustrated in Figure 3-8. The operation of this device is very similar to that of the diac except that it is a unilateral device. The polarity of the voltage required for conduction is shown on the model. With the proper polarity of voltage, the four-layer diode will conduct at the breakover voltage point V_{BO}. However, when the voltage on a four-layer diode is reversed, a very high negative voltage is required to produce a reverse current. The reverse current and reverse voltage conditions are not shown to the same scale on the curve as the forward breakover voltage and current. Usually the zener voltage is much higher numerically than the forward breakover voltage.

In the earlier part of this chapter it was shown that a neon lamp can be used as a relaxation oscillator. As might be expected, both the diac and four-layer diodes can also be used in this type of circuit.

Shockley Diode Oscillator

Figure 3-9 shows a four-layer diode used in place of the neon lamp relaxation oscillator. An important difference between the neon circuit and this Shockley diode circuit is the addition of resistor R_2 in series with the capacitor. It limits the discharge current through the diode. This discharge current would likely exceed the manufacturer's rating for the safe current value.

Thyristors with Three or More Terminals

THE SCR

The silicon-controlled rectifier (SCR) of Figure 3-10 is the solid-state equivalent of a thyratron, but it has several advantages over the tube. It does not

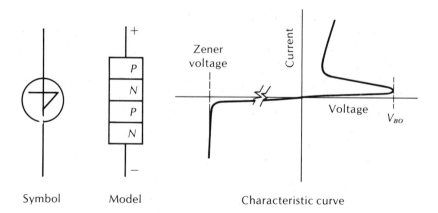

Symbol   Model   Characteristic curve

Figure 3-8.   The four-layer, or Shockley, diode

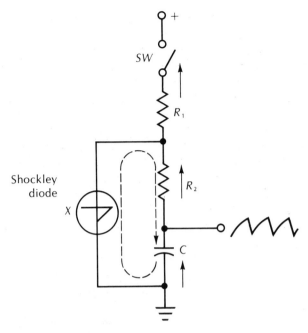

Figure 3-9. Shockley diode relaxation oscillator

require a filament, so its reliability is higher. Also, being a solid-state device, it is more rugged. Another advantage is that the SCR can be operated at lower cathode and anode voltages than most thyratrons.

The symbol for the SCR shows that it has three electrodes: cathode, anode, and gate. When a dc voltage is placed across the SCR with the positive voltage at the anode, the SCR is ready for conduction. However, it cannot conduct until a positive voltage—usually in the form of a pulse—is delivered to the gate. It is more accurate to say that the gate *current* causes the SCR to conduct. The SCR, like the bipolar transistor, is a current-operated device. This is a significant difference between the SCR and the voltage-operated thyratron. However, since gate current is the result of a gate voltage, it is a common practice to refer to the gate pulse voltage.

As with the thyratron, the gate voltage required for conduction is dependent upon the amount of dc voltage on the cathode and anode. Once the gate voltage turns the SCR on, the anode current cannot be controlled with the gate. To stop conduction it is necessary to open the anode or cathode current or reduce the anode-to-cathode voltage to zero volts.

Two different breakover points are illustrated on the characteristic curve of Figure 3-10. As shown, the amount of gate voltage (and hence the amount of gate current) determines the breakover point for the SCR. In this illustra-

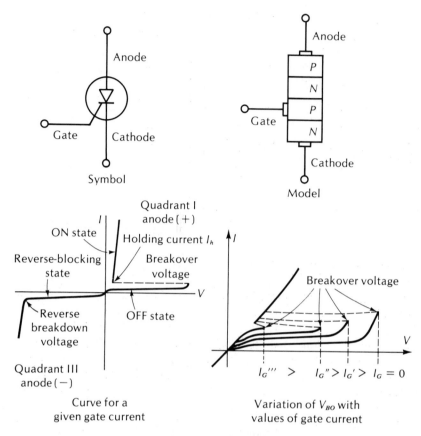

Figure 3-10. The SCR

tion, I''_G is greater than I'_G which is greater than $I_G = 0$. Thus, the higher the pulse amplitude applied to the gate, the lower the breakover point.

When $I_G = 0$, breakover still occurs, but this is an undesired condition. It is related to the temperature of the SCR. Basically, it means that the SCR will break over with a high anode-to-cathode voltage even though there is no gate current. It is not considered a good practice to operate an SCR this way.

Once the SCR fires, its conduction is limited only by the external circuitry, and it is important to be sure that there is a device to limit the current. The model of Figure 3-10 shows that the SCR is similar to the four-layer diode except that the input voltage in one of the layers can be used to trigger the device into operation.

SCR Circuits

Two examples of basic SCR circuits are shown in Figure 3-11. These are *static switching circuits*. As explained in Chapter 1, this term comes from the fact that switching of the load is accomplished without moving mechanical parts. Actually, for the lower circuit the lamp is turning on and off so rapidly that to the eye it appears to be on at all times.

Consider first the static switching circuit with the dc power applied by closing SW_1. No load current can flow because the SCR has not been triggered into conduction. In this illustration a lamp is used to illustrate the load, so the lamp will be OFF. Resistor R is used to limit the amount of gate current when switch SW_2 is closed. When SW_2 is pushed momentarily, a positive pulse is delivered to the gate of the SCR. The SCR is thereby triggered into conduction, and a dc current flows through the load. Once the SCR fires, the lamp will be in the ON condition even though switch SW_2 is opened because the gate has no control over SCR conduction once it occurs. To turn the lamp off it is necessary to open SW_1.

It may seem that not much is accomplished by switching the load current with the SCR. However, it is possible to use a small switch to control a very

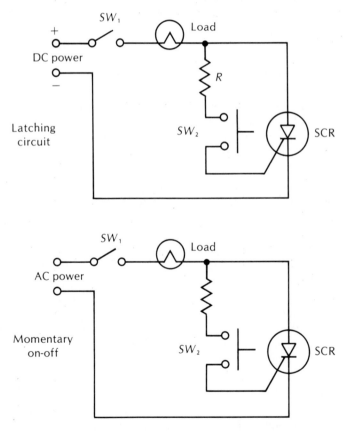

Figure 3-11. Basic SCR circuits for static switching

large current to a load with this circuit. That is the advantage of the SCR circuit over using a switch to directly control the large current.

In the circuit with the ac power applied in Figure 3-11, the operation is not the same. Assume that SW_1 is closed. Momentarily touching the gate switch (SW_2) will turn the SCR on and produce a current flow through the load. Since the applied power is ac, the SCR can only conduct on the half cycle in which the anode is positive with respect to the cathode. The ac power returns to zero volts on each half cycle after closing the switch. However, if the switch is held in the closed position, the lamp will appear to stay ON. As soon as the switch is released, the lamp will go off.

This is a momentary ON/OFF circuit, as opposed to the latching circuit in which case the lamp goes ON and stays ON. The important thing to be learned here is that the SCR can be turned off by using an ac load because of the fact that the ac returns to zero on each half cycle.

The SCR is a controlled *rectifier,* which means that it permits current to flow in one direction only. Manufacturers refer to this type of device as a *reverse blocking thyristor.* Current is flowing for only one-half of each full cycle.

Figure 3-12 shows a timing circuit that uses a diac for triggering an SCR. Switch SW_1 applies dc power to the lamp and SCR. There is no capacitor charging current in the RC circuit with SW_2 open.

When SW_2 is closed, the capacitor charges through R with point a increasing in positive voltage. This positive voltage is not delivered to the gate of the SCR until the breakover voltage of the diac is reached.

The curve shows that the voltage across C follows a normal time constant curve until the diac breakover voltage. This fires the SCR and turns the lamp ON.

The amount of time delay depends upon the time constant of the RC circuit and the value of breakover voltage for the diac.

Summary of SCR Turnoff Methods

Figure 3-13 shows several ways to stop conduction of an SCR.

A manual switch in the anode or cathode circuit will stop conduction of an SCR with a dc load.

The *NPN* transistor between the anode and cathode will stop conduction if two basic requirements are met. First, the positive pulse must saturate the transistor and the emitter-collector forward drop when saturated must be lower than the cathode-anode drop of the SCR. Second, the duration of the turnoff pulse must be sufficiently long to assure that the SCR is stabilized in the OFF condition. This minimum pulse duration is usually about 0.1 millisecond.

The *RLC* turnoff circuits rely upon oscillation to drive the anode negative. When the SCR conducts, it shock excites the inductor-capacitor combination into oscillation. On the first half cycle that the anode side of the coil goes negative, the SCR is turned off.

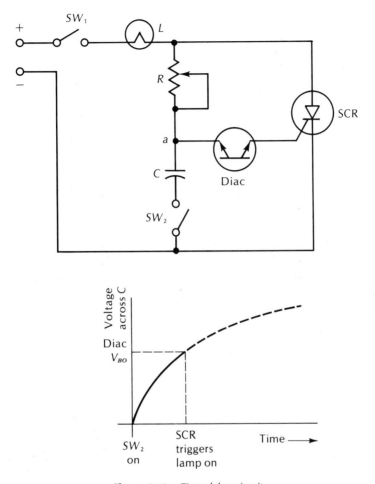

Figure 3-12. Time delay circuit

An SCR in an ac circuit has its anode driven negative on alternate half cycles. This method, which was discussed with the static switch, automatically turns the SCR off for one-half of each full cycle.

TRIACS

Another three-terminal semiconductor thyristor, called a *triac,* is illustrated in Figure 3-14. As shown by the model, it is the equivalent of two SCR's connected back to back. The characteristic curve shows it is bilateral. Once the breakover voltage point is reached on the gate, the triac conducts in either direction. There is no vacuum-tube equivalent for the triac, but two thyratrons could be connected back to back to obtain a similar action. However, they would still have the disadvantages of tube thyratrons.

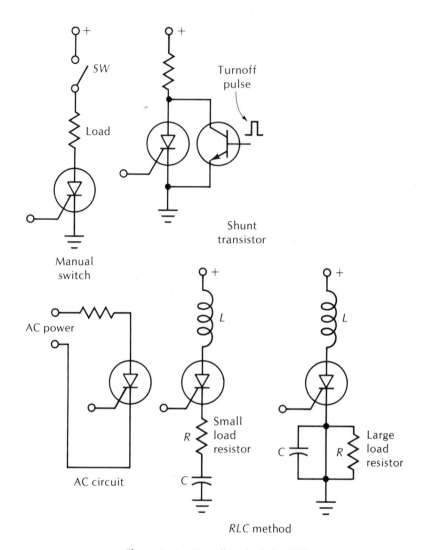

Figure 3-13. Turnoff methods for SCR's

Triac Switch

Figure 3-15 shows a static switching circuit for a triac. It is very similar to the SCR static switching circuit of Figure 3-11. The lamp current waveform shows that the triac conducts on both half cycles. In comparing SCR and triac static switching, the lamp will glow brighter for the triac version because the power is delivered to it on both half cycles of input power.

The time delay circuit of Figure 3-12 can be made with a triac instead of an SCR, but this would be wasteful since the reverse conduction ability is

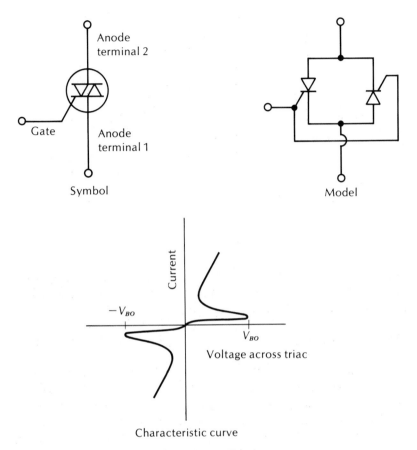

Figure 3-14. The triac

not needed. As a general rule the SCR is used in dc circuits or circuits with dc loads, while the triac is more useful in ac circuits or circuits with ac loads.

Phase Control

The phase relationship between the voltage across a resistor and the voltage across a capacitor can be used to delay the triggering of an SCR or triac. The circuit discussed here will be used with a triac, but an identical circuit can be used with an SCR.

Figure 3-16 shows a series *RC* circuit connected across an ac source of voltage. The voltage across the capacitor lags behind the voltage across the resistor by 90° in this circuit, assuming that the capacitor is perfect— meaning that it has no resistive losses. In practice this condition can be very nearly obtained.

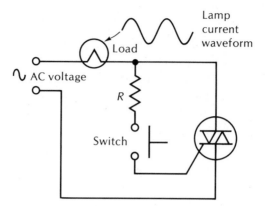

Figure 3-15.   A static switching circuit for a triac

The waveforms of Figure 3-16 show that at any point along the time axis the values of V_C and V_R combine to equal the applied voltage V. Although $V_R = V_C$ in this example, this is not always true. It only happens when the value of resistance is equal to the value of capacitive reactance of the capacitor.

Figure 3-17 shows a better way to represent the voltages in the circuit. They are shown as vectors which are more properly referred to as *phasors*.

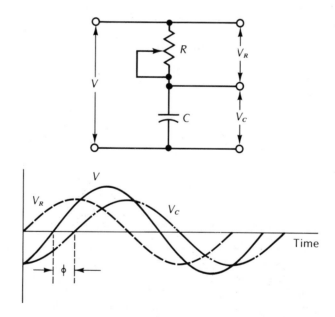

Figure 3-16.   A phase-shift circuit and related waveforms

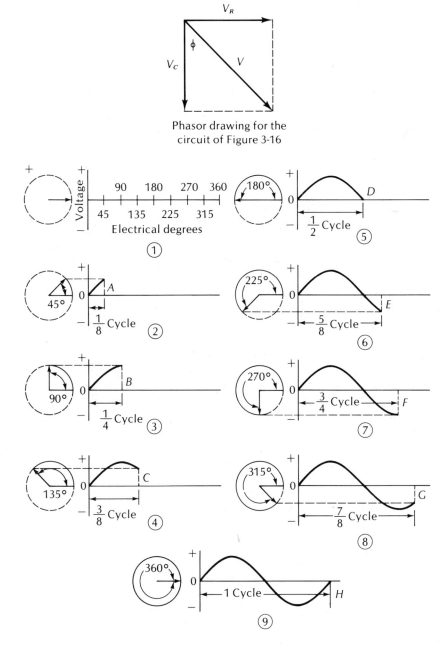

Phasor drawing for the
circuit of Figure 3-16

A rotating phasor generates a sine wave. The angle can be converted to time if the phasor rotates at a constant rate.

Figure 3-17.   Basic phasor diagrams

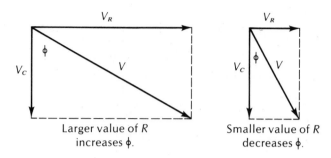

Figure 3-18. Phase angle between V and V_C changing with the resistance of R

A phasor is a line that represents the magnitude (usually the peak value) and the phase angle of a sinusoidal waveform. It is important to understand that the standard direction of rotation for phasors is counterclockwise. Figure 3-17 shows how a rotating phasor generates a sine wave.

If the three phasors of Figure 3-17 are rotated counterclockwise, they will generate the sine waves shown for the circuit of Figure 3-16. Note that V_C lags behind the applied voltage at some angle (ϕ) that is less than 90°.

If the voltage across the resistor is increased (by making R larger) or decreased (by making R smaller), the phase angle between V_C is changed. This condition is shown in Figure 3-18.

Figure 3-19 shows how a phase shift circuit is used to control power to a load through a triac. The phase angle between V_C and V is controlled by varying R as just described. The diac prevents V_C from triggering the triac until the diac breakover voltage is reached. At the instant $V_C = V_{BO}$, a positive voltage is applied to the triac gate and it conducts through the lamp load.

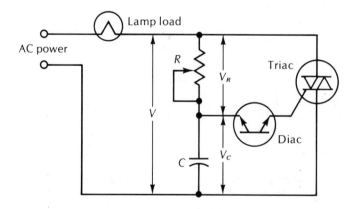

Figure 3-19. Triac circuit with phase control

Figure 3-20 shows two different waveforms for the current through the load. The black portion of these waves represents the portion of the sine wave current that is removed by holding off the conduction of the triac. Note that when R is adjusted to a low resistance value, the sine wave is nearly complete. When R is adjusted to a high value of resistance, the lamp current flows for a much shorter period of time.

The brightness of the lamp is directly related to the wave shape. The greater the amount of time the triac is held off, the lower the rms value of current, and the lower the power $(V \times I)$ delivered to the load.

If an SCR is used for the circuit in place of the triac, one complete half cycle of waveform is removed because an SCR is a reverse blocking triac. Figure 3-21 shows how a diode may be connected across the SCR to permit reverse current flow and increase power to the load.

GATE-CONTROLLED THYRISTORS

There are a number of thyristors in which the gate exercises control over current flow. They are different from SCR's and triacs in which the gate has no control of current once the device is triggered. They have no tube equivalent.

Figure 3-22 shows one example, called a *silicon-controlled switch* (SCS). This switch can be turned either ON or OFF by applying the proper gate voltages.

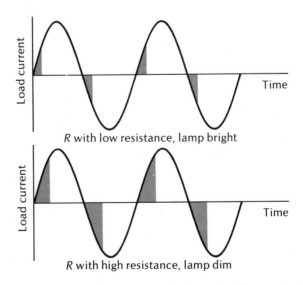

Figure 3-20.   Current waveforms through the lamp for different values of R

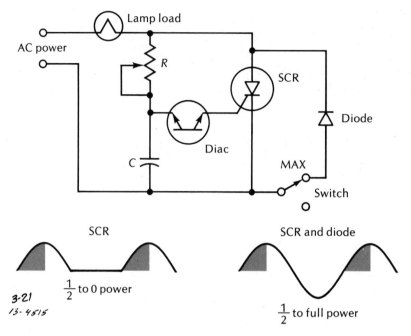

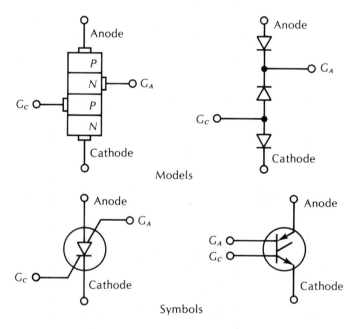

Figure 3-21. When the switch is in the MAX position current flows through load on both half cycles of input voltage

Figure 3-22. Silicon-controlled switch (SCS)

The models and symbols of the SCS show that it is a four-layer device having two gate connections—one to the cathode (G_C) and one to the anode (G_A). The silicon controlled switch is turned on in a manner very similar to the SCR. In fact, it can be used as an SCR with a gate turn-on voltage applied to gate G_C. Generally, silicon-controlled switches do not have the capability of switching the high power available with SCR's, but other than that the action is very similar.

To turn the gate off a positive pulse can be applied to either G_C or G_A, and the connection of circuitry determines which gate is better for turnoff. This is illustrated in Figure 3-23. Note that when the load is in the anode lead the turnoff is best accomplished with a positive pulse applied to gate G_C, but when the load is in the cathode lead the turnoff is best accomplished by applying a positive pulse to gate G_A. In either case a positive pulse turns the SCS off very rapidly, and this type of thyristor can be used to turn the load ON and OFF during one-half cycle of input ac.

Figure 3-24 shows the symbol for a gate-controlled switch. This device is similar in action to an SCR, but has the added feature that it can be turned off with a gate voltage. It is different from the silicon-controlled switch in that there is only one gate terminal for turning anode current ON (with a positive pulse) and OFF (with a negative pulse).

The characteristic curve of Figure 3-24 shows that the amount of negative gate turnoff voltage (V_{GTO}) required depends upon the amount of anode current flowing. For example, it takes a negative gate voltage of 4 V and a negative gate current (I_{GTO}) of 100 mA to turn the device off when the anode current (I_A) is 1 A. For an anode current of 2 A, it requires a negative voltage of 6.5 V and a negative gate current of 140 mA to turn it off.

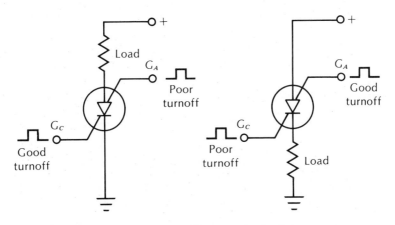

Figure 3-23. Turnoff circuitry for the SCS

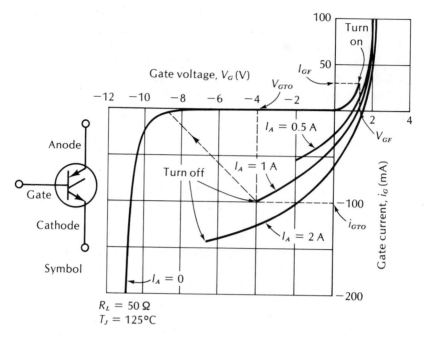

Figure 3-24.   Gate-controlled switch

OPTOELECTRONIC THYRISTORS

Many of the thyristors discussed in this chapter have an optoelectronic equivalent. This means that they can be turned on and off with a light beam.
Two examples are the light-activated SCR and the light-activated SCS. Their
symbols are shown in Figure 3-25. In both cases the light can be used to turn
the devices on, but note that they also have gate input leads. The effect of
the gate is to set the sensitivity of the device and this setting determines in
turn the amount of light required to switch the devices on.

Unijunction Transistors

The *unijunction transistor,* or *UJT,* is a three-terminal switching device.
Like the SCR and triac, it is a breakover device. This is shown by its characteristic curve in Figure 3-26. The symbol shows that the UJT has one emitter
and two base leads (marked B_1 and B_2 on the symbol).

Assume that a voltage is applied across B_1 and B_2. Conduction will not
occur until the emitter voltage reaches a certain predetermined value. Once
the emitter voltage rises to this value, the unijunction conducts and the voltage across it decreases rapidly. This device is not used for controlling high

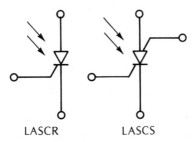

LASCR   LASCS

Figure 3-25.   Optoelectronic thyristors

power loads. It is a very fast switch and convenient for producing pulses that can be used for controlling SCR's, SCS's, and triacs.

Figure 3-27 shows a simple UJT relaxation oscillator circuit that is used to produce both positive and negative pulses. At the instant energy is applied, capacitor C begins to charge through resistors R_1 and R_2. The unijunction transistor cannot conduct at this time because the emitter is below the firing potential.

As the capacitor continues to charge, it will eventually reach the voltage required to start conduction in the UJT. Two things happen when a unijunction conducts. First, the current flow is in the direction of the solid arrow (from base 1 to base 2) shown on the illustration. This current produces a positive-going voltage at base 2 and a negative-going voltage at base 1. Second, once a unijunction fires, conduction occurs between the emitter and base 2. This discharges capacitor C, reducing its voltage to some point below the triggering voltage. The UJT is turned off by the low capacitor voltage, and the next cycle begins.

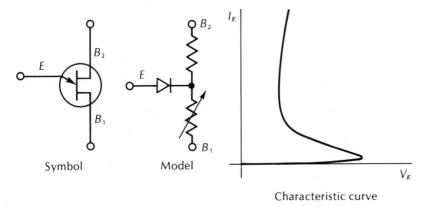

Symbol   Model   Characteristic curve

Figure 3-26.   The unijunction transistor (UJT)

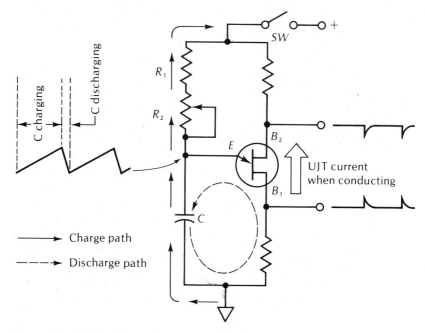

Figure 3-27. UFT oscillator—frequency of an oscillation directly related to time constant of R_1, R_2, and R_3

The UJT only conducts for a brief moment, so the output signals at B_1 and B_2 are pulses. The voltage waveform at the emitter is a sawtooth corresponding to the charge and discharge of capacitor C.

UJT Time Delay Circuit

The use of a UJT in an SCR time delay circuit is shown in Figure 3-28. To start the action, switch SW_1 is closed and capacitor C begins to charge through R_1 and R_2. The SCR is not on at this time because the gate voltage has not received a positive pulse. Therefore, the lamp is off. When the capacitor reaches a sufficiently high voltage to cause the unijunction to conduct, a short-duration positive pulse will occur at base 1. This positive pulse will gate the SCR into conduction, and it will continue to conduct because the applied voltage is dc. The lamp cannot be shut off until switch SW_1 is open. The UJT continues to produce positive pulses at the SCR gate, but they have no effect since the SCR is already conducting.

The amount of delay time in the circuit of Figure 3-28, and the frequency in the circuit of Figure 3-27, both depend upon the RC time constant of R_1, R_2, and capacitor C. However, another factor must be taken into consideration in the unijunction transistor. This factor is called the *intrinsic standoff*

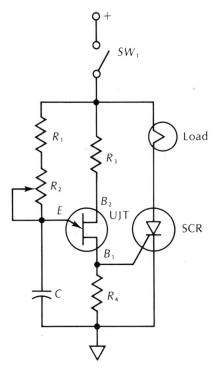

Figure 3-28. Time delay circuit using UJT

ratio. It determines what proportion of the total applied dc voltage must be present at the emitter before the unijunction can fire. Typical values of intrinsic standoff ratio are 0.6 to 0.9. An intrinsic standoff ratio of 0.6 means that the emitter voltage must be six-tenths of the applied dc voltage between B_1 and B_2 before the unijunction can conduct.

PROGRAMMED UNIJUNCTION TRANSISTORS

A disadvantage of the unijunction transistor is that the intrinsic standoff ratio is set by manufacturing, and the circuit must be designed according to its value. Furthermore, the intrinsic standoff ratio for a given type of unijunction may vary (within specified limits) from one device to another.

Both of these undesirable conditions can be eliminated by using a *programmable unijunction transistor (PUT)* of the type shown in Figure 3-29. The difference between this device and the regular UJT is that the gate voltage can now be used to set the intrinsic standoff ratio. Therefore, the gate voltage determines what portion of the applied voltage the gate must reach in order to equal the triggering level.

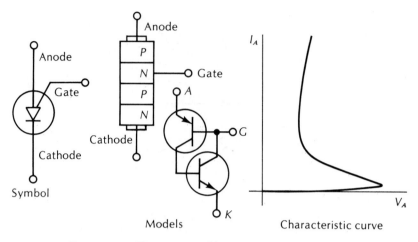

Figure 3-29. The programmable unijunction transistor (PUT)

PUT Oscillator

A simple PUT oscillator circuit is shown in Figure 3-30. Resistors R_2 and R_3 set the gate voltage, and therefore, the intrinsic standoff ratio. Except for this difference, the oscillator circuit is identical to the one shown in Figure 3-27. When energy is first applied, capacitor C begins to charge through resistor R_1. The anode voltage becomes more and more positive as the capacitor charges. When the emitter voltage equals the anode-to-cathode voltage times the intrinsic standoff ratio, the PUT is triggered into conduction. A positive-going pulse is produced at the cathode and a negative-going pulse at the anode.

The mathematical relationship between R_2, R_3, and the intrinsic standoff ratio (η) is

$$\eta = \frac{R_3}{R_2 + R_3}$$

Since the intrinsic standoff ratio sets the voltage required for the PUT to conduct, it also sets the frequency of the oscillator within a narrow range of values.

Table 3-1 is used to summarize the symbolism and characteristics of the two-terminal and three-terminal devices used in industrial electronic control circuits.

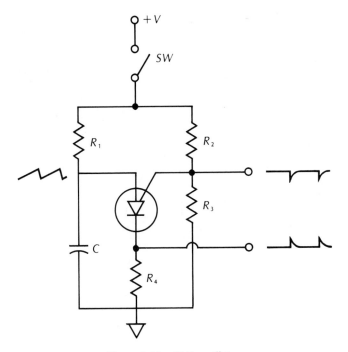

Figure 3-30.   PUT oscillator

Relays

Relays are the electromechanical counterparts of thyristors. They have only two states of operation: ON and OFF. It is interesting to note that when three-terminal and two-terminal thyristors first became popular it was predicted that relays would no longer be needed or used. But, years after the demise of the relay was supposed to have taken place, they are still quite common in many applications.

Relays have some advantages over other devices. One of them is the fact that when the relay operates, its operation can be seen. In terms of troubleshooting, this makes it possible to see what is working and what is not working, and simplifies the problem of finding the trouble. Another advantage relays have is that the *fan-in* and *fan-out* can be very high. In other words, it is possible to control a large number of circuits or have a large number of circuit inputs with one relay.

The disadvantages of relays outnumber their advantages in many applications. Relays are very slow to operate compared to the time for operation of a solid-state thyristor. Also, they are subject to *contact bounce*, meaning that once the relay is energized, switching contacts can bounce several times

TABLE 3-1. Summary of two-terminal and three-terminal devices

Type	Name of Semiconductor Device	Graphical Symbols Used in This Textbook	Main Terminal V-I Characteristic
Diodes	Rectifier diode	A —▷⊢ K	I_A vs V_{A+}
	Breakdown (zener and avalanche) diode Unidirectional	Preferred: A —▷⊢ K / Alternate: A —▷⊢ K	I vs V_{A+}
	Bidirectional (also used for thyrector selenium ac voltage suppressor)	Preferred: / Alternate: B	I vs V
	Tunnel diode	Preferred: A —▷ K / Alternate: A —▷ K	I_A vs V_{A+}
Transistors	PNP	E — C / B	V_{C-}, I_{B1} ... I_{Bn}, $-I_C$
	NPN	E — C / B	$I_{Bn} > I_{B1}$, $+I_C$, I_B, I_{B1}, V_{C+}
Thyristors	Diac (bidirectional diode thyristor)		I vs V
	LAS (light activated switch) light activated reverse blocking diode thyristor	* A —▷⊢ K	I_A vs V_{A+}

A = Anode C = Collector G = Gate
B = Base E = Emitter K = Cathode

Note: circles around graphical symbols are optional except where shown *. In these cases circle denotes an envelope that either encloses a nonaccessible terminal or ties a designator into symbol.

before coming to rest. Contact bounce can cause problems in inductive circuits because it produces rapid on and off voltages. The result is high inductive kickback voltages which cause arcing at the switch contacts and which may destroy other circuit components.

Another disadvantage of relays is the fact that their reliability is often low compared to semiconductor components. However, very highly reliable relays are available that are capable of many millions of operations without breakdown.

Figure 3-31 shows a basic type of relay called the *clapper type,* or *armature relay.* Its operation is very simple. When current flows in the coil, a magnetic field is produced that attracts the armature. The armature thus moves from the position shown to the new (energized) position. Only one set of contacts is shown in the simple illustration, but the inset shows an example of contact pileup. With this arrangement, one motion of the armature can produce many switching functions.

Another type of relay is shown in Figure 3-32. Although this *telephone relay* has a different appearance, its operation is still basically the same. An electric current flows through the coil and the resulting magnetic field attracts the armature. This armature produces a motion which opens or closes the contacts in the pileup. The armature and telephone relays are rugged and can accommodate a large pileup of contacts. A relatively large operating current is needed to produce switching.

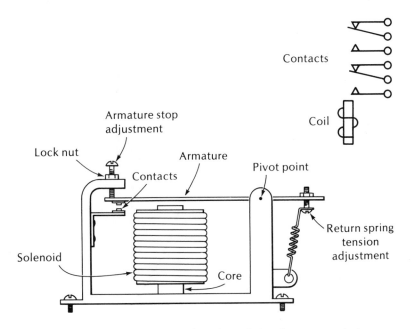

Figure 3-31. An armature relay (also called a clapper-type relay)

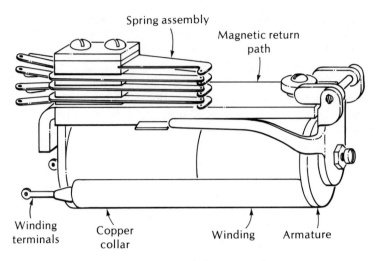

Figure 3-32. A telephone-type relay

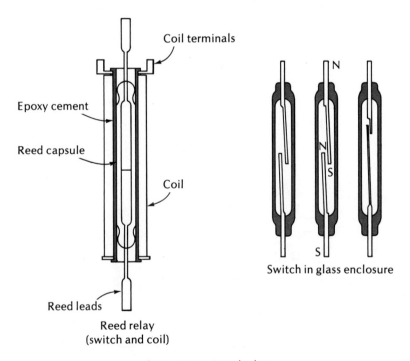

Figure 3-33. A reed relay

A third type of relay called the *reed relay* is shown in Figure 3-33. This relay consists of a glass-enclosed switch contact arrangement with one magnetic movable element and one stationary element. This switch can be closed with a simple magnet. (In another version the switch is normally closed and the magnet causes it to open.) It is the magnetic field that produces the switching action. Enclosing the switch in a bobbin or coil makes it possible to turn it on or off with a coil current.

Reed relay switches are compact and highly reliable, and only a small amount of current is required for their operation. Furthermore, their time for operation is also quite short.

CONTACT FORMS

A relay can cause contacts to either open or close, depending on the form of the relay contacts. These forms are best described by reference to Figure 3-34. In industrial electronic diagrams the simplified symbols are used extensively in drawings, and it is important to memorize these symbols in order to understand the circuitry on schematics. Note that the forms are identified by capital letters *A, B, C, D,* etc. Only four of the forms are shown here, but they are the four most important and most commonly encountered.

With form *A*, the switching contact is normally open (NO). When the relay is energized the contact is closed.

In form *B* the contacts are normally closed (NC). Energizing the relay causes the contacts to open.

In the make-before-break contact arrangement, one of the contacts is normally closed and the other is normally open. When the relay is energized, the NO contacts close and the NC contacts open. This is called a form *C*. It is unique because the NC terminals are opened before the NO terminals are closed.

Make-before-break contacts are called form *D*. Make-before-break contacts are sometimes referred to as *continuity transfer,* which implies that it is possible to switch from one circuit to another circuit without opening the connection to the input.

Figure 3-35 shows a special type of relay called a *stepping switch*. With this device, one impulse causes the stepping switch to go to the first position. The next impulse causes it to go to the second position, and the next to the third position, and so on, until all of the positions have been reached. In many stepping switches, after the last condition is reached a spring returns it to the initial position so that the next pulse brings it to the first position again.

The stepping action is accomplished by a simple relay armature and a small ratchet stop. Each armature action causes the toothed wheel to move one step. This rotates a shaft which turns the switch contacts. Stepping switches can be simulated in semiconductor circuitry.

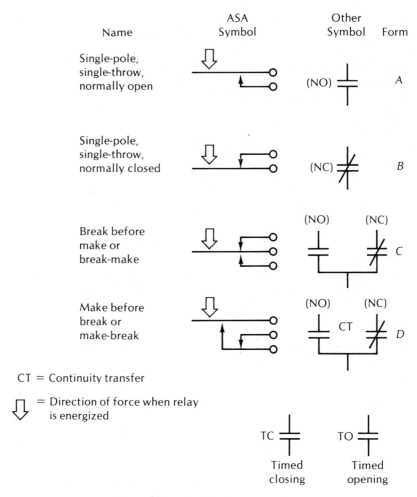

Name	ASA Symbol	Other Symbol	Form

Single-pole, single-throw, normally open — (NO) — A

Single-pole, single-throw, normally closed — (NC) — B

Break before make or break-make — (NO) (NC) — C

Make before break or make-break — (NO) (NC) CT — D

CT = Continuity transfer

⇩ = Direction of force when relay is energized

TC — Timed closing   TO — Timed opening

Figure 3-34. Popular relay contact forms

EXAMPLE OF A RELAY CIRCUIT

Figure 3-36 shows a simple relay circuit with two types of symbols in popular use. Both circuits have identical actions; only the symbols are different. So, it is only necessary to describe the circuit action once.

The applied power for this circuit is battery V. Switch SW supplies current to the relay coil RL. When the switch is closed, current flows through RL and the resulting magnetic field operates the switch contacts. Contact RL_1 is normally open and contact RL_2 is normally closed.

When switch SW is open, lamp L_2 is on. This is because contact RL_2 is normally closed, and the relay is in the deenergized condition or normal condition.

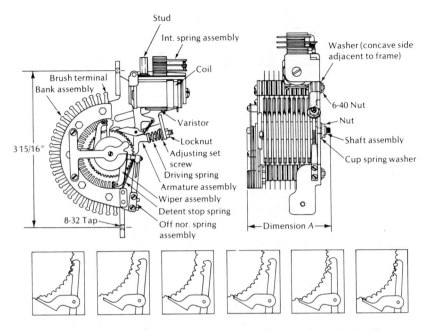

Figure 3-35. Stepping switch with mechanical action shown in detail

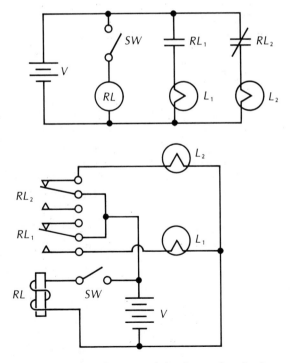

Figure 3-36. Two ways of showing a relay circuit

When switch *SW* is closed, current flows through relay coil *RL* and operates the switch contacts. Now lamp L_1 is on because normally open contact RL_1 is closed in the energized condition. Normally closed contact RL_2 is open and lamp L_2 goes off when the relay is energized.

This circuit diagram should be studied carefully in order to see the relationship between the coils and the contacts, and in order to learn to read both types of schematic symbols.

MAJOR POINTS

1. Thyratrons and neon lamps perform the same functions as semiconductor thyristors. The semiconductors have the advantages of greater reliability, lower operating voltages, and ruggedness.
2. Three-layer and four-layer diodes do not conduct until the voltage across their terminals reaches a minimum breakover value. Once they start to conduct, the voltage across their terminals decreases.
3. A positive pulse at the gate of an SCR will trigger it into conduction. When it is conducting it cannot be turned off with a gate voltage.
4. When connected into an ac circuit, an SCR will turn off each time the ac voltage goes to zero volts. An SCR can also be shut off by connecting a short circuit across its anode and cathode terminals.
5. A triac can be triggered into conduction with a gate pulse, but cannot be turned off with a gate voltage.
6. The SCR and the triac are both three-terminal thyristors. A triac will conduct in two directions, but an SCR is a reverse blocking device.
7. A silicon-controlled switch can be turned on and turned off with positive pulses delivered to the gates. The location of the load (in the anode or cathode circuit) determines the best gate to use for turnoff.
8. A UJT is a three-terminal switching device with one emitter and two bases. It is in the OFF condition until the emitter is raised to the correct positive voltage (with respect to base 1).
9. A programmable unijunction transistor has a cathode, an anode, and an anode gate. The anode gate voltage determines the value of voltage required to trigger the PUT.
10. A relay is an electromagnetic switch. It serves the same function as thyristors when they are used for switching power.

PROGRAMMED REVIEW

(Instructions for using this programmed section are given in Chapter 1.)

The important concepts of this chapter are reviewed here. If you have understood the material, you will progress easily through this section. Do not skip this material, because some additional theory is presented.

---

1. The firing potential of inexpensive neon lamps is lowered when exposed to an increase in heat or light. What effect would an increase in light have on the frequency of the oscillator in Figure 3-2?

   A. It will decrease the frequency. (Go to block 9.)
   B. It will increase the frequency. (Go to block 17.)

---

2. The correct answer to the question in block 17 is B. The difference between the diac and the Shockley diode is that the diac is bilateral and the Shockley diode is unilateral. Placing a diode in series with a diac produces a characteristic curve similar to the one shown in Figure 3-8.
   Here is your next question.
   In the circuit shown the lamp is

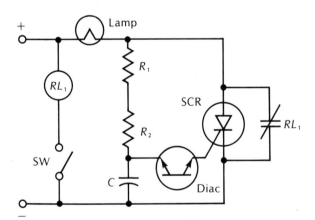

   A. ON. (Go to block 3.)
   B. OFF. (Go to block 14.)

---

3. The correct answer to the question in block 2 is A. When switch SW is open in the circuit of the figure shown above, there is a complete dc path through the lamp and through the NC contact of RL_1.
   Here is your next question.

In the circuit of block 2 capacitor C is
A. charged. (Go to block 22.)
B. discharged. (Go to block 5.)

---

4. Your answer to the question in block 27 is wrong. Read the question again, then go to block 7.

---

5. The correct answer to the question in block 3 is B. The discharge path is through the relay contact, then through R_1 and R_2.
Here is your next question.
Closing switch SW in the circuit of block 2 will

A. not affect the lamp circuit. (Go to block 19.)
B. cause the lamp to go off, then come back on after a period of time. (Go to block 27.)

---

6. Your answer to the question in block 11 is wrong. Read the question again, then go to block 12.

---

7. The correct answer to the question in block 27 is A. Thyratrons and SCR's are both reverse blocking components. Another way of saying this is that they are unilateral. Triacs are bilateral devices.
Here is your next question.
The *NPN* transistor in the circuit shown here controls the rate of capacitor charge. Base bias is obtained with the R_1-R_2-R_3 voltage divider. Adjusting the arm of R_2 to point a should

A. increase the frequency of the oscillator. (Go to block 10.)
B. decrease the frequency of the oscillator. (Go to block 26.)

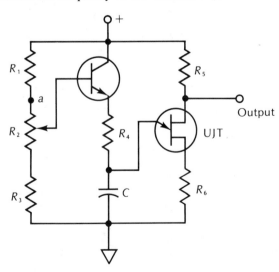

8. Your answer to the question in block 15 is wrong. Read the question again, then go to block 16.

9. Your answer to the question in block 1 is wrong. Read the question again, then go to block 17.

10. The correct answer to the question in block 7 is A. Moving the arm of R_2 toward point a in the circuit shown in block 7 will make the base of the *NPN* transistor more positive. This increases the current through the transistor. The lower opposition to current flow allows the capacitor to charge more quickly, and decreases the time for one cycle of waveform.

Decreasing the time for a cycle of waveform is the same as increasing the frequency.

Here is your next question.

The output signal of the circuit in block 7 should have

A. a sawtooth waveform. (Go to block 18.)
B. a waveform of negative-going pulses. (Go to block 20.)

11. The correct answer to the question in block 20 is B. The voltage on the gate of the UJT sets the intrinsic standoff ratio.

Here is your next question.

Which of the following best describes the waveform of the voltage across resistor R_2 in the circuit of Figure 3-9?

A. Short-duration pulses. (Go to block 12.)
B. Sawtooth. (Go to block 6.)

12. The correct answer to the question in block 11 is A. The waveform is shown below. The voltage across R_2 has the same waveform as the capacitor current. Assuming that the ramp is linear, when the capacitor is charging the current is constant. This current produces the sawtooth ramp. When the capacitor discharges, the current reverses momentarily and a voltage pulse appears across R_2.

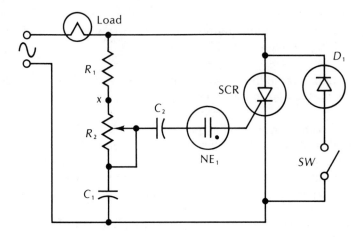

Here is your next question.

Moving the arm of resistor R_2 in the circuit of block 7 toward point x in the circuit shown here should

A. increase the lamp brightness. (Go to block 13.)
B. decrease the lamp brightness. (Go to block 15.)

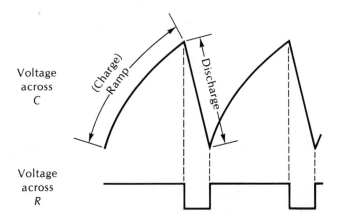

---

13. Your answer to the question in block 12 is wrong. Read the question again, then go to block 15.

---

14. Your answer to the question in block 2 is wrong. Read the question again, then go to block 3.

---

15. The correct answer to the question in block 12 is B. As the arm of the variable resistor in block 12 is moved toward point a, the resistance of the R_1-R_2-C_1 branch is decreased. The circuit becomes more capacitive, and the phase angle of the voltage across C becomes greater with respect to the voltage across the SCR. The overall result is that the SCR fires later.

Here is your next question.

Closing switch *SW* in the circuit for the question in block 12 will cause the lamp brightness to

A. increase. (Go to block 16.)
B. decrease. (Go to block 8.)

---

16. The correct answer to the question in block 15 is A. Diode D_1 in block 12 permits reverse current to flow through the lamp.

Here is your next question.

In the relay circuit shown below the lamp should be ON when

A. the motor is running. (Go to block 21.)
B. the motor is not running. (Go to block 25.)

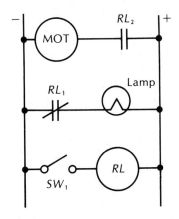

---

17. The correct answer to the question in block 1 is B. When the increase in light lowers the firing potential, the capacitor voltage reaches the firing potential more quickly on each cycle. The time for one cycle is thereby reduced. The relationship between the time for one cycle (*T*) and the frequency (*f*) is given by the equation:

$$f = \frac{1}{T}$$

It is obvious from this equation that a decrease in time *T* will correspond to an increase in frequency *f*.

Here is your next question.

What type of device is imitated (electrically) by placing a diode in series with a diac as shown here?

A. A neon lamp. (Go to block 23.)
B. A Shockley diode. (Go to block 2.)

---

18. Your answer to the question in block 10 is wrong. Read the question again, then go to block 20.

---

19. Your answer to the question in block 5 is wrong. Read the question again, then go to block 27.

---

20. The correct answer to the question in block 10 is B. The waveforms for a UJT are shown in Figure 3-27.
Here is your next question.
Which of the following has an adjustable intrinsic standoff ratio?

A. A UJT. (Go to block 24.)
B. A PUT. (Go to block 11.)

---

21. Your answer to the question in block 16 is wrong. Read the question again, then go to block 25.

---

22. Your answer to the question in block 3 is wrong. Read the question again, then go to block 5.

---

23. Your answer to the question in block 17 is wrong. Read the question again, then go to block 2.

---

24. Your answer to the question in block 20 is wrong. Read the question again, then go to block 11.

---

25. The correct answer to the question in block 16 is B. The motor shown in block 16 is in series with a normally open contact. When switch SW_1 is closed, the relay coil is energized and the motor starts. The lamp is in series with normally closed contacts. The lamp is ON when the relay is not energized.

Here is your next question.

In this chapter it was shown that three-terminal thyristors and thyratron tubes can be used in circuits to control power to a load. For example, if the load is a lamp these circuits control its brightness. Why are these circuits better than those using a variable resistor to control power to a load? (Go to block 28.)

---

26. Your answer to the question in block 7 is wrong. Read the question again, then go to block 10.

---

27. The correct answer to the question in block 5 is B. When the switch in the circuit of block 2 is closed, the short circuit across the SCR is removed and the lamp is turned OFF. This short circuit, which is due to the NC contacts of RL_1, was also across R_1, R_2, and C. The instant the NC contacts open, the capacitor begins to charge. The lamp is OFF until the voltage across C triggers the SCR into conduction. Current through the SCR also flows through the lamp and turns it ON.

Here is your next question.

The thyristor equivalent of a thyratron is the

A. SCR. (Go to block 7.)
B. triac. (Go to block 4.)

---

28. The thyristor controls are more efficient because they deliver power to the load in bursts. They do not conduct during periods between these bursts. Variable resistor controls dissipate more heat, and they conduct at all times.

You have now completed the programmed section.

SELF-TEST WITH ANSWERS

(Answers to this test are given at the end of the chapter.)

1. An advantage of using a triac to control power to a load, compared to using a variable resistor is (a) greater efficiency of the triac circuit; (b) less danger of fire.

2. The four connections to an SCS are _____, _____, _____, and _____.

3. Which of the devices studied in this chapter is called a reverse blocking controlled thyristor?

4. Which of the following devices can be turned off with a gate pulse: SCR, triac, SCS, diac?

5. The tube that does the same basic job as an SCR is called a _____ .

6. The semiconductor counterpart of a neon lamp is called a _____ .

7. The Shockley diode is also known as the _____ .

8. A nonsinusoidal oscillator that depends upon the charge and discharge of a capacitor for its operation is called a _____ oscillator.

9. The frequency of a unijunction transistor oscillator is determined by _____ .

10. A PUT and a UJT have similar operating characteristics. The triggering potential is determined by the intrinsic standoff ratio. In which of these devices is the intrinsic standoff ratio set by the manufacturing process?

Answers to Self-Test

1. The correct answer is (a). The triac uses very little power, and normally conducts for only part of a cycle. A variable resistor dissipates power at all times when controlling load power.
2. Anode, cathode, anode gate (G_A), and cathode gate (G_C).
3. The SCR.
4. SCS.
5. Thyratron.
6. Diac.
7. Four-layer diode.
8. Relaxation.
9. RC time constant of the emitter circuit.
10. UJT.

CHAPTER 4

TRANSDUCERS

In Chapter 1 the use of a transducer in a closed-loop system was discussed. It performs the task of sensing some quantity to be measured or controlled.

The terms *sensor* and *transducer* are used to mean the same thing in industrial electronics. A unit that permits the energy of one system to control the energy of another system is called a *transducer*. A well-known example is a loudspeaker in which the electrical energy input controls the sound energy output.

A transducer is sometimes defined as *a component that converts energy from one form to another*. Actually, this definition is not technically accurate, but it is a convenient way to think of these devices. So, you can think of a loudspeaker as being a transducer that converts electrical energy to mechanical energy for producing sounds. You can think of a microphone as being a transducer that converts mechanical energy to electrical energy.

Transducers are used extensively in making measurements. For that application a transducer is defined as *a unit that converts a measurand into an electrical quantity*. A *measurand* is something that is being measured, such as the amount of fuel in a tank, or the distance from one point to another. This definition of a transducer is better because it applies to cases where the transducer is being used to sense energy in one form or another; it can also be applied to cases where the measurand is not energy.

There are two kinds of transducers: *active* and *passive*. Active transducers generate a voltage that is related to (actually, controlled by) some form of energy. For example, some types of *accelerometers* generate a voltage that is proportional to acceleration. The voltage causes a current to flow in a circuit. For convenience, it can be said that the accelerometer converts mechanical energy to electrical energy.

Passive transducers undergo a change in a circuit characteristic such as resistance, inductance, or capacitance when operated upon by a form of energy. An example is a light-dependent resistor (LDR) in which light energy controls the amount of resistance in an electric circuit. Again, for convenience, the system is said to convert light energy into electrical energy. Both active and passive transducers are used extensively in industrial electronic systems.

A good place to start the study of the operation of active transducers is to review the six basic methods of generating a voltage. All of these methods have applications in industrial electronics, although not all are used directly as transducers.

The study of passive transducers will be introduced by first reviewing the methods of varying resistance, capacitance, and inductance. These are the main properties utilized in the construction of passive transducers.

Once the operation and variety of applications for the transducers are understood, the next step is to put them in circuits. That will be the concluding part of this chapter.

THEORY

Six Methods of Generating Voltage

There are six methods of generating a voltage: *electrostatic, or frictional; chemical; piezoelectric, or pressure; photoelectric, or light; thermoelectric, or heat; and electromagnetic.*

THE ELECTROSTATIC METHOD

Static electricity is generated by rubbing two insulating materials together. An easy way of thinking of this kind of generator is that rubbing the material scrapes charge carriers off one surface and deposits them on the other surface. Since the materials are insulators, the charges are isolated and cannot move. Classic examples of the electrostatic method of generating electricity are rubbing glass with a piece of silk, moving a comb through hair, or scuffing shoes across a carpet. In each case a rubbing or frictional action is required to produce the electric charge.

One important characteristic of static generators is the very large amount of voltage that can be produced by this method. When a person walks across a carpet and reaches for a door knob a spark may jump between his fingers and the metal knob. To produce such a spark in air requires many hundreds of volts, but the person is not electrocuted because the amount of current is very small.

PROBLEM: How much voltage is required to produce a spark through the air between two electrodes 0.16 cm apart?

This type of problem is not easily solved. Many factors, such as the moisture in the air and the shape of the electrodes, can cause the answers to vary. The approximate value of voltage is given by the equation:

$$V = A + Bpl$$

where A and B = constants equal to 1700 and 39 respectively in dry air
p = atmospheric pressure of the air measured in millimeters of mercury. A typical value is 760 mm.
l = distance between the electrodes in centimeters. In this problem $l = 0.16$ cm (about 1/16 inch).

Therefore, $V = 1700 + (39 \times 760 \times 0.16)$
$= 1700 + 4742.2$
$= 6442.4$ volts   (Answer)

The large voltages produced by static electricity can be a nuisance. Technicians sitting at a bench may generate static charges as they move around on the seat. If they pick up a MOSFET or a small-signal diode, the static voltages can immediately destroy the component. In fact, the manufacturers ship these components in metal foil or conductive foam to prevent accidental destruction by electric fields or static charges.

Measuring probes of older types of test instruments and soldering guns caused considerable damage to semiconductor components because of static charges. Newer designs have nearly eliminated the problem.

Another example of the difficulty that static electricity presents is in the field of electronographic printing. This is an extremely rapid printing process in which electronics is employed to form the letters on paper. The greatest limitation to the speed of electronographic printing is in the ability to remove the static charges from the paper. Many techniques have been employed, including wires and foil rubbing across the paper. Another method tried was atomic radiation directed at the paper to neutralize the charge. However, none of these methods have proven to be successful, and static electricity is still a great problem in industrial electronic systems.

Static electricity is seldom employed in generating useful power in electronic systems. The reason is that friction is needed to produce the electricity, and this friction makes the generators inefficient. Despite this drawback, there are a few isolated applications. For example, in x-ray systems extremely high voltages (millions of volts) are needed for accelerating the electrons. There is practically no current required, and static generators are a convenient method of obtaining the required high voltage values. A popular way of generating such high voltages is the Van de Graaff generator illustrated in Figure 4-1.

While the electrostatic method of generating voltage is not used for transducers, electrostatic charges can create problems in industrial electronic systems.

THE CHEMICAL METHOD

The chemical method of generating a voltage is not used extensively for making transducers, but this method has extensive application in industrial electronics. Batteries are used in portable equipment for making tests and measurements. They are also used for standard reference voltages. Electroplating is an electrochemical process indirectly related to the chemical method of generating voltages.

Because of the importance of the chemical method for generating voltage, it will be discussed here. Any time that two dissimilar conductors are immersed in an alkali or acid solution, a voltage is generated between the two

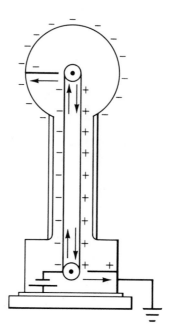

Figure 4-1. Van de Graff generator

metals. A simple experiment, illustrated in Figure 4-2, can be used to demonstrate this fact. A piece of coathanger wire and a piece of copper wire are inserted into an acid-type fruit such as a lemon or grapefruit. This forms a simple voltaic cell. The amount of voltage produced is quite small, but the principle is the same as that employed in commercial cells and batteries.

The voltage source obtained by inserting two dissimilar conductors in an alkali or acid solution is called a *cell*. The amount of voltage generated across the terminals of a cell depends upon the kinds of conductors (or semiconductors) used. They are called the *electrodes*. The amount of voltage also depends upon the type of chemical, which is called the *electrolyte*. The physical condition of the electrolyte and of the conductors is a third factor in determining the amount of cell voltage.

All cells have a certain amount of internal resistance. As shown in Figure 4-3, the terminal voltage of a battery is reduced when a load current flows through the internal resistance. The greater the internal resistance the lower the terminal voltage. As a battery ages, its internal resistance increases. The condition of a battery can be indirectly determined by measuring its terminal voltage while it is delivering current to a load. When the voltage is below a limited value set by the manufacturer, the battery is said to be in a discharged condition.

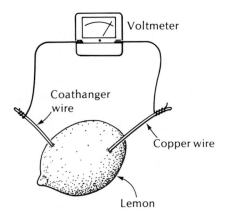

Figure 4-2. A simple battery

Cells may be connected in series to obtain a larger voltage. A *battery* is simply a combination of cells in series and parallel to obtain the desired output voltage and current characteristics.

Cells are connected in combinations to obtain a larger current-delivering ability, or a larger voltage. When they are connected in parallel the voltage rating of the cells must be equal. The total amount of current that a battery can deliver is increased by the parallel connection. However, if there is a difference in the terminal voltage between cells, then when they are connected in parallel those cells with the larger voltage will deliver current to the cells with the lower voltage.

All cells (and batteries) can be divided into two classes: *primary* and *secondary.* A primary cell cannot be recharged and a secondary cell can be

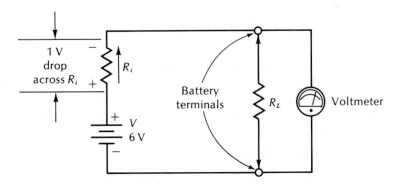

Figure 4-3. Internal resistance causes a reduction in terminal voltage

recharged. It is important to understand what happens when a battery is recharged. A reverse current through the battery reestablishes the chemical properties of the electrolyte and electrodes. A lead-acid battery, like the kind used in cars, is an example of a secondary cell.

The carbon-zinc cells, such as those used in flashlights, cannot be recharged. They are examples of primary cells. Despite this fact, "chargers" are being sold that supposedly recharge carbon-zinc cells. Actually, they do not recharge the cell because this is impossible. What they do is *rejuvenate* the cell. Rejuvenation is a method of improving the electrolyte by eliminating some of the gaseous impurities from it. Heating a carbon-zinc cell will rejuvenate it. This is actually what the so-called rechargers for primary cells accomplish.

Some small cells such as silver cadmium and mercury cells can be recharged. They are, in every sense of the word, secondary cells that can be recharged many times.

A high-impedance voltmeter connected across a battery will not draw sufficient current through the internal resistance to cause a voltage drop. For this reason, a battery in a discharged condition will have almost the full terminal voltage that it has when charged. To determine the condition of a battery with a voltmeter, its terminal voltage must be measured *under load,* that is, when it is delivering a reasonable amount of current through a resistor.

THE PIEZOELECTRIC METHOD

Certain crystalline materials, such as quartz, will generate a voltage across a surface when their shape is distorted. These are called *piezoelectric materials*. They are used extensively in electronic systems.

A typical feature of these crystals is that they also have a property called *electrostriction*. This is the reverse of the piezoelectric effect. It means that an electric field will alter the shape of the crystal. When describing the use of piezoelectric components in electronic circuitry, it is common practice to combine the effects of piezoelectric generated voltage and electrostriction into one single property. For that reason crystals which control the frequency of transmitters employing both the piezoelectric effect and electrostriction are generally referred to as *piezoelectric devices*.

One application is in the ceramic crystal pickup used for disk record playback. The principle is illustrated in Figure 4-4. The piezoelectric crystal material flexes as the stylus moves back and forth in the groove of the record. The flexing causes the crystal to generate a voltage that is proportional to the amount of motion. This voltage is amplified as the output voltage in the system. A form of disk recording is used for memory in industrial systems.

THE PHOTOELECTRIC METHOD

There are several important photoelectric effects of importance in electronics.

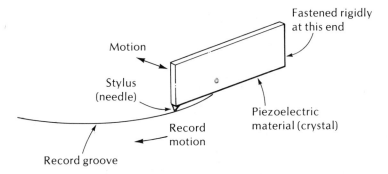

Figure 4-4. One application of a piezoelectic transducer

The *photoconductive effect* refers to the fact that some materials change their electrical resistance when exposed to light. Light-sensitive resistors (LSR's) are made on the basis of this property.

Some materials emit electrons when exposed to light. This is the *photoemissive effect*.

When the junction of two dissimilar materials is exposed to light, a voltage may be produced. This is called the *photovoltaic effect*. The amount of voltage is a function of the amount of light present.

Light-emitting diodes and many other optoelectronic components operate on the basis of two important principles:

1. When an electron passes from a higher energy level to a lower one it gives off energy. The energy may be in the form of visible light, and the LED operates on this principle.
2. When light passes through a semiconductor material, charge carriers are released. The result is a lowering of the material resistance.

Optoelectronic components are used for displays in industrial systems. (Examples will be discussed in Chapter 8.) Among other uses are counters on assembly lines, intrusion alarms, and position transducers.

THE THERMOELECTRIC METHOD

Two important effects relate electricity to temperature: the *Seebeck effect* and the *Peltier effect*. Whenever the junction of two dissimilar materials is heated, a voltage is created across their ends. This is the Seebeck effect, and it is illustrated in Figure 4-5. A well-known application of the Seebeck effect is in the thermocouple. It produces a voltage proportional to the amount of heat present at the junction. The thermocouple ammeter of Figure 4-6 works on this principle. Current flowing through the wire heats the thermocouple

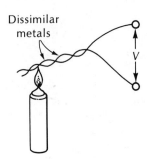

Figure 4-5.   Voltage generated when junction of two metals is heated

and produces a voltage that deflects the meter. The amount of meter deflection is proportional to the amount of current flowing through it, which in turn is directly related to the amount of heat.

Thermocouple ammeters are very useful for measuring high rf frequencies where ordinary meters cannot function because of their inductive and capacitive properties.

If the ends of the heated junction shown in Figure 4-5 are twisted together, heat will be absorbed at the second junction. See Figure 4-7. In other words, heat applied to point *a* produces a cooling at point *b*. At one time this amount of cooling was only a lab curiosity, but new developments in technology, especially in metallurgy, have made it possible to produce a measurable amount of cooling. In fact, small refrigerators have been made by this process.

If the current is reversed through junction *b,* then instead of absorbing heat, it will radiate heat. This is also part of the Peltier effect.

The thermoelectric method of generating voltage is employed in transducers of instruments designed for measuring temperature.

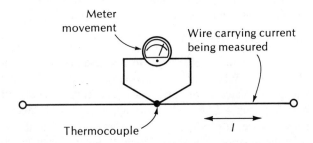

Figure 4-6.   Principle of the thermocouple ammeter

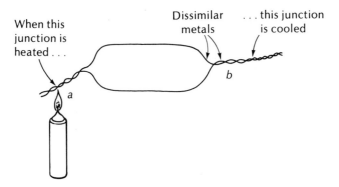

When this junction is heated . . .

Dissimilar metals

. . . this junction is cooled

a

b

Figure 4-7. When one junction is heated the other is cooled

THE ELECTROMECHANICAL METHOD

Whenever there is relative motion between a magnetic field and a conductor, a voltage is produced across the ends of the conductor. This is a statement of *Faraday's law of induction*. It does not matter if the inductor is moving through a magnetic field or if the magnetic field is cutting across the conductor. In fact, both the conductor and the magnetic field can be moving. The only requirement is that they are moving relative to one another. In other words, a person standing on the conductor would be experiencing a change in magnetic flux even though the field and conductor are both moving.

The amount of voltage produced in a conductor or a number of conductors with relation to the magnetic field is expressed by the equation:

$$V \propto N \, \frac{d\phi}{dt}$$

This mathematical statement simply means that the amount of voltage *(V)* induced increases as the number *(N)* of conductors moving through the field increases. The expression *dφ/dt* means that the amount of voltage induced is proportional to the rate at which the flux is cutting across the conductor. The symbol ∝ means "is proportional to." It means that *V* will be increased by increasing *N* or by increasing *dφ/dt*.

The expression for Faraday's law can be made into an equation by introducing a constant of proportionality. This constant is needed for making the units constant.

So far only the amount of induced voltage has been discussed. If the conductor is connected to a closed circuit, then current will flow. The magnetic field produced by this current is always such that it opposes the

relative motion between the conductor and the magnetic field. This is called
Lenz's law.

The principle of Lenz's law can be demonstrated by the simple experiment illustrated in Figure 4-8. When the generator is turned without a load
(that is, without current flowing), it turns quite easily because there is no
opposition other than friction on its bearings. When a resistor is connected
across the output terminals of the generator so that a current flows, then the
generator is more difficult to turn. The reason for this is that the current
flowing through the coils of the generator produces a magnetic field that
opposes the motion that produces the generated voltage.

Lenz's law is the reason for the negative sign normally accompanying the
mathematical statement of Faraday's law. The law is generally written then
as:

$$V = -N \frac{d\phi}{dt}$$

Motors and generators both make use of Faraday's law. Analog current
measuring instruments and accelerometers (transducers that produce a voltage proportional to acceleration) are also applications for the electromechanical method of generating voltage.

Methods of Varying Resistance, Inductance, and Capacitance

Passive transducers permit the energy being sensed or the measurand to control either the resistance, inductance, or capacitance of a circuit. The study

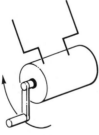

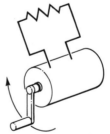

The generator
turns easily
when there is
no load current.

When a load current
is flowing, the
generator is
more difficult
to turn.

Figure 4-8. Effect of load current on a generator

of passive transducers will be started by reviewing the factors that control the amount of resistance, capacitance, or inductance of components.

FACTORS THAT AFFECT RESISTANCE

Figure 4-9 summarizes the factors that control the amount of resistance of a resistor or a conductor. The resistance of a wire is obtained by using the following equation:

$$R = \rho \, \frac{l}{A}$$

where R = resistance
ρ = resistivity of the wire material
l = length
A = cross-sectional area

The dimensions of ρ, l, and A are chosen so that the resistance (R) is in ohms. The equation indicates that changing the area or length of a given wire will change its resistance. The resistance is also dependent upon the type of material. This is reflected by the resistivity (ρ) of the material that the wire is made of.

The equation gives the resistance of a wire at a certain temperature— usually 20°C. When the temperature coefficient (α_t) and the change in temperature (Δt) are known, then the change in resistance (ΔR) can be determined from the following equation:

$$\Delta R = \alpha_t R_0 \Delta t$$

where R_0 is the initial resistance before the temperature change.

A resistor may have a positive temperature coefficient, which means that its resistance increases with an increase in temperature. A negative temperature coefficient means that the resistance decreases with an increase in temperature. For most resistors operated within their power rating, the amount of change in resistance is relatively small over a wide range of temperatures.

However, special resistors called *thermistors* are manufactured which are very sensitive to changes in temperature. With a thermistor, a small change in temperature causes a relatively large change in thermistor resistance. Thermistors are made with both negative and positive temperature coefficients.

Materials used in electronics can be divided into categories: insulators, semiconductors, and conductors. Among the conductors, silver will pass electricity with the least resistance. Copper has the least resistance of any material with a reasonable cost. Copper is a better conductor than aluminum, but aluminum has been used in many wiring circuits because of its light weight and low cost.

The resistance of a conductor is inversely proportional to its cross-sectional area. The area of a conductor is proportional to the square of the diameter, so it follows that the resistance of the conductor varies inversely with the square of the diameter. This means that if you double the diameter of a conductor, its resistance is only one-fourth as great.

Finally, the resistance of a conductor is directly proportional to its length. Doubling the length of a wire will double its resistance, assuming that the cross-sectional area is constant along the complete length and that the temperature is constant along its length.

All of the methods of varying resistance as illustrated in Figure 4-9 can be utilized in making passive transducers. There are other factors that indirectly control the resistance of a material. An example is illustrated in the Hall effect device of Figure 4-10. The magnetic flux passes through a semiconductor material at a right angle and causes the charge carrier (usually electrons) to move away from the center of the material. Effectively, this decreases the cross-sectional area over which the charge carriers can move, and therefore the resistance of the material. Hall devices are used to sense magnetic fields.

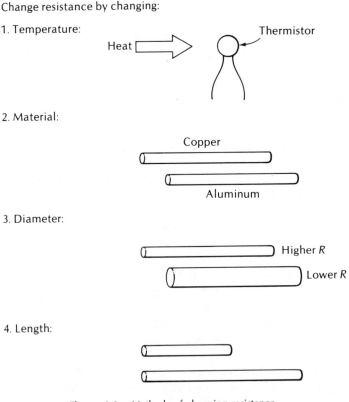

Change resistance by changing:

1. Temperature:

    Heat Thermistor

2. Material:

    Copper

    Aluminum

3. Diameter:

    Higher R

    Lower R

4. Length:

Figure 4-9. Methods of changing resistance

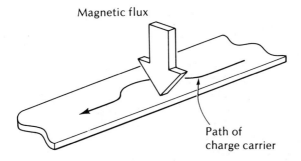

Figure 4-10. The Hall effect

Another factor that may be used to control the resistance of a material is light. Certain materials have a resistance that is indirectly related to the amount of light falling on them. This example is shown in Figure 4-11. Such devices are sometimes erroneously referred to as *photocells,* but in reality they are light-dependent resistors (LDR's). (To be a true photocell it must generate a voltage that is proportional to the amount of light.)

FACTORS THAT AFFECT INDUCTANCE

The most important factors that affect inductance are shown in Figure 4-12.

The inductance is directly affected by the type of material and the amount of magnetic material used for the core. Soft iron and ferrite materials have lower reluctance than air. In other words, they are more magnetic than air. Closing the shell in the inductance or moving the ferrite core into the coil are both ways of increasing the inductance and both are illustrated in Figure 4-12.

The equation for the inductance of a single-layer coil is

$$L = \frac{(Nr)^2}{9r + 10l}$$

where N is the number of turns, r is the radius of the coil, and l is the length of the coil.

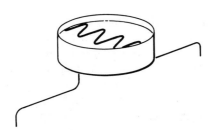

Figure 4-11. Light-dependent resistor

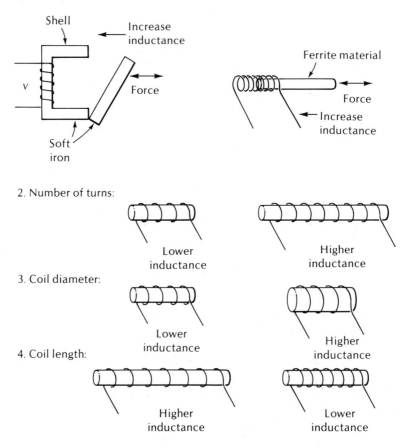

Figure 4-12.   Methods of changing inductance

   This equation presumes that the coil is wound on a nonmagnetic material (relative permeability equal to 1.0). The units for r and l are chosen so as to give L in microhenries (μH).

   There are many equations for inductance for coils of various sizes and shapes. This equation shows how inductance can be varied in the design of transducers. It shows that inductance is directly proportional to the number of turns of the coil, and to the radius, and therefore to the cross-sectional area of the core material.

   It should be remembered that inductors work from the principle of induced voltages. Each turn of the inductor carries an expanding and contracting current that cuts across adjacent turns. This property accounts for the countervoltage established in a coil. Winding the turns closer together improves the flux linkages between windings, and therefore increases the inductance.

The amount of inductance is also affected by the temperature and the shape of the coil, but these factors are not used extensively in making transducers.

Transformers also work on the principle of induced voltages and they are sometimes used as transducers. Figure 4-13 shows the most important factors utilized in varying coupling between the primary and the secondary. Although the symbols for transformers show the windings separately, this is not a typical method of making transformers. Normally, they are wound on the same core and the windings are piled on top of each other. However, in the case of transducers, separate cores may be used so that the windings can be moved with relation to one another.

The secondary voltage can be affected by turning the secondary at an angle to the primary. Another way of affecting the secondary voltage is to move the secondary winding closer or farther away from the primary. Both methods are shown in Figure 4-13.

FACTORS THAT AFFECT CAPACITANCE

The equation for capacitance of a parallel plate capacitor is

$$C = \frac{kA}{d}$$

where C is the capacitance, A is the area of plates opposite to each other, d

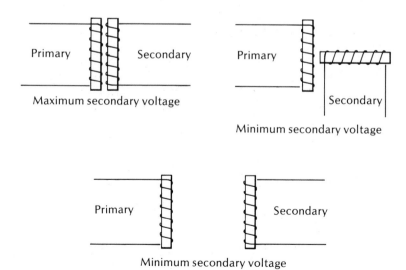

Figure 4-13. Methods of changing transformer secondary voltage

is the distance between the plates, and k is a constant that is determined by the material in the dielectric.

This equation indicates that the capacitance of a capacitor can be varied by changing the type of dielectric material, the area of the plates facing one another, or the distance between the plates. All three of these methods of varying capacitance are utilized in the making of passive transducers (see Figure 4-14).

Capacitors are also affected by temperature changes. In some circuits this can be a serious problem, and ceramic capacitors may be marked to show their temperature coefficient. As an example a capacitor marked N750 has a *negative* temperature coefficient. It will *decrease* by 750 parts per million for an increase of 1°C. A capacitor marked NP0 has a negative-positive temperature coefficient of zero. In other words, the capacitance value does not change with a change in temperature.

Capacitive transducers and inductive transducers require ac voltages to be applied to the output terminal. Changing the capacitance or inductance changes the reactance, and therefore changes the amount of ac current that can flow in the circuit. The current is measured, and its value reflects the magnitude of the quantity being sensed. For resistive transducers either a dc or an ac voltage may be applied to make the transducer operable.

Change capacitance by changing:

1. Dielectric

Dielectric material

2. Area of facing plates

Air

3. Distance between plates

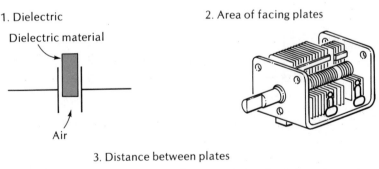

Capacitor plates

Insulation

Plastic screw

Figure 4-14. Methods of changing capacitance

Examples of Transducers

It would not be possible in one volume to include all of the different methods of utilizing generated voltages and passive parameters for making transducers. However, a few examples will illustrate the importance of the chapter material just studied.

ACTIVE TRANSDUCERS

Figure 4-15 shows four examples of active transducers. An *accelerometer* produces an output that is proportional to the amount of acceleration that the unit is subjected to. In the particular case shown in Figure 4-15, a heavy weight is mechanically connected to a piezoelectric crystal and held in place by a heavy spring. When the unit is accelerated, the weight pushes against the piezoelectric transducer. The transducer produces a voltage that is directly related to the amount of acceleration.

The *tachometer* shown in Figure 4-15 is one of many types in use. In this particular application magnets are attached to the outside of a rotating shaft. A soft-iron core with a coil wrapped around it is placed near the shaft. The moving magnets induce a flux in the soft-iron core, and the induced flux produces a voltage across the coil. The amount of voltage is related to the rate of flux change. (You will remember that this is in accordance with Faraday's law: $v = N(d\phi/dt)$.) The faster the shaft turns, the more quickly the magnets move past the soft-iron core and the larger the voltage induced. The induced voltage is processed and delivered to a display so that the angular velocity—usually in revolutions per minute (r/min)—can be read directly.

The event counter uses a photocell that normally receives light from a lamp on the assembly line. The purpose of this arrangement is to count the boxes on a conveyor belt as they move past the photocell. The boxes interrupt the light from the lamp and reduce the voltage output from the photocell. An inverter and amplifier process the pulses from the photocell so that they can operate a counter that displays the number of boxes.

Most ammeters are calibrated to measure only pure sinusoidal waveforms. A nonsinusoidal waveform gives an erroneous reading on such meters. On a thermocouple ammeter, as shown in Figure 4-15, the waveform is used to heat a wire which is connected to a junction of a thermocouple. The heating effect of the current (that is, the rms value) determines the current in the sensitive galvanometer. The meter is calibrated to indicate rms current. The thermocouple ammeter can be used in applications where nonsinusoidal currents are to be measured. This type of ammeter is also very useful for measuring high frequency rf currents.

PASSIVE TRANSDUCERS

Figure 4-16 shows some examples of how passive circuit parameters can be used in transducers. Two liquid level sensors are shown. The variable

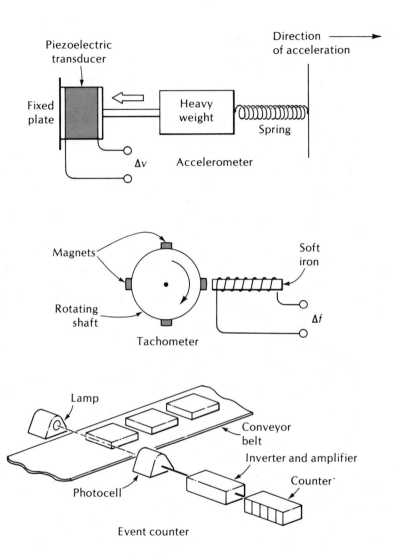

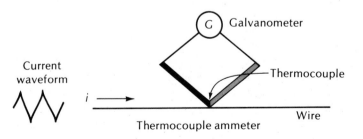

Figure 4-15. Examples of active transducer uses

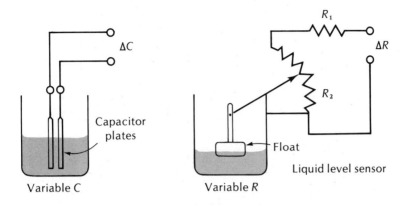

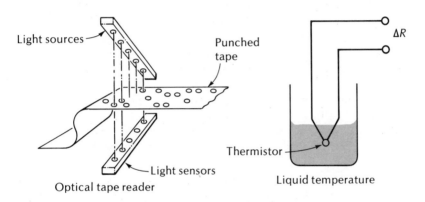

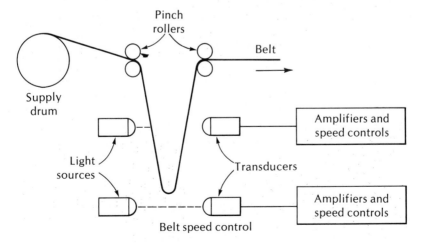

Figure 4-16. Examples of passive transducer controls

capacitor liquid type works on the principle that capacitance is directly dependent upon the type of dielectric material between the plates. The dielectric in this case is partly air and partly the liquid being sensed. When the tank is full the capacitance dielectric is liquid; and when the tank is empty the dielectric is air. The capacitance changes gradually as the liquid level decreases, and this changing capacitance is used to control current in an ac circuit. The amount of current is measured and related to the amount of liquid level.

The second liquid level sensor shown in Figure 4-16 uses a variable resistor. A float is connected to the arm of the variable resistor. When the liquid level is high, the arm shorts across R_2 and the circuit resistance consists only of R_1. (This resistor is necessary in order to prevent a short circuit from occurring under this condition.) When the liquid level is low, then all of R_2 is in the circuit and the circuit resistance is maximum. The overall result is that the resistance depends directly upon the level of the liquid. An ac or a dc voltage can be applied to this system, and the resulting measured current is related to the liquid level.

The optical tape reader consists of a source of light on one side of the punched tape, and light sensors on the other side. At the particular instance shown only three holes are in position, and three beams of light pass the sensors. The three holes in position permit three of the light sensors to produce an output. These light sensors may be active or passive, and the output is a resistance that depends upon the amount of light. Since there are only two possibilities, *light* and *no light,* the output of the sensor is either high or low, depending on whether there is a hole in position.

The overall result is that the holes in the punched tape can be used to produce electrical signals that are directly related to the position of the holes. These electrical signals are processed in logic circuitry, then used to control machinery. An alternative to the punched tape system shown in Figure 4-16 is a magnetic tape system. In such a system magnetic flux is established on the surface of the tape. As the flux passes an electromagnetic pickup, it produces a small voltage that is processed in logic circuitry and used to control machinery. Both optical and magnetic tapes are used extensively in industrial control systems.

The liquid temperature measuring device in Figure 4-16 consists of a thermistor immersed in the liquid. For most thermistors an increase in temperature will result in a reduction in thermistor resistance. In other words, a thermistor usually has a negative temperature coefficient.

The output terminals of a thermistor are normally connected into a bridge circuit. The purpose of the bridge circuit is discussed in the next section.

The belt speed system shown in Figure 4-16 regulates the speed of a conveyor belt or of some material being wound on a drum. An example is a paper mill in which the paper is wound on huge rolls. When the end of the loop is between the two light sources as shown, the belt speed is correct. If the loop dips to the lower light source, the speed is too low; and if it goes above the upper light source, the speed is too fast. The transducers connect to control circuits that increase or decrease the speed as required for maintaining the correct loop.

Circuits for Transducer Output

Figure 4-17 shows one method of reading the output of an active transducer. The transducer produces a voltage that can be read directly by a voltmeter. An example of this method is the thermocouple ammeter shown in Figure 4-15, where the output of the active transducer is read by a sensitive current meter. The scale for the meter can be *calibrated,* that is, marked to indicate the measurand. For example, if the active transducer is a thermocouple, the voltmeter (or ammeter) can be marked to read temperature directly.

A disadvantage of the circuit in Figure 4-17 is the fact that the voltmeter must be constantly calibrated for accuracy. A small change in the voltmeter indication can represent a relatively large change in the output of the transducer. It would be better to have a system that did not depend upon voltmeter calibration for accuracy. Such a system is shown in Figure 4-18. A known voltage (V) is placed across three resistors $(R_1, R_2,$ and $R_3)$ in a voltage divider network. The active transducer is connected across R_1 and part of R_2 to the arm of variable resistor R_2.

The output of the transducer (V) is to be determined. When the arm of R_2 is set so that the voltage across the transducer equals that portion of the voltage across R_1 and R_2, there will be no current flow in the *galvanometer.* A galvanometer is a very sensitive current indicator.

With the method shown in Figure 4-18 the voltages are being balanced out so that no current flows in the galvanometer. Resistor R_2 is connected to a dial that is calibrated to read the measurand directly. For example, if the transducer is a thermocouple, then the dial will be calibrated to read temperature. When the dial is adjusted until the galvanometer reading is zero, the temperature value is read off the dial.

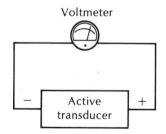

Voltmeter

Figure 4-17.   One method of reading an active transducer output

Known voltage

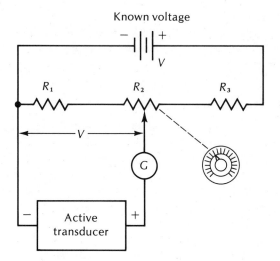

Figure 4-18. Active transducer circuit

An advantage of the system in Figure 4-18 is that the active transducer does not supply current at the time the reading is taken. This is due to the fact that the voltage across the divider to the arm of R_2 is exactly equal to the transducer voltage. The two voltages are equal so there is no resultant voltage to produce a current. If the active transducer had to supply current, it would be heated internally and this could make calibration difficult.

Another advantage is that the meter scale is unimportant since it is only necessary that it be accurate at the zero point. Usually the meter reads zero at center scale to allow for deflection in the + or − direction.

Figure 4-19 shows two circuits for connecting a passive transducer. In the basic circuit the passive transducer is connected through resistor R (a

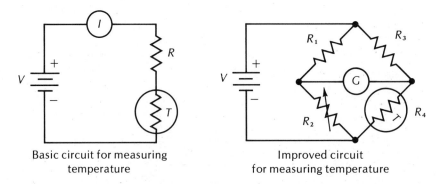

Basic circuit for measuring
temperature

Improved circuit
for measuring temperature

Figure 4-19. Thermister circuits

current-limiting resistor) and through a current indicator to a voltage source. The resistance of the thermistor varies with the temperature being measured, so the current flow will also vary with temperature. The idea behind this basic circuit is to calibrate the current indicator to read temperature directly.

An important problem related to the basic circuit for measuring temperature is the fact that the current going through the thermistor causes it to heat. Therefore, an increase in temperature must add to the temperature of the thermistor that is due to current flow. Any small change in the applied voltage V will change the current through the thermistor and change its temperature significantly. This, in turn, will make it necessary to recalibrate the current meter.

An improved circuit for measuring temperature is also shown in Figure 4-19. It involves placing the thermistor in a bridge circuit. This bridge circuit is *balanced* when

$$\frac{R_1}{R_2} = \frac{R_3}{R_4}$$

When the bridge is balanced, there is no current flow in the galvanometer. The advantage of this circuit over the basic circuit is the fact that a change in applied voltage *(V)* will not seriously affect the calibration of the system. This is true because the ratio of R_1 to R_2 and R_3 to R_4 will be the same regardless of the value of applied voltage.

The galvanometer in the bridge circuit could be calibrated to read temperature directly. Changes in thermistor temperature cause the bridge to become unbalanced, and the result is a current through the galvanometer that is related to the temperature.

A better way to use the bridge circuit is to calibrate the variable resistor (R_2) to indicate temperature. The resistor is adjusted to balance the bridge, and the temperature (or other measurand) is read directly from the dial. This method has the advantage that the current through the galvanometer is zero when the reading is taken, so calibration of the meter is not a significant factor in taking a reading. (Of course, the meter must indicate zero current accurately.)

The bridge circuit can be used for any kind of passive transducer. The disadvantage of the improved circuit of Figure 4-19 is that some current still must flow through the thermistor, and this causes it to be heated. Also, any change in ambient (surrounding) temperature of the bridge will slightly affect the resistance of R_2, and that could also unbalance a sensitive bridge.

A better bridge circuit is shown in Figure 4-20. In this circuit R_2 and R_4 are both thermistors. One of the thermistors (R_2) is maintained at a steady temperature so that its resistance is independent of any changes in ambient temperature. Resistor R_4 is a thermistor that is connected to a point where the temperature is being measured. Any unbalance in the thermistor bridge must be almost entirely due to a change in the temperature of R_4.

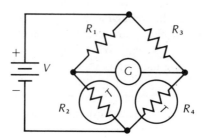

Figure 4-20.   A temperature-compensated
bridge

If the applied voltage (V) changes, the current through both thermistors will change by the same amount and this will not cause any unbalance of the bridge. A change in ambient temperature will affect the resistance of R_1 and R_3 by the same amount, so this will not unbalance the bridge.

An important advantage of the bridge circuit for measuring the outputs of passive transducers is that a very sensitive current meter can be used as an indicator. Very small changes in the resistance in one of the legs of the bridge will produce a relatively large change in current indication. The bridge can be made even more sensitive by connecting an amplifier in place of the galvanometer (Figure 4-21). The amplifier used is a special type called a *differential amplifier*. It produces an output voltage that is proportional to the difference in voltage between points a and b. Operational amplifiers, which will be discussed in Chapter 6, are ideal for this application. The output of the amplifier is fed to an analog/digital (A/D) device. It converts the small changes in output dc voltage to a signal that can operate a digital display.

Telemetering

There are many applications where the quantity being sensed (either energy or a measurand) is located some distance from where it is to be observed and

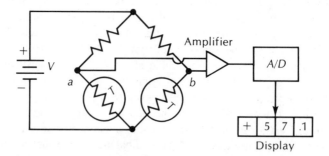

Figure 4-21.   A very sensitive temperature-measuring circuit

recorded. A few examples are intrusion alarms, measurement of extremely high temperature, measurement of atomic radiation, x-ray input, and assembly line counter controls.

When the output of a transducer is to be transmitted over a considerable distance, the procedure used is called *telemetering*. Figure 4-22 shows the basic principle. The quantity being sensed is either some form of energy or some type of measurand. In either case the sensor produces an output which is proportional to the quantity being sensed. This output is amplified and delivered to a transmitter. The transmitter conditions the signal for transmission over a distance.

Signal conditioning involves changing the dc signal from the sensor into a radio signal. Mixing a signal to be transmitted with a transmitter signal is called *modulation*.

The output of the transmitter may go to an antenna where it is transmitted to the receiving antenna. This transmitter-receiver combination is similar to those used in two-way radio systems. In other applications the output of the transmitter may go to a transmission line where it is transmitted over great distances. In either case the receiver processes the transmitted signal and produces an output that is directly related to the sensor output.

The signal processing that occurs in a receiver is called *demodulation*. It converts the radio signal into a usable voltage.

Figure 4-23 shows two popular types of modulation that can be used in telemetering systems. *Amplitude modulation* produces a signal amplitude output that is directly related to the transducer output. For example, a high amplitude corresponds to the high output of the transducer. For *frequency modulation* the frequency of the output signal is directly related to the output of the signal. A high output voltage corresponds to the high frequency, and a low output voltage corresponds to a low frequency.

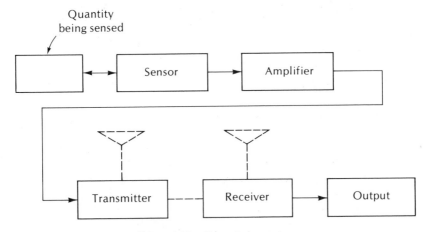

Figure 4-22. Telemetering system

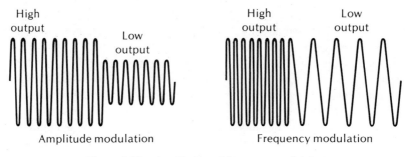

High output

Low output

High output

Low output

Amplitude modulation

Frequency modulation

Figure 4-23. Amplitude and frequency modulation

Figure 4-24 shows the types of pulse modulation frequently used in tele-metering. The modulation signal shown in this illustration is a sine wave, but variations in dc level from a transducer are normally used. The timing signal is a row of pulses to be modulated.

One way, called *pulse amplitude modulation* (PAM), modulates the pulses to change their amplitude. Another way is to change the width of the output

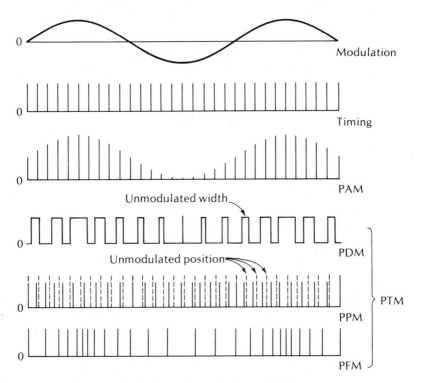

Modulation

Timing

PAM

Unmodulated width

PDM

Unmodulated position

PTM

PPM

PFM

Figure 4-24. Types of pulse modulation

pulses in correspondence with the input signal. This is called *pulse duration modulation* or *pulse width modulation*. Note that the wider pulses correspond to the higher amplitudes. The position of the pulses can be varied in correspondence with the amplitude of the signal. This is called *pulse position modulation* (PPM). Finally, the frequency of the pulses can be varied. This is *pulse frequency modulation* or PFM. These types of modulation are all used in telemetering systems.

MAJOR POINTS

1. Transducers are also called sensors. They permit the energy of one system to control the energy in another system, or they convert a measurand into an electrical quantity.
2. An active transducer produces an output voltage; and a passive transducer produces a change in a circuit parameter (resistance, capacitance, or inductance). The output of active and passive transducers is related to the quantity being sensed.
3. Active transducers generate a voltage by one of the six basic methods: electrostatic, chemical, piezoelectric, photoelectric, thermoelectric, and electromechanical.
4. Optoelectronic components, such as photodiodes and phototransistors, are also used in transducer systems.
5. Of the six methods of generating a voltage, the electrostatic and chemical methods are seldom used in transducer systems.
6. A passive transducer may vary resistance in relation to an input. Resistance is affected by temperature, type of material, and dimensions (length and cross-sectional area) of the material.
7. A passive transducer may vary capacitance in relation to an input. Capacitance is affected primarily by the type of dielectric, the area of the plates, and the distance between the plates.
8. A passive transducer may vary inductance in relation to an input. Inductance is affected primarily by the type of core material and by the dimensions (length and diameter) of the coil.
9. By connecting a passive transducer into a bridge circuit, its output can be made independent of the voltage supply.
10. Telemetering is used when a transducer output is to be transmitted over a distance. A number of different types of modulation are used in telemetering.

PROGRAMMED REVIEW

(Instructions for using this programmed section are given in Chapter 1.)

The important concepts of this chapter are reviewed here. If you have understood the material, you will progress easily through this section. Do not skip this material, because some additional theory is presented.

---

1.   When pressure is exerted on certain crystalline materials, a voltage is generated across their surfaces. This is called

A. Faraday's law. (Go to block 9.)
B. the piezoelectric effect. (Go to block 17.)

---

2.   The correct answer to the question in block 25 is A. In addition to telemetering, the two-way radio systems used in industrial applications also employ various methods of modulation. AM and FM are most commonly used for voice communications.

Here is your next question.
An instrument that measures r/min is the

A. tachometer. (Go to block 11.)
B. ratemeter. (Go to block 18.)

---

3.   Your answer to the question in block 24 is wrong. Read the question again, then go to block 13.

---

4.   The correct answer to the question in block 17 is A. Thermocouple action is based on the Seebeck effect. The thermocouple consists of two dissimilar metals which, when heated, generate a low value of dc voltage.

Here is your next question.
Which of the inductors shown here has the higher value of inductance?

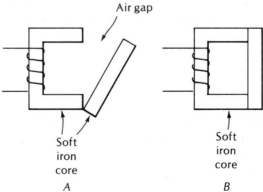

A

B

A. The one marked *A*. (Go to block 16.)
B. The one marked *B*. (Go to block 24.)

5. Your answer to the question in block 11 is wrong. Read the question again, then go to block 23.

6. Your answer to the question in block 8 is wrong. Read the question again, then go to block 21.

7. Your answer to the question in block 19 is wrong. Read the question again, then go to block 26.

8. The correct answer to the question in block 23 is B. Both primary and secondary cells are available in very small packages. They are used in portable instruments.

Here is your next question.
Moving the plates of a capacitor closer together will cause

A. an increase in capacitance. (Go to block 21.)
B. a decrease in capacitance. (Go to block 6.)

9. Your answer to the question in block 1 is wrong. Read the question again, then go to block 17.

10. Your answer to the question in block 17 is wrong. Read the question again, then go to block 4.

11. The correct answer to the question in block 2 is A. Not all tachometers are electronic. Mechanical tachometers that connect to a rotating shaft are also used extensively.

Here is your next question.
A spark is caused to jump across a 0.15 cm gap. The voltage required to do this is

A. over 6000 V. (Go to block 23.)
B. less than 1000 V. (Go to block 5.)

12. Your answer to the question in block 21 is wrong. Read the question again, then go to block 19.

13. The correct to the question in block 24 is B. Here is another definition of a transducer: A device that is actuated by one system and supplies signals to another system.

Here is your next question.
The two wires in this illustration are made of different materials. When junction *a* is heated, junction *b* cools. This is called

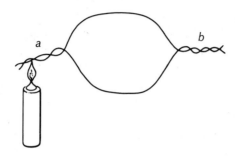

A. the Thompson effect. (Go to block 20.)
B. the Peltier effect. (Go to block 25.)

---

14. Your answer to the question in block 26 is wrong. Read the question again, then go to block 22.

---

15. Your answer to the question in block 23 is wrong. Read the question again, then go to block 8.

---

16. Your answer to the question in block 4 is wrong. Read the question again, then go to block 24.

---

17. The correct answer to the question in block 1 is B. Piezoelectric transducers are used for generating ultrahigh frequency sound waves (usually 35 to 65 kHz). These sound waves are used in ultrasonic cleaners. They vibrate the cleaning fluid at the ultrasonic rate, which produces a scrubbing action on small parts.

Ultrasonic waves are also used for checking large metal castings to see if there are any internal bubbles or cracks. As shown in the illustration, the waves enter the casting through a small transmitting transducer. If there are no bubbles or cracks, the waves are reflected from the opposite surface. Any imperfection in the casting will reflect the wave sooner, and the shorter time is monitored on the oscilloscope.

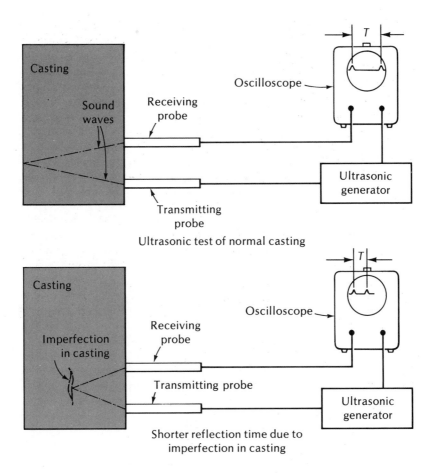

Ultrasonic test of normal casting

Shorter reflection time due to
imperfection in casting

Here is your next question.
A thermocouple is

A. an active transducer. (Go to block 4.)
B. a passive transducer. (Go to block 10.)

---

18. Your answer to the question in block 2 is wrong. Read the question again, then go to block 11.

---

19. The correct answer to the question in block 21 is B. Hall devices are used to sense the presence of magnetic fields.
   Varactors are diodes that act like capacitors.

Here is your next question.
Another name for PDM is

A. pulse width modulation (PWM). (Go to block 26.)
B. pulse amplitude modulation (PAM). (Go to block 7.)

---

20. Your answer to the question in block 13 is wrong. Read the question again, then go to block 25.

---

21. The correct answer to the question in block 8 is A. If the capacitor is charged, the voltage across it will decrease as the plates move closer. The voltage will increase as the plates are moved apart.

The most important application of capacitors is to store energy.

Here is your next question.
As the strength of a transverse magnetic field is varied, the amount of current through a certain device changes. This device is a

A. varactor. (Go to block 12.)
B. Hall device. (Go to block 19.)

---

22. The correct answer to the question in block 26 is B. The calculation is performed as follows:

$$\frac{500}{200} = \frac{750}{R_x}$$

Set the product of the means (200 and 750) equal to the product of the extremes (500 and R_x):

$$500 \times R_x = 200 \times 750$$

Divide both sides of the equation by 500:

$$R_x = \frac{200 \times 750}{500}$$

$$= 300 \text{ ohms}$$

The bridge circuit shown in block 26 is called a Wheatstone bridge. In addition to passive transducer inputs, this type of bridge is used for accurately measuring unknown resistance values.

Here is your next question.
A component that generates a voltage which is proportional to acceleration is called _____ . (Go to block 28.)

---

23. The correct answer to the question in block 11 is A. A sample problem was worked in the chapter. Static charges can destroy electronic components, and they can produce arcs that interfere with signals.

Here is your next question.
A chemical cell that can be recharged is called

A. a primary cell. (Go to block 15.)
B. a secondary cell. (Go to block 8.)

---

24. The correct answer to the question in block 4 is B. The inductance of an inductor is dependent upon the number of flux linkages. By closing the iron path, the flux lines are confined to the iron and fewer lines are lost. Thus, the inductance is increased.

Here is your next question.
When a transducer is being used to measure a quantity, that quantity is called the

A. measurable. (Go to block 3.)
B. measurand. (Go to block 13.)

---

25. The correct answer to the question in block 13 is B. Small refrigerators and cooling devices for components have been built upon the Peltier principle.

Here is your next question.
The output of a transducer controls the signal from a telemetering transmitter by

A. modulation (Go to block 2.)
B. demodulation. (Go to block 27.)

---

26. The correct answer to the question in block 19 is A. The initials PDM stand for pulse duration modulation. This is also called pulse width modulation.

Here is your next question.
This bridge is balanced when the galvanometer indicates zero current. In order to be balanced, the resistance of R_x must be

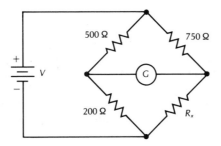

A. 275 ohms. (Go to block 14.)
B. 300 ohms. (Go to block 22.)

---

27.   Your answer to the question in block 25 is wrong. Read the question again, then go to block 2.

---

28.   An accelerometer.

You have now completed the programmed section.

SELF-TEST WITH ANSWERS

(Answers to this test are given at the end of the chapter.)

1.   What is another name for a transducer?

2.   What is the name for a transducer that does not generate a voltage?

3.   What is the name for a transducer that generates a voltage?

4.   Of the six methods used for generating voltage, which two are least likely to be used as transducers?

5.   Certain materials generate a voltage when a pressure is exerted on them. This is the _____ effect.

6.   When a battery is delivering current to a load, the terminal voltage of the battery drops as the load current increases. This drop in voltage is due to _____ .

7.   A voltage is generated when the junction of two wires made of different materials is heated. This thermoelectric effect is called the _____ .

8.   Name four factors that determine the resistance of wires.

9.   Name four factors, excluding temperature, that influence the inductance of an inductor.

10.   Name three factors, excluding temperature, that influence the capacitance of a capacitor.

11.   Transmitting transducer output over a distance is called _____ .

12.   A device that changes resistance when placed in a varying magnetic field is the _____ .

Answers to Self-Test

1.   Sensor
2.   Passive transducer
3.   Active transducer
4.   Frictional
      Chemical
5.   Piezoelectric
6.   Battery internal resistance
7.   Seebeck effect
8.   Temperature
      Type of material
      Length
      Cross-sectional area
9.   Type of core material
      Number of turns of wire
      Coil diameter
      Coil length
10.   Type of dielectric material
      Area of plates
      Distance between plates
11.   Telemetering
12.   Hall device

CHAPTER 5

THEORY OF MOTORS AND GENERATORS

Many years ago when industrial electronics was in its infancy, it was necessary for an industrial electronics technician to be a combination of an electronics and an electrical equipment expert. In those days the amount of electronics used in industry was very limited. The questionable reliability of tubes and other electronic components was one reason for the lack of interest in electronic control systems.

Today, an industrial electronics technician is primarily a specialist in industrial electronics. The electrical and electronics fields have each become so specialized that each has its expert technicians for troubleshooting and repairing equipment. Be that as it may, it is desirable for you, as an electronics technician, to have a basic understanding of some of the electrical devices that will interface with the electronic equipment. This knowledge will give you insight into the output signals that are necessary in control systems. It will also make your work more interesting because it will enable you to understand the complete system rather than just the electronic sections.

Obviously a complete in-depth study of such electromagnetic devices as motors, generators, transformers, and even relays is not possible in one chapter of a book. You should not expect to become an expert in these areas on the basis of this study. You *should* get a sound fundamental understanding of the principles of the electrical devices—especially as they relate to electronic systems.

The special characteristics of the devices are emphasized in this chapter. For example, the factors that control the direction of rotation, the speed, and the torque of a motor directly influence the operation of electronic motor control systems. These are the factors that you should concentrate upon. Other factors, such as the motor construction and methods of winding motors, have very little practical value or interest to electronics technicians. These subjects are not included in the chapter.

You should keep in mind that it is absolutely necessary to start with a summary of basic principles in many chapters in this book. Certainly some of the readers will already be well educated in these basic fundamentals. This material will then serve as a quick review. Other readers will find it informative and helpful in understanding the devices to be discussed. For these reasons, this chapter, as in previous chapters, begins with a quick survey of the basic fundamental principles that affect or determine the operation of the devices discussed in the chapter.

After the review material, the devices to be studied in this chapter include dc and ac generators, dc and ac motors, magnetic amplifiers, synchros and servos, and resolvers.

THEORY
Principles that Affect the Operation of Electromagnetic Components

THE UNIT NORTH POLE

The concept of a unit magnetic north pole is useful in understanding the magnetic terms used in the study of electromagnetic devices. A *unit north pole* is a very small isolated north pole of a magnet that has no corresponding south pole. This is only a concept. So far, no one has ever been able to produce a north pole that does not have a south pole.

If you were to place the unit north pole near the north pole of a magnet, it would follow a path similar to the one shown in Figure 5-1. Note that it moves away from the north pole (like poles of a magnet repel) and it moves toward the south pole (unlike poles attract).

An interesting thing about the unit pole moving around the path is that if you started it at a slightly different position, it would take another path

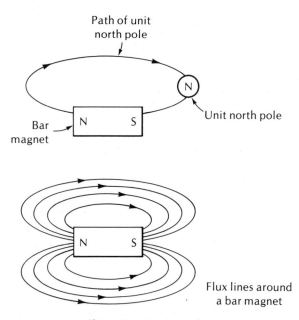

Figure 5-1. Magnetic flux

which is similar in shape but either closer to or farther away from the bar magnet. If you drew all of the paths that the unit north pole could follow, you would find that they trace lines around the bar magnet. These lines, called *flux lines,* are also illustrated in Figure 5-1. The term *flux,* then, means a path of magnetic intensity that a unit north pole would follow.

With a strong magnet there are more flux lines (more paths), and with a weak magnet there are only a few flux lines. Now this is a very important point. *The arrows that are usually drawn on these flux lines do not mean that the lines themselves are moving in that direction!* The arrows show the direction that a unit north pole will go if placed near the bar magnet.

ELECTROMAGNETIC FIELDS

Whenever an electron moves, regardless of whether it is moving in space or if it is moving in a conductor, there is always a magnetic field surrounding that electron.

When an electron current flows through a conductor, the magnetic fields surrounding all of the moving electrons combine to produce a magnetic field around the conductor. This magnetic field is identical in every way to the magnetic field that surrounds a permanent magnet. Its *direction* is the direction that the unit north pole would follow if it were placed in the flux of the field. The direction of the magnetic field around an electron current can be determined with the left-hand rule as shown in Figure 5-2.*

The strength of the magnetic field is directly determined by the amount of current flowing in a conductor. Two examples are shown in Figure 5-3. Note that in both cases the strength of the field decreases as you move farther and farther away from the conductor. Another way of saying this is that a north pole would have many paths to follow if it was close to a conductor but the paths are farther apart as you increase the distance from the conductor.

FORCE ON A CONDUCTOR

If a conductor that is carrying electron current is placed in a permanent magnetic (pm) field, the field surrounding the conductor will interact with the pm field. The conductor moves as shown in Figure 5-4. If the conductor is free to move, it will be forced to follow a path that is dependent upon two things: (1) The direction of the permanent magnetic field, and (2) the direction of the electron current flowing through the conductor.

Although a permanent magnetic field is used in the example of Figure 5-4, the conductor would also move if the field was produced by an electromagnet. Keep in mind that the electromagnetic fields are no different than the permanent magnetic fields.

---

*The left-hand and right-hand rules in this chapter are based on the assumption of electron flow. For conventional current the rules have to be reversed. So, the left-hand rule for electron current becomes the right-hand rule if conventional current is assumed.

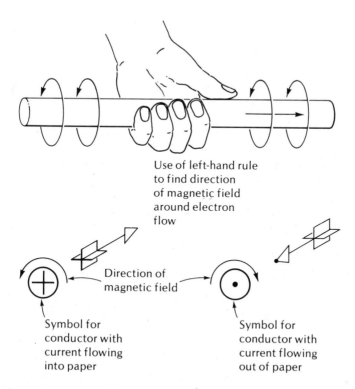

Use of left-hand rule to find direction of magnetic field around electron flow

Direction of magnetic field

Symbol for conductor with current flowing into paper

Symbol for conductor with current flowing out of paper

Figure 5-2.   Flux around current

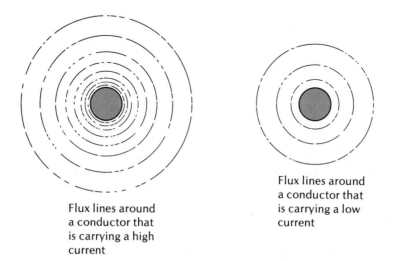

Flux lines around a conductor that is carrying a high current

Flux lines around a conductor that is carrying a low current

Figure 5-3.   Flux lines around a conductor with high current and a conductor with low current

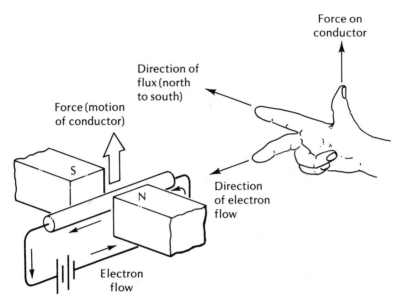

Figure 5-4. Right-hand rule for motors

The direction of force, and resulting motion, of a current-carrying conductor in a magnetic field is determined by the right-hand rule as illustrated in Figure 5-4. The thumb and first two fingers of the right hand are held at right angles to one another as shown. If the first finger points in the direction of the magnetic field, and the second finger points in the direction of the electron current through the conductor, then the direction of force and motion will be indicated by the direction of the thumb.

The importance of this rule is that it shows you the factors that determine the direction of motion. These same factors determine the direction of rotation of an electric motor.

GENERATED VOLTAGE

Whenever a conductor is moved through a magnetic field, a voltage is always generated. This is sometimes referred to as *Faraday's law of induction*. The amount of voltage *(v)* is directly proportional to the number of conductors being moved *(N)* and the rate at which they are being moved *(dϕ/dt)*. The expression *dϕ/dt* is simply a mathematical shorthand way of saying *the rate of change of flux with respect to time,* that is, the rate at which the conductors are moving through the field.

Faraday's law shows you that you can increase the amount of voltage by increasing the number of conductors or by increasing the rate at which the conductors and the magnetic field are moving with relation to one another.

Even though a voltage is generated, it does not mean a current flows. In order for current to flow there must be an external connection to the moving conductor. Figure 5-5 shows how to tell which direction the generated electron current will flow when the conductor is moved through a magnetic field. Although this current is a result of a generated voltage, it is sometimes referred to as a *generated current*.

A generated current has an associated magnetic field since electron flow *always* has a magnetic field around it. The magnetic field due to generated current will try to push the conductor in the opposite direction to the direction it is being moved. This is called *Lenz's law*. Because of Lenz's law, the equation for generated voltage is usually written

$$v = -N\frac{d\phi}{dt}$$

The negative sign shows that the generated voltage will produce a current with a magnetic field that opposes the motion.

With these basic principles of electricity and magnetism in mind, it is now possible to review the principle of operation for dc motors and generators.

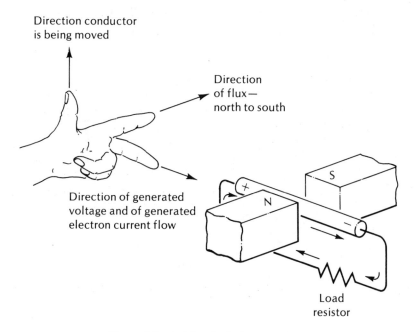

Figure 5-5.   Left-hand rule for generators

Generators

Dc motors and generators are constructed much in the same way. As a matter of fact, a dc generator will behave as a motor under proper conditions. (Likewise, a dc motor will behave like a generator under proper conditions.) It is necessary to make provisions for preventing a dc generator from acting like a motor when it is being turned slowly. In cars when generators were used for charging the battery, a cutoff removed current to the generator field winding when the engine was idling. This prevented the generator from acting like a motor and turning against the fan belt.

Figure 5-6 shows the basic principle of operation for a generator. Here, a loop of wire called the *armature* is suspended between the poles of two electromagnets. The magnetic field in the generator of Figure 5-6 is obtained with a coil wound around a soft-iron core. In pm generators, usually called *magnetos,* a field winding would not be needed. Some magnetos are used for low-power applications, but for larger powers it is the practice to use an electromagnetic field.

When a separate dc source is used to supply the current for the field winding, it is called a *separately excited generator.* By using an external dc source, the field current, and therefore the generator output voltage and current, can be precisely controlled.

When the current for the generator field winding comes from the generator itself, it is said to be *self-excited.* This type is convenient to use because no external power supply is needed. Self-excited and separately excited generators are compared in Figure 5-7.

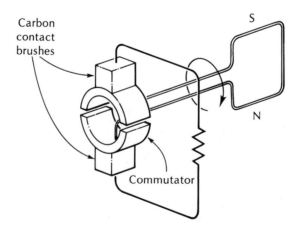

Figure 5-6. Basic principle of operation for a generator

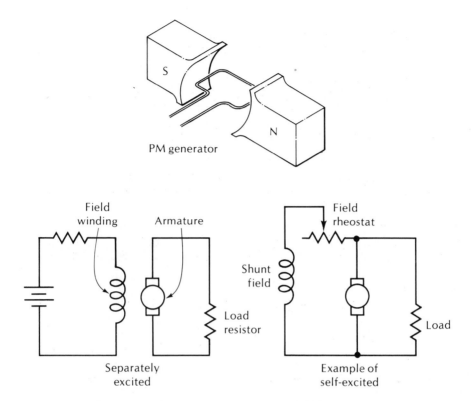

Figure 5-7. Three ways to get a magnetic field for a generator

The armature is caused to rotate so the conductors cut through magnetic flux lines. Since voltage is generated any time a conductor moves through a flux line, the arrangement of Figure 5-6 can be used as a generator.

In Figure 5-8 the armature is being turned through the magnetic field. To simplify the drawing a permanent magnet is used in this group of illustrations. The arrows point to the direction of generated voltage, which is also the direction of generated current when an external load is connected. To distinguish between the two halves of the conductor, it is drawn with a black side and a white side.

In the first drawing the conductors are moving parallel to the magnetic flux lines and no voltage is generating. Remember that the conductors must cut *through* the flux lines in order to generate a voltage.

In the second drawing of Figure 5-8 the conductors are beginning to move through the flux lines and a voltage is generated in the direction shown by

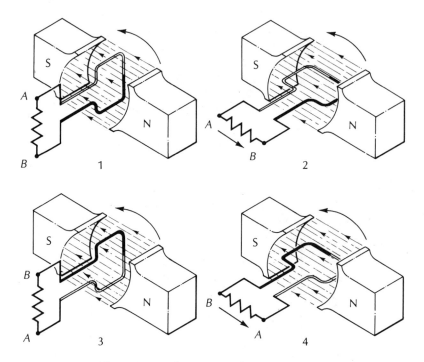

Figure 5-8. Voltage generated as armature turns

the arrows. Note carefully that the direction of current is into the paper on the black side and out of the paper on the white side. This can be verified by the generator rule shown in Figure 5-5. The voltage and maximum current will occur when the rate of cut of flux lines is maximum. This will happen when the conductors are halfway between the ends of the poles.

In the third drawing of Figure 5-8 the conductors have moved to a point where they are again parallel to the flux lines and no voltage is being generated.

In the fourth drawing of Figure 5-8 the black conductor is moving down through the flux lines instead of up as in the second drawing, thus causing the generated voltage and current to reverse through the load resistor.

Clearly, then, if this action continues, an ac voltage waveform will be generated by the rotating conductor.

To obtain a dc output it is necessary to have some form of rectification for the ac current flow. In a dc generator the rectification is accomplished by a commutator as shown in Figure 5-9. Now the rotating conductors are connected to metal rings, called a *commutator*. It rotates with the conductors. As the conductors reach the zero-voltage point, where they are moving parallel with the flux lines, the commutator segments are in a position to

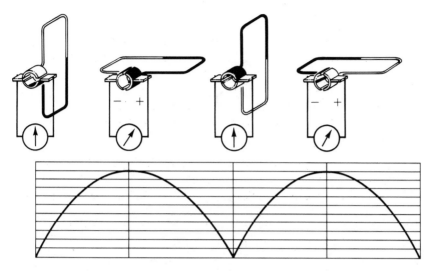

Figure 5-9. Commutator converting output to pulsating dc

reverse the connections to the outside load. The result is that the conductor that moves up through the flux lines is always connected to one of the fixed position brushes and the conductor that moves down through the flux lines is always connected to the other fixed brush. This result is true regardless of whether the conductor is black or white.

The resulting output waveform is shown in Figure 5-9. The disadvantage of this waveform is that it goes to zero on each half rotation. This disadvantage can be eliminated by using a greater number of segments and a greater number of conductors rotating through the field as shown in Figure 5-10. With this arrangement there is no instant of time when the output is at zero because there are always conductors moving through the field. Since the output waveform never drops to zero volts, it more nearly represents a dc voltage.

In the generators just discussed, a permanent magnet was used to supply the magnetic flux. In practice it is more common to use an electromagnetic field obtained with a field winding as shown in Figure 5-6. The field winding requires a dc current to produce the magnetic flux. The current may be supplied from an outside source, in which case the generator is said to be *separately excited;* or the dc current may come from the dc voltage being generated by the generator, in which case it is called a *self-excited generator.*

SELF-EXCITED GENERATORS

Self-excited generators are usually classified by the way the field winding is connnected. Three possible arrangements are illustrated in Figure 5-11.

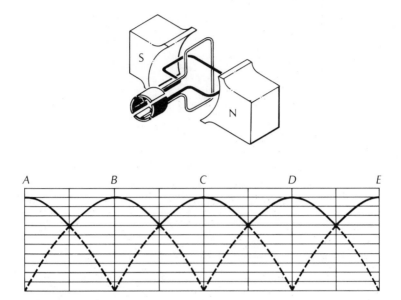

Figure 5-10.   Segmented commutator prevents zero-volt points

Their output characteristics are directly dependent upon the internal field and armature connections. Before discussing these characteristics, the term *regulation* must be defined.

Regulation is a measure of how well the generator maintains a constant output voltage when the output current changes from zero (no load) to the rated current for the generator (full load). Mathematically,

$$\text{Percent regulation} = \frac{V_{NL} - V_{FL}}{V_{FL}} \times 100$$

This is actually a percentage decrease in generator voltage when the full load is applied. For an ideal generator the value of percent regulation would be zero percent.

The schematic drawings of Figure 5-12 show various connections for self-excited dc generators, and the characteristic curve associated with each type. The first curve is a straight line parallel to the load current axis. This would be the output of a generator that has perfect regulation, that is, zero percent regulation. The graph shows that the voltage across the load does not change regardless of how much current flows through it. Electromechanical generators cannot be made with this type of curve. However, this curve can be closely approached electronically by using regulated power supplies.

In the series-wound generator the same current that flows through the armature also flows through the field and through the load resistor. When the

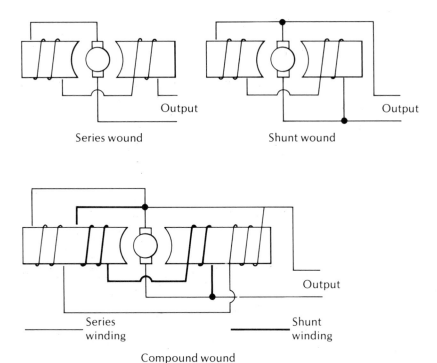

Series wound

Shunt wound

Series
winding

Shunt
winding

Compound wound

Figure 5-11. Three types of generator windings

armature rotation is first started, a small amount of residual magnetism in the soft-iron cores of the field provides enough of the flux to generate a small amount of voltage. This voltage causes a low value of current through the load and field winding. The greater flux, resulting from the field winding, causes a higher value of generator voltage.

The characteristic curve shows that the voltage and current in the series generator rise together until a peak value is reached. At that point the iron in the generator is saturated so no further magnetic flux increase is possible. The voltage drops off very rapidly if the current is increased beyond the saturation point.

In a shunt-wound generator the field coil is across the armature. In this case there is an output voltage regardless of whether or not a load current is flowing.

As shown by the characteristic curve, the shunt-generator voltage is nearly constant over a wide range of load current values. However, there is a slight drop in voltage as the current increases, so this is not a generator with zero percent regulation.

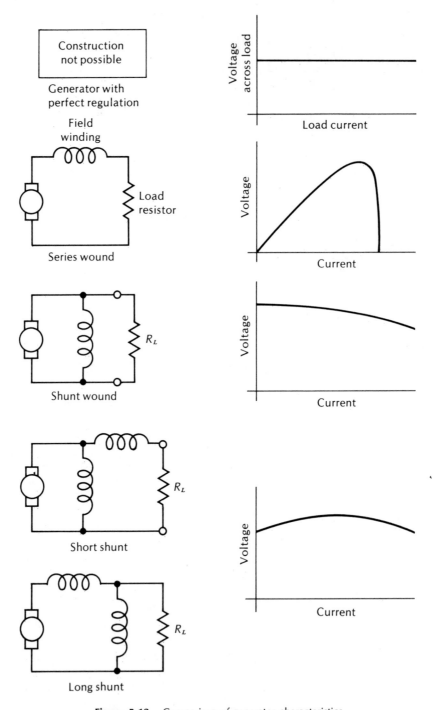

Figure 5-12. Comparison of generator characteristics

Two examples of compound-wound generators are shown in Figure 5-12. If the armature winding is directly across the shunt winding, the connection is called a *short shunt*. If the armature is in series with a parallel combination of the field winding and load resistance, it is a *long shunt*.

Regardless of whether the generator connection is a long shunt or short shunt, it is obviously a combination of series- and shunt-wound generator circuits. Looking back at the characteristics for the series- and shunt-wound generators, it is seen that with the series wound there is an increase in voltage as the current increases, whereas with the shunt wound there is a slight decrease in voltage as the current increases.

By combining these two into a single generator it is possible to get a relatively flat compound-generator characteristic like the one shown in Figure 5-12. In that case a generator is said to be *flat compounded*. Depending on the size of the series and shunt coils, it is possible to design a generator having almost any characteristic curve shape. Examples are shown in Figure 5-13.

The solid curves of Figure 5-13 represent the possibilities when the design is a *cumulative compound*. This simply means that the magnetic fields of the series-field winding and the shunt-field winding combine in the generator to produce the total magnetic field. It is also possible to connect the compound generator in such a way that the fields oppose one another. This is called a *differential compound*. The dashed line in Figure 5-13 shows a typical characteristic curve for the differentially compounded generator.

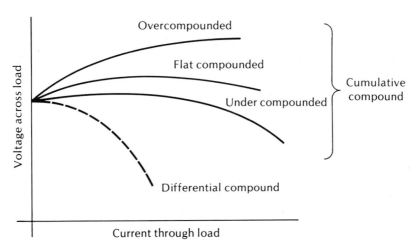

Figure 5-13. Comparison of characteristics for cumulative and differential compound generators

ARMATURE REACTION

An important consideration in dc generators is the effect that the current flow in the armature has on the magnetic field between the poles. This effect, which is called *armature reaction,* is illustrated in Figure 5-14. Note that the magnetic flux is no longer in a straight line between the north and south poles. Instead, it tends to curve around the moving armature. This presents a problem at the commutator. Remember that the commutator was designed in such a way that the armature connections were reversed on each half rotation, thus producing a pulsating dc output. If the brushes are set for proper commutation at a low r/min, the armature links will be switched at the instant the wires are running parallel with the original field. The point where ideal switching occurs is called the *neutral plane.* (See Figure 5-14.) As the generator output current increases, the flux lines change their position, and this changes the neutral plane. So, the original point of commutation is no longer ideal. As a matter of fact, there will be a great amount of arcing at the brushes when the load becomes high.

One way to eliminate this arcing would be to turn the brushes with the neutral plane. This is not a convenient method, however, because of the mechanical problems involved. Another way is to use *interpoles*—electromagnets placed between the north and south poles of the generator field. The interpoles react with the flux lines between the north and south poles in such a way that they prevent them from curling around the armature as the load current increases.

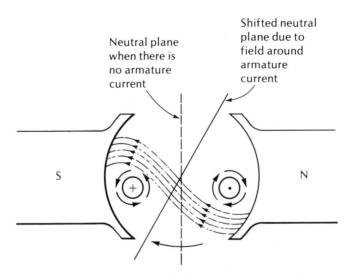

Neutral plane when there is no armature current

Shifted neutral plane due to field around armature current

S

N

Figure 5-14. Shift of neutral plane

Dc Motors

It was stated earlier in this chapter that you could make a dc generator behave like a dc motor by applying a voltage to its armature, and conversely, a dc motor will behave like a dc generator if its shaft is turned. Dc motors and generators are similarly constructed. Therefore, some motors have a permanent magnetic field and others have an electromagnetic field.

As a technician you should be aware of the fact that the characteristics of the motor vary greatly with series, shunt, and compound connections.

Figure 5-15 shows a characteristic curve for the different dc motor connections. In the series-wound motor, the torque increases and the speed decreases as armature current becomes greater and greater. (*Torque* is a measure of the turning force of a motor.) Series-wound motors are capable of extremely high torque at low speeds because armature countervoltage is low. The armature countervoltage reduces the armature current. When the motor is first starting there is no countervoltage opposition to current flow and the starting torque is very high. As the motor speed increases, the countervoltage increases and the armature current decreases.

You can see from the graph that when the armature current decreases the speed begins to increase. As a matter of fact, if there is no load on the motor, the speed will just continually increase until a *runaway condition* occurs. This means the motor destroys itself from the enormous speed. In some cases the friction of the motor may be great enough to prevent it from a runaway condition, but in most cases it will occur in series-wound motors. For this reason series-wound motors should never be used in applications where a *mechanical load* (opposition to rotation) is removed.

An interesting feature of the series-wound motor is the fact that it will run on both ac and dc. To understand this, suppose the dc source for a motor is reversed. The direction of rotation of a motor depends upon the use of the right-hand rule. You can easily demonstrate from Figure 5-4 that the direction can be reversed by reversing the direction of *either* the field *or* the armature current. However, in a series-wound motor when you reverse the polarity of applied voltage, you reverse both the field and the armature. Therefore the rotation of direction does not change. Furthermore, if you put an ac voltage across the series-wound motor, the motor will run. Hence, it is often called a *universal motor*.

Figure 5-15 shows that the torque of a shunt-wound dc motor is directly dependent upon the armature current and they both increase linearly. The speed of a shunt motor decreases slightly.

The speed of the shunt-wound motor is almost independent of the armature current up to its full mechanical load. For that reason, a shunt-wound motor is sometimes referred to as a *constant speed* type.

Compound motors are wound in such a way that they take advantage of the high starting torque of a series-wound motor and the constant speed of the shunt-wound motor combined. As with the generator, the motor can be

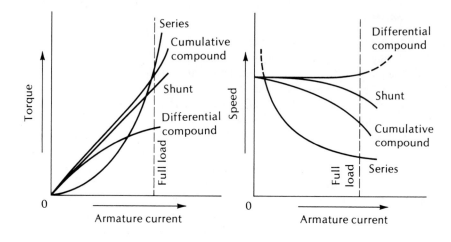

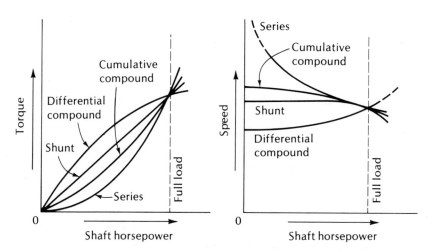

Figure 5-15. Characteristic curves of series, shunt, and dc compound motors

differentially compounded or accumulative compounded. The design of a compound motor is always a compromise between the starting torque and the constant speed.

SUMMARY OF DC MOTOR CHARACTERISTICS

Dc motors are used in control systems because they are easily reversible and their speed can be accurately controlled. Table 5-1 summarizes some of the important features of dc motors.

TABLE 5.1.   Characteristics of dc motors

Characteristic	Type of dc Motor			
	Series Wound	Shunt Wound	Cumulative Compound	Differential Compound
Starting torque	Very high	Fair	Very high	Very poor
Speed	Varies with load	Speed is easily controlled	Constant speed under changing load	Within set limits speed is constant
Runaway	Will run away without a load	Will run away if field winding is opened	Won't run away without a load	Won't run away

Alternators

Ac generators, called *alternators,* can be divided into two classes: those with rotating armatures and those with rotating fields. Figure 5-16 shows a simplified drawing of an ac generator with a rotating armature. A conductor is rotated between the fields of the permanent magnet or an electromagnet —usually an electromagnet—and the ac voltage is generated. The power is removed by two slip rings.

On one half cycle the black conductor is moving up past the north pole, and on the next half rotation it is moving down past the south pole. The generator rule shows that the voltages of the two half rotations will be reversed. (The same is true for the white conductor of the armature.) The result is that the ac voltage periodically reverses, that is, it is an ac voltage. If the flux lines between the north and south poles are equally distributed, the voltage generated would be a pure sine wave.

If you try to get a very high voltage and/or current from the rotating armature of Figure 5-16 you will soon find that an important limiting factor is in the physical size of the armature. High voltages require thick insulation to prevent internal arcing, and high currents require large diameter conductors. These two things combine to make a bulky armature that is difficult to rotate at high speeds.

A better arrangement for high output power is shown in Figure 5-17. In this case the field is rotated and the armature—where the voltage is taken away from the generator—is stationary. In this illustration the rotating field is produced by a permanent magnet. Earlier it was stated that a voltage is induced whenever a conductor is moved through a magnetic field. The same

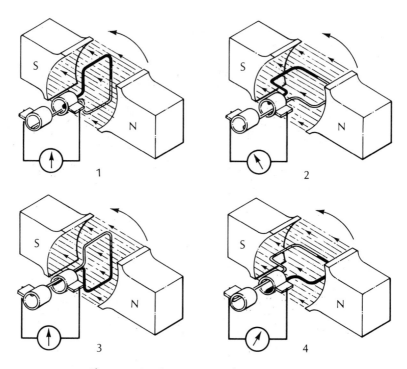

Figure 5-16.   Ac generator with rotating armature

is true if the magnetic field is moved through a conductor. It does not make any difference which is moving—the field or the conductor—provided there is relative motion between the two.

The rotating field produces a changing flux in the core. This changing flux, in turn, induces a voltage in the windings. The output voltage and power of this generator do not have to be delivered through the slip rings. This additional advantage of the rotating field arrangement means that there

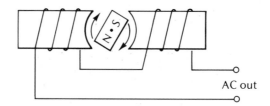

Figure 5-17.   Ac generator with a rotating
permanent magnet

is no armature reaction, no loss of power, or maintenance problems that occur with arcing at slip rings.

The power output from the alternator with a rotating and permanent magnet is still somewhat limited. More power can be obtained by using a rotating electromagnetic field. This is shown in Figure 5-18. In this type of generator the dc current for the electromagnet is introduced through the slip rings. At first you might think this gets us back to the same problem we had with slip rings in Figure 5-16. However, the amount of power necessary to produce the rotating electromagnetic field is very small compared to the output power of the ac generator. So the generator efficiency is still very high.

Returning again to the construction of the simple generator in Figure 5-17, the output ac voltage will be single phase. By adding two more sets of coils for armatures, it is possible to generate three-phase ac voltage readily. Three-phase voltage is more efficient and the idea behind this arrangement is shown in Figure 5-19. For greater efficiency, a rotating electromagnet would normally be used. The permanent magnet is shown here to simplify the drawing.

Three output voltages from the 3ϕ (three phase) generator are shown in the illustration. Three-phase alternating current is more efficient for operating three-phase motors and power supplies. It is usually used for industrial equipment. Note that the 3ϕ waveform has no point where the voltage drops to zero as it does in a single-phase ac voltage.

Ac Motors

For efficient operation with low maintenance requirements the ac motors may be chosen over dc motors. They have a disadvantage, however; unlike dc motors, the direction of rotation of ac motors is difficult to control. Also,

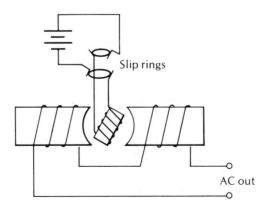

Figure 5-18. Ac generator with rotating
electromagnetic field

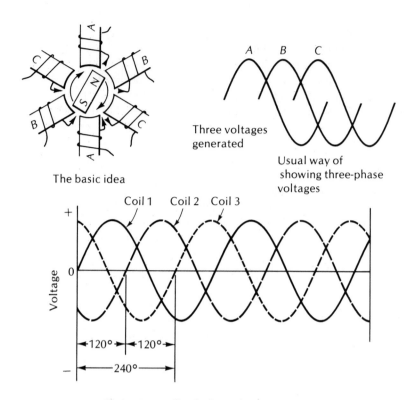

Three voltages generated

Usual way of showing three-phase voltages

The basic idea

Figure 5-19. Simple three-phase generator

it is more difficult to get a high torque at low motor speeds with ac motors. Before discussing the ac motor, a few additional basic principles will be reviewed.

BASIC PRINCIPLES OF OPERATION

Magnetic induction occurs when the magnetic field around a permanent magnet or electromagnet is induced into a soft-iron material. Figure 5-20 shows an example. The soft-iron material has a low *reluctance,* that is, a low opposition to the establishment of flux lines. The field lines tend to crowd together and pass through the soft iron because the reluctance of iron is lower than the reluctance of air. For all practical purposes, the soft iron becomes a magnet with a north and south pole.

The direction of the flux lines through the soft iron is the same as the direction of the flux lines through the core of the electromagnet. If the switch is opened, the flux in an electromagnet goes to zero. Since the field around the electromagnet is lost, the field around the induced magnet is also lost. If the battery is reversed, and the switch is closed, the soft iron will again

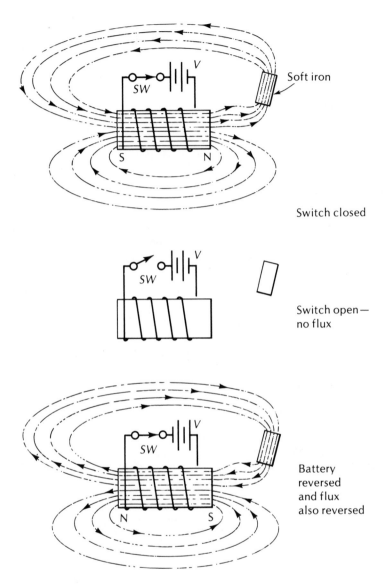

Soft iron

Switch closed

Switch open—
no flux

Battery
reversed
and flux
also reversed

Figure 5-20. Flux line crowding through low-resistance iron

become a permanent magnet. This effect is also shown in Figure 5-20. Note, again, that the induced magnet has north and south poles in the same relative position as an inducing field. This means that the induced magnet will be attracted to the electromagnetic inducing field for both circuits.

The principle of magnetic induction is used in the synchronous motors that are described later in this chapter.

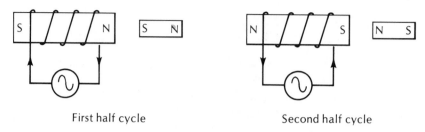

First half cycle   Second half cycle

Figure 5-21.   Alternating field periodically reversing induced magnetism

The second basic principle to be reviewed is shown in Figure 5-21. In this case an alternating electromagnetic field is produced with an alternating current. On one half cycle the north pole will be on one end, and on the next half cycle the north pole will be at the opposite end.

The north and south poles of an electromagnet can be determined by the left-hand rule for induced magnetism. This rule is illustrated in Figure 5-22. If the coil is grasped with the left hand so that the fingers circle the coil in the same direction as the electron current flows, then the thumb will point toward the north pole of the electromagnetic field. With conventional current the right-hand rule is used.

Using the left-hand rule, you can determine for yourself that alternating current in the coil of Figure 5-21 produces a continually reversing field. The expanding and contracting field around the electromagnet will induce a voltage in any conductor in the vicinity. In the illustration a copper ring is placed near one end of the electromagnet. The expanding and contracting magnetic field causes a voltage to be induced in the copper ring. Since the

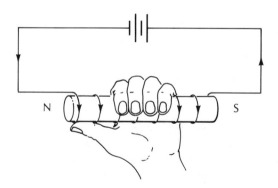

Figure 5-22.   Left-hand rule for direction of electromagnetic field

copper ring is a short circuit, a circulating current will flow. This current will have a magnetic field of its own which will react with the electromagnet. The two magnetic fields will always be such that they attract; that is, they will always be in a north-to-south arrangement as shown in the two half cycles of the illustration. Thus, the copper ring will be attracted to the electromagnet. This principle is used in some ac relays. It is also used in electromagnetic induction motors.

The third basic principle to be reviewed is that of repulsion. The principle is shown in Figure 5-23. In this case an ac field is expanding and contracting near a coil that is wound around a soft-iron material. The coil is wound in such a way that the magnetic field produced in the soft iron opposes the inducing field. The left-hand rule can be used to verify the positions of the north and south poles in other illustrations.

Since like poles repel, the movable coil and its core will be repelled away from the electromagnet. This principle is used in repulsion motors.

Having discussed the basic principles of operation, it is an easy matter to understand how the ac motors work.

Synchronous Motors

The principle of the synchronous motor is illustrated in Figure 5-24. In this case, two sets of electromagnets are arranged around a compass. (The coils are not shown.) On one-half cycle, the magnetic fields are arranged in such a way that the south poles of the electromagnet are attracting the north poles of the permanent magnet (the compass needle). At the same time the north poles are attracting the south poles of the compass needle. The attraction of the electromagnets causes the compass needle to rotate.

On the next half cycle, the currents through the coils have reversed. By this time, however, the compass needle has moved to a new position. Once

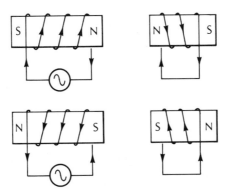

Figure 5-23. Principle of repulsion motor

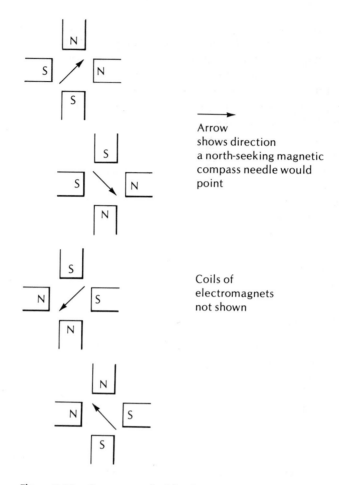

Arrow
shows direction
a north-seeking magnetic
compass needle would
point

Coils of
electromagnets
not shown

Figure 5-24.   Compass needle following a rotating magnetic field

again, the north poles of the electromagnet are attracting the south poles of the permanent magnet, and the south poles of the electromagnet are attracting the north poles of the permanent magnet.

Note that on each half cycle of operation, shown in Figure 5-24, the like poles of the electromagnet and permanent magnet are in a position that allows their repelling forces to aid in the rotating force. The combined attractions of unlike poles and repulsions of like poles produce the turning of the magnet. If, instead of using a compass as shown in Figure 5-24, a soft-iron rotor is used, the action will still be the same. This is the way synchronous motors are made; their basic construction is shown in Figure 5-25. To understand this operation, return again to Figure 5-21. Note that if the soft iron is

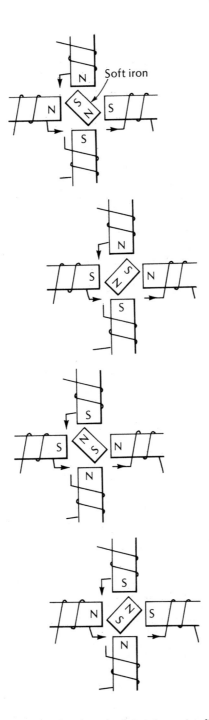

Figure 5-25.   Rotation of soft iron in varying field

placed in a magnetic field it always becomes magnetized in such a way that it is attracted by the electromagnet. The soft-iron rotor will continually be magnetized by the magnetic fields of the coils in such a way as to cause it to rotate. Once the soft iron becomes magnetized, its magnetism does not reverse on each half cycle. It behaves like the rotating permanent magnet discussed earlier in this section.

Although only a few positions are described in Figure 5-24, it can be shown that north and south poles of the electromagnets are continually rotating and dragging the soft-iron core in the center with them. The rate at which the core rotates is directly dependent upon the frequency of the alternating current, because the rate at which the field rotates depends upon the power line frequency. And, as the rotor turns with the field, its speed also depends upon the frequency. This is the principle of operation for the synchronous motor.

A nonexcited synchronous motor can be made with one set of electromagnetic coils by using a trick called a *shaded pole,* which is illustrated in Figure 5-26. The shading pole has an electromagnetic field produced by the current induced in the copper ring. The ring around the shading pole has a large rotating current due to the induction of the varying magnetic field. This current reverses with each alternation. The overall effect is that the copper ring produces a field that is very intense and behaves like the second set of poles in Figure 5-24. One problem with the shaded pole synchronous motor in Figure 5-26 is that the rotor must become magnetized by the field

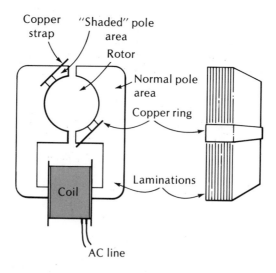

Figure 5-26. Shading pole for producing rotating field

when it is first energized. However, when the armature is stationary, the field in the soft iron continually reverses with each half cycle. See Figure 5-27. Since opposite fields occur in the armature and rotor coils, the induced field holds the shaft steady. To start such a motor it is necessary to spin the shaft mechanically.

To get around this problem some synchronous motors have a separate small damper winding on the rotor that produces a stator-rotor interaction to get the motor started. This is a self-starting synchronous motor. Large synchronous motors may be started with another smaller motor.

Induction Motors

Instead of placing a soft-iron material at the rotor, copper coils can be used. This makes it an *induction motor*. The principle of operation is shown in Figure 5-28. The induced current flows in the copper coils of the rotor and produces a magnetic field which is attracted to the rotating magnetic field of the stationary poles.

Figure 5-29 shows an example of a rotor that can be used in this type of motor. This *squirrel cage* arrangement consists of copper bars that are shorted at the end. Each copper bar then forms the closed conducting loop. When an induced voltage is produced by a varying magnetic field, the short circuit wire will have a high induced current. The magnetic field of the induced current rotates with the electromagnetic field of the stationary windings.

The squirrel cage is only one of many ways to make the rotor. Individual copper wires wound on a slotted rotor are also used.

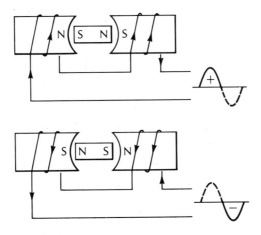

Figure 5-27. Soft iron rotor cannot move

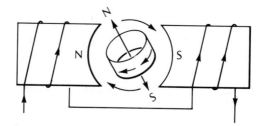

Figure 5-28. Basic principle of induction motor

Once the rotor begins to turn, its speed increases until it is almost equal to the speed at which the flux is revolving. Obviously, if the speed of the rotor and flux were the same, no voltage would be induced in the rotor. Without induction, however, this motor will not run, so the rotor goes slower than the revolving flux. The difference between the flux and rotor speeds is called the *slip*.

It takes four poles to start the motor. Once it is going, however, only two poles are necessary, provided the fields reverse at exactly the right moment. The *synchronous speed* is the speed that the rotor must turn so that the poles reverse at the right time to keep it going.

The rotating field can be produced by a single coil with a shaded pole, or by using two sets of coils with voltages 90° out of phase. (Two-phase power consists of two voltages that are 90° out of phase.) The out-of-phase voltages can be obtained by making use of the 90° phase difference between voltage and current in a capacitor.

Figure 5-30 shows this arrangement used in a capacitor start motor. When energy is first applied, the capacitor switch is closed, and the ac current flowing in a capacitor circuit is 90° out of phase with the original current.

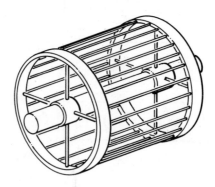

Figure 5-29. Squirrel-cage rotor

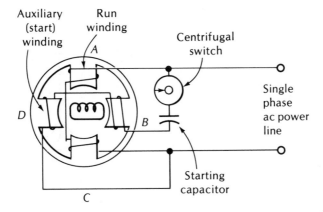

Figure 5-30. Capacitor-start induction motor

This puts the current in poles B and D 90° out of phase with the current in poles A and C. The magnetic field will rotate as the ac current reverses direction. The rotor increases its speed until it approaches synchronous speed, at which time the capacitor switch is opened. Remember that only one set of coils is needed to operate the motor at synchronous speed. The speed is maintained by the alternating current in coils A and C. The switch is usually designed so that it is opened automatically by the centrifugal action of the rotating shaft of the motor.

The rotating field for the three-phase induction motor is obtained by reversing the method shown in Figure 5-19. The supply produces a rotating field in the three sets of coils and the rotor follows the field.

Repulsion Motors

The principle of the repulsion motor is illustrated in Figure 5-31. The short-circuited coil produces a magnetic field in the core material that is repelled by the magnetic field of the stator. This repulsion causes the rotor to turn as the rotor field tries to get away from the light poles of the stator field.

The construction of the repulsion motor is obviously similar to that of a dc motor. In fact, it has some of the same characteristics: high starting torque, high operating speeds, and its direction is easily reversed. To reverse the direction of rotation on the repulsion motor, the position of its brushes is changed from one side of the neutral plane to the opposite side. This is also shown in Figure 5-31.

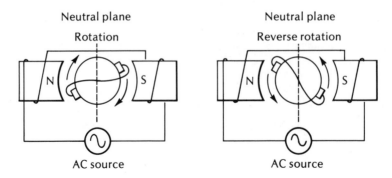

Figure 5-31. Repulsion motor

Motor Combinations

You will remember that in the dc motors the series-wound and the shunt-wound devices have different characteristics, and it is possible for the designer to combine these in a compound-wound generator that produces a compromise between starting torque and good speed regulation.

Essentially the same thing happens in ac motors. Instead of making a pure induction motor or a pure synchronous motor, it is possible to put induction windings in a synchronous motor or repulsion windings in an induction motor to get some specific desired characteristics. Such motors are usually identified according to the types of characteristics that predominate. For example, if it is primarily a synchronous motor but has an induction winding to get it started, it would still be called a synchronous motor.

The important characteristics of ac motors are summarized in Table 5-2.

Synchros, Servos, and Resolvers

DEFINITIONS

Synchros, servos, and resolvers are electrical devices used in control systems. Again, even though they are electrical, it is a good idea for electronics technicians to have a basic understanding of what these devices do and how they operate.

By definition a *synchro* is a remote positioning circuit that is made with two devices. One is called the *synchro transmitter* and the other is called the *synchro receiver*.

A *servo* is an automatic system for controlling power to a mechanical load. A servo system may control the position, direction of rotation, or the speed of the servo motor. The servo system is comprised of a feedback loop that senses the output with a transducer and uses the sensed factor in the control loop.

TABLE 5.2. Characteristics of ac motors

Characteristic	Type of ac motor		
	Synchronous	Induction	Repulsion
Starting torque	0 (started mechanically)	Good	Very high
Speed	Lock to powerline frequency	Dependent on number of poles and load	Poor speed regulation

In Chapter 1 you were introduced to a system as an example of a feedback control system. The term *servo* is sometimes used to refer to any closed-loop feedback system.

A *resolver* is a transformer with a rotating primary that senses the angle of rotation. The output of the resolver consists of one or two secondaries, and the voltage across the secondary is proportional to the sine or cosine of the angle through which the primary has been rotated.

SYNCHROS

Figure 5-32 shows how a synchro is used. As a matter of fact, this is one of the first important applications of synchros. An antenna that rotates is located outside of the ship. It is desired for someone inside to be able to tell which direction the antenna is pointing at all times. (The person inside cannot see the antenna.)

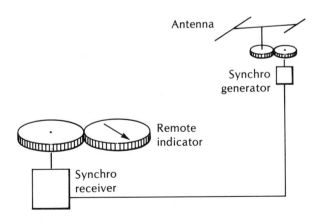

Figure 5-32. Synchro used for remote indication

The synchro transmitter senses the direction that the antenna is being turned. The synchro signal positions the synchro receiver so that the arrow on the dial points exactly to the same direction as the antenna. The transmitter is sometimes thought of as being a transducer that converts position into electrical signals. The synchro receiver is a transducer that converts the electrical signal back to a position indicator.

Figure 5-33 shows the basic construction of the synchro. This could be either a transmitter or a receiver. Winding *AB* can be rotated on a shaft, and the device gives very much the appearance of a motor.

Consider the case of the synchro transmitter. In the position shown in Figure 5-33 the primary coil will induce a maximum voltage into the winding marked S_1, and lower values of induced voltage values into windings S_2 and S_3. The voltage and polarity relationships between the windings are shown.

If you wind the primary winding counterclockwise 60°, it will be in line with S_3. Winding S_3 would receive the maximum voltage, in which case S_1 and S_2 will have reduced voltage values. The amount of voltage and the phase of the voltage induced in each winding depend upon the position of the primary with reference to secondaries S_1, S_2, and S_3.

Now assume that this is a synchro receiver. Windings S_1, S_2, and S_3 have applied voltages as shown. The coil connected to terminals *A* and *B* is single-phase ac. Remember that this winding is free to rotate. With the voltages shown in Figure 5-33, the movable winding will rotate until it is aligned with coil S_1 as shown. The positive side of the movable coil is attracted to the negative voltage at the bottom of S_1, and the negative side of the movable coil is attracted to a point halfway between S_2 and S_3.

The connections for the synchro transmitter and synchro receiver are shown in Figure 5-34. The movable coil transmitter is aligned with S_1. The movable winding in the receiver has exactly the same phase and amplitude as the primary winding in the transmitter. Also, the stationary coils have the same voltage and phase as S_1, S_2, and S_3 in the primary. So, the movable coil in the receiver will align itself in exactly the same position as the movable coil in the transmitter. If you change the position of the movable coil in the transmitter, the movable coil in the receiver will turn to exactly the same position so that its magnetic field aligns with the fields produced by S_4, S_5, and S_6.

Normally, the receiver is connected to some type of indicating device, or it may be used in an analog computer, but because of its construction it is not usually able to drive a large mechanical load.

SERVO SYSTEMS

Figure 5-35 shows the basic principle of operation for a servo system. This is not unlike the closed-loop system that was previously discussed. The servo motor is capable of driving a heavy mechanical load, and it can be controlled in direction or speed or position. The servo motor is mechanically

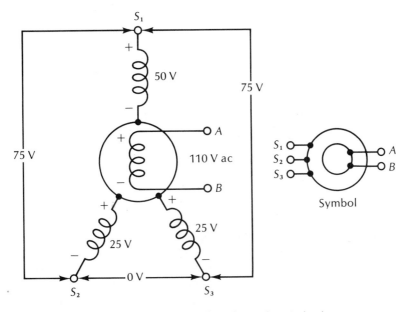

Figure 5-33.   Construction of synchro with typical voltages

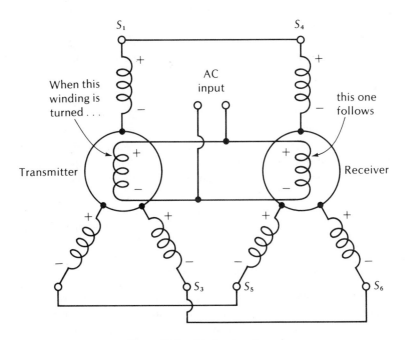

Figure 5-34.   Mechanics of synchro

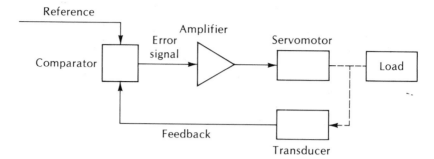

Figure 5-35. Closed-loop servo system

connected to the load, as indicated by dashed lines, and also to the sensing transducer. The type of transducer depends upon the type of servo system used. The feedback signal is compared with the reference signal in the comparator. If there is an error (difference between signals), it is amplified and delivered to the servo motor to correct the servo motor's speed (or position or direction). Correction is made continually until there is no output error signal from the comparator.

The servo system may use a synchro as a transducer to sense the position of the servo motor. The feedback control system may be continuous, or it may be an OFF/ON device (like the control circuit of the furnace in a home).

In modern electronic systems the transducer, comparator, and amplifier are all electronic, but in the earlier servo systems these functions were accomplished by mechanical or electromechanical methods. In the chapter on power supplies you will study examples of electronic feedback control.

RESOLVERS

A *resolver* has an output voltage that is proportional to the angle to which the primary winding is rotated. Figure 5-36 shows an example of a resolver with the two secondary windings. The amount of voltage induced by the primary is rotated from one position to another. The actual output voltage is proportional to the sine or the cosine of the angle to which the primary is rotated.

Resolvers can be used as a position transducer in a servo system. They are also used extensively in analog computers when it is desired to multiply some factor by the sine or cosine of an angle.

SATURABLE REACTORS AND MAGNETIC AMPLIFIERS

Figure 5-37 illustrates the basic principle of a saturable reactor. To begin the discussion, assume that switch *SW* is open so that no dc current flows through coil L_2. The lamp load (or any other kind of load) is in series with

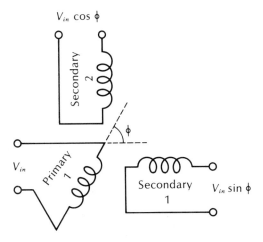

Figure 5-36. Resolver with two secondary windings

coil L_3 in the ac line. The inductive reactance of coil L_3 will cause a voltage drop and reduce the amount of voltage across the load.

If switch SW is closed, it will cause dc current to flow in winding L_2. The amount of current is controlled by a rheostat R. The dc current through L_2 will produce a flux in the soft-iron core that combines with the flux due to current through L_3. On one half cycle the two fluxes add, and on the next half cycle of ac the two fluxes subtract, or oppose.

Remember that the inductance of a coil depends upon the type of core material and the amount of flux in that core. If the dc current in L_2 is sufficiently large, then on one half cycle it combines with the flux due to L_3, and

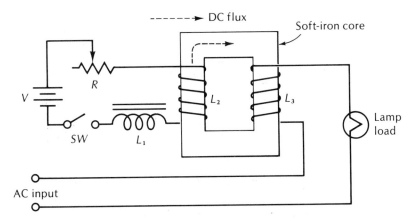

Figure 5-37. A saturable reactor controlling a lamp load

the two will produce a saturation in the soft-iron core. When the core is saturated there can be no further increase in flux, and therefore there is no inductive reactance. The effect of saturation, then, is to eliminate the effect of the iron core and reduce the inductive reactance.

The overall result is that R controls the inductive reactance of L_3, so it controls the power to the load.

Since an ac current is flowing in L_3, it will induce an ac voltage in coil L_2. Normally, this flow would cause an ac current to flow in the battery circuit. To prevent this, a choke is placed in the dc line to oppose any alternating current flow.

Figure 5-38 shows an arrangement of a saturable reactor that will not produce an ac current in the dc line. In this circuit the flux due to the ac coil tends to flow around the outside of the core, while the flux from the dc tends to divide into the two halves. The overall result is that the ac flux does not induce a voltage in the dc control winding.

The circuit of Figure 5-38 is modified slightly in the circuit of Figure 5-39. Two diodes have been added in series with each coil and in series with the load. When an ac voltage is applied across terminals A and B, the voltage alternates between positive and negative half cycles. On the half cycle when A is positive with respect to B, the two diodes conduct; and when B is positive with respect to A, both diodes are cut off. The importance of this

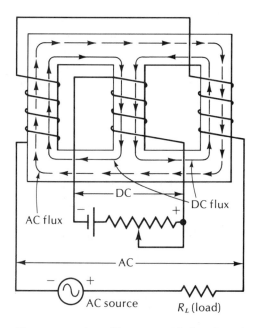

Figure 5-38. Saturable reactor with three-legged core

arrangement is that the current only flows in one direction in each of the two ac coils.

To see why this single direction of flow is important, return again to the simple saturable reactor of Figure 5-37. On one half cycle the fluxes add to reduce the inductive reactance, but on the opposite half cycle the fluxes oppose. This means that the inductive reactance is higher than normal because it is harder for the core to establish a flux. The result is a nonsymmetrical waveshape and a waste of energy.

In the circuit of Figure 5-39 the situation is avoided. This combination of saturable reactor and diodes is called a *magnetic amplifier*. It is an amplifier in the true sense of the word because a small amount of dc power is used to control a large amount of ac power. The coils are arranged in such a way that the current through them only flows in one direction and only on one half cycle. The dc current in the dc winding will always have the same effect on both half cycles of current in the ac coils. If the transformer is designed in such a way that it saturates on alternate half cycles of ac, then the dc winding is adjusted in such a way that it reduces the flux. That way the dc winding will have the greatest control. By adjusting variable resistor R_a, it is possible to set the transformer to various levels of saturation and therefore obtain the maximum possible control output voltage.

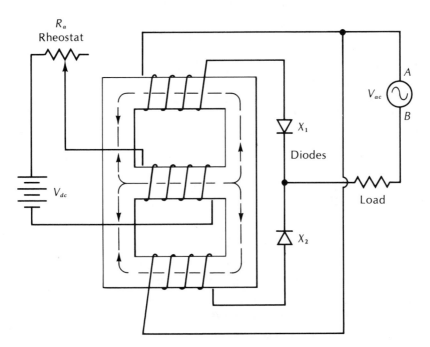

Figure 5-39. Modification of Figure 5-38

MAJOR POINTS

1. Generators operate on the principle that relative motion between a conductor and a magnetic field causes a voltage to be generated in the conductor.
2. In dc generators a commutator is used for converting the ac voltage generated in the armature to a pulsating dc.
3. In smaller low-power ac generators the armature may be rotated and the field stationary. In larger ac generators the field is rotated and the armature is stationary.
4. Self-excited dc motors are classified by the way the field winding is connected. The field winding is in *series* with the load, parallel *(shunt)* with the load, or there is a series-parallel *(compound)* connection.
5. Ac motors are classified by the method used to turn the rotor. The rotor is *synchronized* with the rotating field, or a magnetic field is *induced* in the rotor, or *repulsion* occurs in the rotor due to opposing magnetic fields.
6. A series-wound motor is sometimes called a *universal motor* because it will run on ac or dc power. (If it is designed to run on ac, there will be a few minor changes in construction.)
7. A synchro system was originally used for remote indicating. Synchro transmitters sense the amount of rotation. Synchro receivers follow the rotation of the transmitter.
8. A closed-loop servo system is used to control the speed, position, or output power of a motor.
9. A saturable reactor is an inductor in which a dc current controls the amount of inductive reactance.
10. A magnetic amplifier uses diodes to assure that the saturation is controlled on both half cycles of ac. As in a saturable reactor, a dc current controls a large amount of ac power in a magnetic amplifier.

PROGRAMMED REVIEW

(Instructions for using this programmed section are given in Chapter 1.)

The important concepts of this chapter are reviewed here. If you have understood the material, you will progress easily through this section. Do not skip this material, because some additional theory is presented.

---

1. In the simple generator shown here the rotating conductor is called the

A. armature. (Go to block 17.)
B. field. (Go to block 9.)

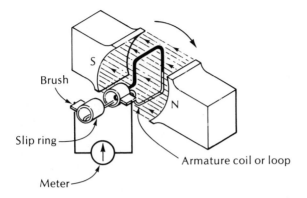

Brush

Slip ring

Meter

S

N

Armature coil or loop

2. Your answer to the question in block 24 is wrong. Read the question again, then go to block 22.

3. Your answer to the question in block 11 is wrong. Read the question again, then go to block 28.

4. Your answer to the question in block 19 is wrong. Read the question again, then go to block 10.

5. Your answer to the question in block 18 is wrong. Read the question again, then go to block 27.

6. The correct answer to the question in block 20 is A. Perfect regulation would be zero percent.

   Here is your next question.
   Which of the following is a type of ac motor that has a speed which is directly dependent upon the frequency of the ac power?

   A. Universal. (Go to block 15.)
   B. Synchronous. (Go to block 11.)

7. Your answer to the question in block 21 is wrong. Read the question again, then go to block 13.

8. Your answer to the question in block 13 is wrong. Read the question again, then go to block 24.

9. Your answer to the question in block 1 is wrong. Read the question again, then go to block 17.

10. The correct answer to the question in block 19 is A. A series-wound dc motor is sometimes called a universal motor because it will run on ac power or dc power.

Here is your next question.
Which type of ac motor can have its direction of rotation reversed by changing the position of its brushes? (Go to block 30.)

11. The correct answer to the question in block 6 is B. Synchronous motors are used for clocks and other applications where a constant speed is desirable.

Here is your next question.
A closed-loop system is used to turn a heavy antenna and hold it in position. This is an example of

A. a synchro system. (Go to block 3.)
B. a servo system. (Go to block 28.)
C. a resolver system. (Go to block 29.)

12. Your answer to the question in block 22 is wrong. Read the question again, then go to block 18.

13. The correct answer to the question in block 21 is A. There is a rotating field in a synchronous motor and in induction motors. One way to get the required rotating field is to use a shading pole. Another way is to use two-phase or three-phase power. The start capacitor provides the phase shift necessary for converting single phase power to two-phase power for split-phase starting.

Here is your next question.
In the simple generator shown below the voltage across terminals X and Y is

A. ac. (Go to block 24.)
B. dc. (Go to block 8.)
C. pulsating dc. (Go to block 16.)

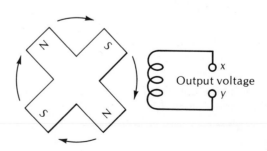

14. Your answer to the question in block 27 is wrong. Read the question again, then go to block 20.

15. Your answer to the question in block 6 is wrong. Read the question again, then go to block 11.

16. Your answer to the question in block 13 is wrong. Read the question again, then go to block 24.

17. The correct answer to the question in block 1 is A. You should not think of the armature as being the rotating part of a generator. In many alternators the field is rotated and the armature is stationary.

   The armature is the part of the generator in which the voltage is generated. Its output is delivered to the load resistance.

Here is your next question.
Dc motors may be used in control systems because

A. of their low maintenance problems. (Go to block 26.)
B. their speed and direction of rotation are easy to control. (Go to block 21.)

18. The correct answer to the question in block 22 is A. If you reverse both the field and the armature windings the direction of rotation will not be changed. You can demonstrate this principle by applying the generator rule to the simple generator of Figure 5-5.

Here is your next question.
A closed-loop control system that holds the speed of a motor to some desired value is called

A. a synchro. (Go to block 5.)
B. a servo. (Go to block 27.)

19. The correct answer to the question in block 28 is B. Flux lines do not move. The arrows drawn on flux lines show the direction that an isolated unit north pole would move in the field.

Here is your next question.
Which of the following types of motors will run on either a dc or an ac voltage?

A. Series wound. (Go to block 10.)
B. Shunt wound. (Go to block 4.)

20. The correct answer to the question in block 27 is A. There is very little difference between the construction of dc generators and dc motors. One difference in some applications is the size of the wires used.

Here is your next question.
The regulation of a generator is a measure of how well it holds its output voltage when the load current changes. Is a high percent regulation more desirable than a low percent regulation? (Assume that it is desired to have a steady voltage under varying load current conditions.)

A. A low percent regulation is better. (Go to block 6.)
B. A high percent regulation is better. (Go to block 23.)

21. The correct answer to the question in block 17 is B. The speed of a dc motor is controlled by adjusting the armature or field current. The direction can be reversed by reversing either the field or armature connection.

Here is your next question.
In a capacitor start induction motor, the purpose of the capacitor is to obtain a

A. two-phase voltage for operating the field windings during startup. (Go to block 13.)
B. three-phase voltage for operating the field windings during startup. (Go to block 7.)

22. The correct answer to the question in block 24 is B. A compound-wound generator can be designed with better regulation than a shunt-wound generator. The poorest regulation is obtained with a series-wound generator.

Here is your next question.
Reverse the direction of rotation of a dc motor by

A. reversing either the field winding or the armature winding connections. (Go to block 18.)
B. reversing both the field winding and the armature winding connections. (Go to block 12.)

23. Your answer to the question in block 20 is wrong. Read the question again, then go to block 6.

24. The correct answer to the question in block 13 is A. A rotating permanent magnet is only used in low-power ac generators.

Here is your next question.
In comparing the series and shunt generators, which has the better regulation?

A. Series. (Go to block 2.)
B. Shunt. (Go to block 22.)

25. Your answer to the question in block 28 is wrong. Read the question again, then go to block 19.

26. Your answer to the question in block 17 is wrong. Read the question again, then go to block 21.

27. The correct answer to the question in block 18 is B. A servo is a system that controls power to a load.

Here is your next question.
Is this statement correct? If a dc voltage is applied to a dc generator it will run like a motor.

A. Correct. (Go to block 20.)
B. Not correct. (Go to block 14.)

28. The correct answer to the question in block 11 is B. As mentioned before, servos can be used to control heavy loads. Resolvers produce an output voltage, not torque.

Here is your next question.
Is this statement correct? Flux lines move from the north pole to the south pole in a magnet.

A. Correct. (Go to block 25.)
B. Not correct. (Go to block 19.)

29. Your answer to the question in block 11 is wrong. Read the question again, then go to block 28.

30. The correct answer to the question in block 10 is repulsion type.

You have now completed the programmed section.

SELF-TEST WITH ANSWERS

(Answers to this test are given at the end of the chapter.)

1. Which of the following types of generators can be designed so that it operates nearly like a constant-voltage generator? (1) Series wound; (2) Compound wound.

2. When conductors are moved through a magnetic field a voltage is generated. The amount of voltage depends upon the strength of the magnetic field, the number of conductors moving through the field, and (1) the speed at which the conductors are moved; (2) the direction of motion.

3. Which of the following is an advantage of a repulsion motor? (1) Its direction of rotation can be easily reversed; (2) Its rotor is locked to a speed by the power frequency.

4. A device that produces an output voltage which is proportional to the amount of rotation of its shaft is the (1) synchro; (2) resolver.

5. Which of the following is correct for comparing types of ac generators? (1) A low-power compact construction is used for the rotating armature type, and a high-power three-phase construction is used for the rotating field type; (2) A low-power compact construction is used for the rotating field type, and a high-power three-phrase construction is used for the rotating armature type.

6. An advantage of using dc motors in control systems is that their speed is easily controlled by electronic means. Another advantage is (1) very low maintenance cost; (2) easy reversal of direction by electronic methods.

7. An inductor is designed so that the flux in its core can be set by a dc current in a separate winding. This is called (1) a swing choke; (2) a saturable reactor.

8. Which of the following is a motor that will eventually destroy itself with high speed unless it is connected to a mechanical load? (1) A series-wound motor; (2) A shunt-wound motor.

9. Another name for an ac generator is (1) alternator; (2) magneto.

10. A closed-loop motor control system is called a (1) synchro system; (2) servo system.

Answers to Self-Test

1. (2)
2. (1)
3. (1)
4. (2)
5. (1)
6. (2)
7. (2)
8. (1)
9. (1)
10. (2)

CHAPTER 6

LINEAR AND DIGITAL INTEGRATED CIRCUITS

Integrated circuits, or IC's, are usually placed in one of two broad categories: *linear* or *digital*. A linear integrated circuit is one that has an output which is directly related to the input in some ways. A good example is a Class A amplifier. It has an output signal that is an amplified version of the input signal. The word *analog* is sometimes used to mean linear in integrated circuit technology.

A digital integrated circuit is one that functions like a switch. It has only two states: ON and OFF. These states are often called 1 (for on) and 0 (for off).

The operational amplifier (op amp) is probably the most popular of the linear integrated circuits. The term *operational amplifier* is misleading. It is based on the fact that these amplifiers were used in early analog computers to perform arithmetic operations such as addition, subtraction, multiplication, and division. The modern integrated circuit version of the op amp can still be used for these purposes, but that is only a small part of their overall use.

The secret of the op amp's success in modern circuits is its versatility. One type of amplifier can be used for a wide variety of applications. Furthermore, now that op amps are made as integrated circuits, their reliability is very high and their cost is low. You will study the use of op amps in this chapter.

Logic gates are digital circuits that perform basic functions. Digital integrated circuit designs are usually made by combining these logic gates. The six basic gates are covered in Chapter 1.

The complete operation of the digital gate can be described in a *truth table*. The truth table is really a simplified way of explaining all of the possible inputs and outputs for a particular gate. If you learn to use the truth table properly, you can readily predict the input and output of more complicated circuits that are combinations of basic gates.

Integrated circuit logic developed quite rapidly starting in the early 1960s. Unfortunately, different manufacturers produce different versions of the logic gates. These versions are not necessarily compatible—that is, they cannot be readily combined in a single system. For this reason it is important to be able to identify logic families and some of their characteristics. In this chapter you will study some of the more popular digital IC families, and you will learn how interfacing can be accomplished between incompatible families. The term *interfacing* simply means the connection between one circuit and another.

LINEAR INTEGRATED CIRCUITS

Although the op amp is the most popular of the linear integrated circuits, there are other types that you should also be familiar with.

The simplest and most versatile linear integrated circuit is an *array*. An array is simply a combination of semiconductor components on an integrated circuit. It is distinguished from other integrated circuits by the fact that the IC components are not connected in any way to perform a particular function. Figure 6-1 shows several examples of manufactured linear IC arrays.

Many linear integrated circuits are designed to perform complete functions in a system. A good example is a linear IC audio amplifier. Another example is the regulator in a power supply. These integrated circuits are designed to perform one specific type of function. They are different from op amps, which can be connected to perform a wide variety of functions.

Integrated circuits are normally operated at low power levels. They do not have provisions for radiating the heat generated in power amplifiers. However, special integrated circuit combinations have been designed to produce large output powers. These combinations are referred to as *hybrids*. They are hybrid in the sense that they contain two different breeds of circuits. One is the integrated circuit that performs the voltage amplification in the system, and the other is the power amplifier, which is capable of operating a transducer. Hybrid circuits are contained in one single package just like op amps and arrays.

The Operational Amplifier (Op Amp)

Figure 6-2 shows a block diagram of a typical *differential operational amplifier*. The term differential operational amplifier comes from the fact that the input circuit is a differential amplifier. There are op amps that use other types of input circuitries, but this is by far the most popular type.

The two signal input terminals are marked – and +. When the input signal is delivered to the terminal marked –, the output is inverted. In other words, the output is 180° out of phase with the input. When the input signal is at the + terminal, the output signal is not inverted; that is, it is in phase with the input signal.

The output of the differential amplifier goes to one or more stages of high-gain voltage amplification. Usually direct coupling is used between the amplifier stages. However, this is not always true, as you will see when you study the chopper-stabilized amplifier that is discussed later in this chapter.

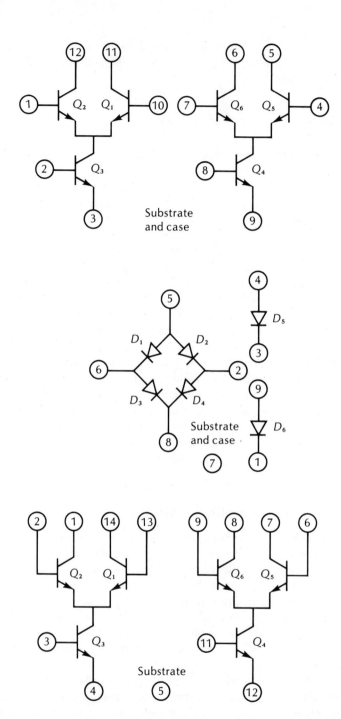

Figure 6-1. Integrated circuit arrays

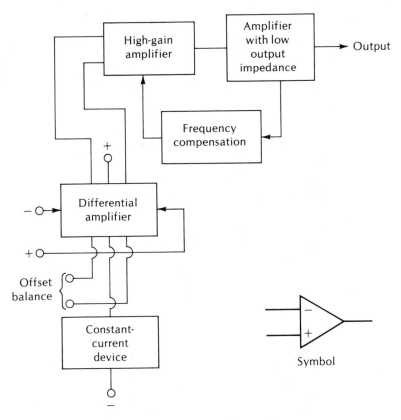

Figure 6-2. Block diagram of differential op amp

The output of the high-gain amplifier goes to an amplifier stage that has a low output impedance. This stage is sometimes referred to as the *power output stage,* but the term should be used with reservation because it does not produce a sufficient amount of power to drive most transducers.

In the block diagram of Figure 6-2 is a feedback circuit marked *frequency compensation.* This feedback circuit determines the high-frequency response of the op amp. For some types of op amps the frequency compensation circuit is added externally to the op amp package. In others it is internal and part of the monolithic construction.

An important characteristic of differential amplifiers is that they must be balanced in order to work properly for some applications. This is accomplished by an *offset balance* input, which requires the connection of external circuitry to control the amount of conduction through the individual components. In some types of op amp circuits, however, an offset balance adjustment is not required and these terminals are not connected to an external circuit.

A differential amplifier works best when it is connected through a constant-current source (also shown in the block diagram of Figure 6-2).

The operational amplifier symbol does not usually show the power supply connections. Nor is it necessary to show the balance terminals and the frequency compensation terminals unless they are actually required for circuit operation. The symbol for the differential op amp is shown in the inset of Figure 6-2.

Characteristics of a Typical Op Amp

Some of the most important characteristics of an op amp in work applications are gain, rolloff, impedance, and coupling.

GAIN

Early tube-type op amps had gains of about 20,000, with a minimum value of 10,000. Today the integrated circuit op amps have gains as high as 3,000,000, but a typical value is over 100,000. These values are known as *open-loop* gains. When the op amp is connected into a circuit with negative feedback, the gain is greatly reduced.

LINEAR ROLLOFF

In order for the op amp to work properly, it must have a linear *rolloff* at the high frequency end of its response curve, or *Bode plot*. Linear rolloff refers to the drop in gain with increase in frequency as shown in Figure 6-3. Any plot that shows the gain of an amplifier or the phase of the output signal with

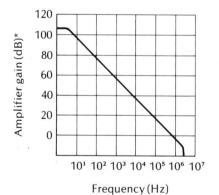

*Amplifier gain in dB = 20 log (voltage gain)
= 20 log R_f/R_i for inverting input

Figure 6-3. Typical op amp Bode plot

reference to the input vs the frequency is called a Bode plot. Note that the response is flat up to about 7 or 8 Hz, and from that point the response drops off linearly up to a frequency of 1 MHz.

AMPLIFIER IMPEDANCE

It is desirable for the op amp to have a high input impedance so that it does not load the preceding circuit. Also, it should have a low output impedance so that it can be matched with almost any circuit or follower.

DIRECT COUPLING

The overall amplifier must act like a direct-coupled device so that any change in dc voltage at the input terminal will produce a change in the output voltage. In most circuits this effect is accomplished simply by direct coupling all of the amplifiers within the op amp package.

Additional Requirements

In addition to the requirements just mentioned, it is also desirable that the op amp have low power consumption and low internal noise. Another desirable feature is low *temperature drift*. This simply means that the output voltage will not change over long periods of time due to changes in temperature of the IC.

OFFSET

Since the input stage to the differential op amp contains a differential amplifier, you can expect to see such specifications as common-mode rejection and common-mode rejection ratio on their spec sheets. You can also expect to see op amps connected into circuits that have both positive and negative power supply voltages. Figure 6-4 shows some packages used for the popular 741 op amp.

A differential amplifier is made with two amplifying devices—such as tubes or transistors—connected in parallel, usually through a common constant-current amplifier. Figure 6-5 shows an example. In normal operation the two amplifiers must conduct equal amounts of current when there is no input signal. When they do, they are said to be *balanced*. To determine if they are balanced, identical voltages are delivered to the two input terminals. If two identical sine waves are delivered to the input terminals, there should be no sine wave voltage across the output terminals. This is called a *common-mode rejection*.

When the two amplifiers are unbalanced, an external adjustment can be added. The adjustment is usually made so that there is no output signal (or minimum output signal) when the two input terminals have identical input sine wave voltages. This is called an *offset adjustment*.

14 Pin DIP

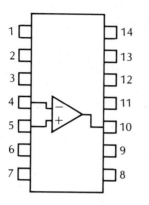

1. No connection
2. No connection
3. Offset null
4. Inverting input
5. Noninverting input
6. V−
7. No connection
8. No connection
9. Offset null
10. Output
11. V+
12. No connection
13. No connection
14. No connection

TO5

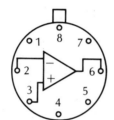

1. Offset null
2. Inverting input
3. Noninverting input
4. V−
5. Offset null
6. Output
7. V+
8. No connection

8 Pin DIP

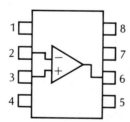

1. Offset null
2. Inverting input
3. Noninverting input
4. V−
5. Offset null
6. Output
7. V+
8. No connection

Figure 6-4. Examples of op amp pin connections

There are several ways to accomplish an offset adjustment with the op amp. Naturally, this adjustment must be made external to the op amp differentiating amplifier because the integrated circuit package is sealed. Figure 6-6 shows three possibilities.

When the op amp is used as an amplifier, the input signal usually goes to only one input terminal. If the input signal is to the inverting terminal, then

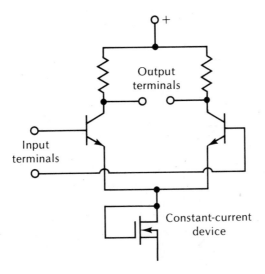

Figure 6-5. A differential amplifier circuit

the offset adjustment is normally made at the noninverting terminal. The adjustment is accomplished by using a variable resistor between the + and − power supply terminals and adjusting this resistor to produce a dc voltage on the noninverting input. If the input signal is to the noninverting terminal, then the offset adjustment is usually made at the inverting terminal. (See Figure 6-6.)

Some op amps are made with special input terminals for making the offset adjustment. The popular 741 op amp is an example. The method of adjusting the 741 for balance is shown in Figure 6-6. What you are actually doing when you make this adjustment is setting the zero volt point between the two legs of the input differential amplifier.

FREQUENCY COMPENSATION

The linear rolloff shown in the Bode plot of Figure 6-3 is accomplished by connecting a capacitive circuit between the output amplifier and some point in the voltage amplifier. Figure 6-7 shows three types of frequency compensation. Some amplifiers do not require external frequency compensation. The 741 is an example.

LATCHUP

Some of the earlier IC op amps had a problem called *latchup*. When the input signal to the op amp is greater than a certain predetermined value, it drives the op amp output into saturation. If there is no provision for preventing latchup, it remains in saturation even though the signal is removed. No change in the input signal can affect the output during latchup.

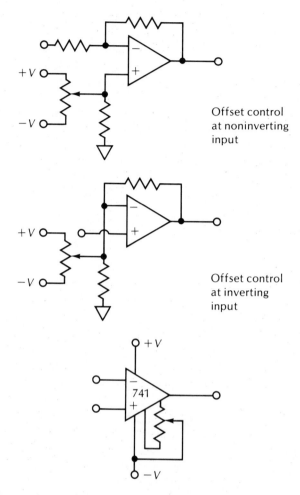

Figure 6-6. Examples of offset and adjustment

The specification sheet for most op amps today carry a provision marked "No latchup." Prevention circuits (Figure 6-8) can be used for earlier circuits which are still being sold. The operation of this circuit is simple. Diode X_1 will conduct when the positive amplitude of the signal voltage is greater than the positive supply voltage. Diode X_2 conducts when the negative amplitude of the signal voltage is more negative than the negative supply voltage. Conduction of either diode prevents any further increase in signal voltage from reaching the input terminal and thus latchup is prevented.

Other latchup circuits use zener diodes to limit the input signal amplitude. Latchup is prevented in the newer designs by providing for a much larger voltage swing at the input terminals.

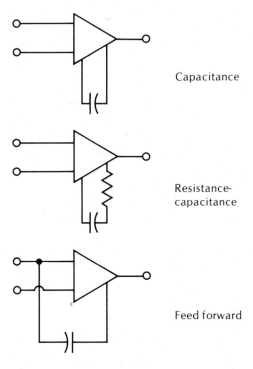

Capacitance

Resistance-
capacitance

Feed forward

Figure 6-7. Examples of frequency compensation

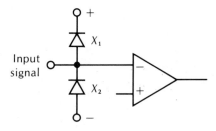

Figure 6-8. Circuit for preventing latch-up

Examples of Op Amp Circuitry

The versatility of the op amp can be best understood by studying examples of actual circuits.

INVERTING AMPLIFIER

Figure 6-9 shows an op amp connected as an inverting amplifier. The input signal is delivered through resistor R_i to the inverting terminal, and a feedback resistor (R_f) is connected from the output terminal (V_{out}) to the input. Resistor R_a is connected between the noninverting terminal and ground. However, in many circuits you will not see this resistor. It is necessary only when the balance of the differential amplifier stage is an important factor in the circuit design.

It can be shown mathematically that when the open-loop gain of an amplifier is very high, such as for an op amp, the output signal is almost entirely a function of the feedback resistor (R_f) and the input resistance (R_i). Mathematically, the voltage gain (A_v) for the inverting amplifier is equal to the ratio of R_f to R_i (see Figure 6-9). The negative sign in the equation indicates that the output is an inverted version of the input.

NONINVERTING AMPLIFIER

Figure 6-10 shows the op amp connected as a noninverting amplifier. Again, the gain of the amplifier is due almost entirely to the ratio of R_f to R_i. Note

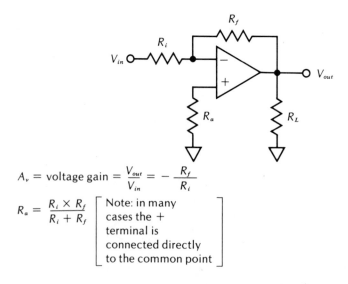

$$A_v = \text{voltage gain} = \frac{V_{out}}{V_{in}} = -\frac{R_f}{R_i}$$

$$R_a = \frac{R_i \times R_f}{R_i + R_f} \quad \left[\begin{array}{l} \text{Note: in many} \\ \text{cases the } + \\ \text{terminal is} \\ \text{connected directly} \\ \text{to the common point} \end{array} \right]$$

Figure 6-9. Op amp inverting amplifier

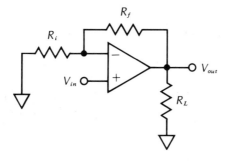

$$A_v = \text{voltage gain} = \frac{V_{out}}{V_{in}} = 1 + \left(\frac{R_f}{R_i}\right)$$

Figure 6-10. Op amp noninverting amplifier

that the feedback circuit is connected between the output and the inverting input, but the input signal is delivered to the noninverting input. The reason for this is that the feedback must be negative. In other words, the feedback signal must be 180° out of phase with the input signal to produce the degeneration required for limiting gain.

Before proceeding with the types of op amp circuitry, a few sample problems will be worked to show the relationship between the feedback circuitry of the op amp and the voltage gain. These example problems will also show the importance of the linear rolloff on the Bode plot.

SAMPLE PROBLEMS

Sample Problem 6-1

Design an inverting op amp circuit with a gain of 200.

Solution

To obtain a gain of 200, the ratio R_f/R_i must be numerically equal to 200. An input resistance value is chosen first. This choice is somewhat arbitrary, but the resistance should not be so low that it loads the previous stage. The input impedance to the op amp is numerically equal to R_i, and values of about 1 kΩ are typical.

There is one additional limit to the choice of R_i. Its value determines the value of R_f for a particular gain. It is undesirable to use a feedback resistance value greater than 1 MΩ because the shunting resistance effect of the amplifier begins to have an effect on dc stability and actual gain value for values of $R_f > 1$ MΩ. Somewhat higher values can be tolerated in op amps that use FET's rather than bipolar transistors.

$$\text{Let } R_i = 3.3 \text{ k}\Omega$$

$$\text{Then } A_v = 200 \text{ (given)}$$

$$A_v = (-)\frac{R_f}{R_i} \qquad \text{(Disregard the minus sign in gain calculation)}$$

$$200 = \frac{R_f}{3.3 \text{ k}\Omega}$$

$$R_f = 660 \text{ k}\Omega$$

The nearest preferred value is 680 kΩ. Using this value rather than the 660 kΩ value calculated gives a gain, in kilohms, of 680/3.3, or 206.

Sample Problem 6-2

Assume the op amp in Sample Problem 1 is a 741 type. What is the bandwidth of the amplifier?

Solution

First, the gain of 200 must be converted to dB gain for the voltage ratio. The mathematical procedure is given here. If you are not familiar with logarithms, you can convert dB gain to voltage gain by using the graph of Figure 6-11.

$$\text{dB gain} = 20 \text{ log} 200$$
$$= 20 \times 2.3$$
$$= 46 \text{ dB}$$

Find the value 46 dB on the graph of Figure 6-3, which is a Bode plot for the 741 op amp. Move to the right until you intercept the curve, then move down to the frequency value. You should get a value of approximately 50 kHz. The amplifier response is flat from approximately 0 Hz (dc) to 50 kHz. You could not predict the bandwidth this way if the rolloff of the response curve were not in a straight line.

Sample Problem 6-3

Using the same resistance values found in Sample Problem 6-1, find the gain of a 741 op amp connected as a noninverting amplifier.

Solution

From Figure 6-10,

$$A_v = 1 + \left(\frac{R_f}{R_i}\right)$$
$$= 1 + 200$$
$$= 201$$

Gain

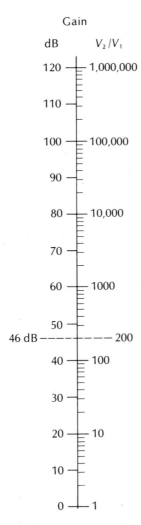

Figure 6-11. Graph for converting voltage ratio to dB

The gain and bandwidth values are almost identical for ratios R_f to R_i greater than 10.

FOLLOWER

Figure 6-12 shows an op amp connected in a follower circuit. The characteristics of this circuit are similar to those of cathode followers, emitter followers, and source followers. The input impedance is high and the output impedance is low. The output signal is in phase with the input signal.

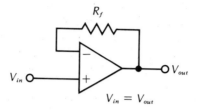

Figure 6-12. Op amp follower

Followers are used primarily as impedance-matching components and also as *buffers* between stages. A buffer is an amplifier that connects the output of one signal source to the input of another circuit. It connects the signal from one point to another, but it acts to isolate the two electrically.

It is not advisable to use an amplifier that is capable of latching for the follower circuit of Figure 6-12. The gain of this circuit is unity, and latchup is more likely to occur in amplifiers with low voltage amplification than in those with high voltage amplification.

SUMMING AND DIFFERENCING

Figure 6-13 shows the op amp used in a circuit that performs adding and subtracting operations. As shown by the equation in Figure 6-13, the output signal is dependent upon the amount of feedback resistance and the sum of the inputs to R_3 and R_4 minus the inputs to R_1 and R_2. In normal operation, resistors R_1, R_2, R_3, and R_4 would be made equal so that the output voltage is

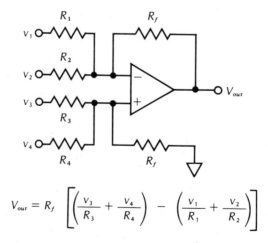

$$V_{out} = R_f \left[\left(\frac{v_3}{R_3} + \frac{v_4}{R_4} \right) - \left(\frac{v_1}{R_1} + \frac{v_2}{R_2} \right) \right]$$

Figure 6-13. Adding and subtracting op amp circuit

simply the ratio of R_f to the input resistance and the sum and differences of the voltages.

The input signals for the circuit of Figure 6-13 can be either ac or dc, depending on the desired application. When the circuit is used without R_1 and R_2, it is referred to as a *summing amplifier*. When it is used without R_3 and R_4, it is called a *difference amplifier*.

The simplicity of the adding and subtracting circuit makes the job of designing an analog computer quite simple, especially if adding and subtracting have to be accomplished in many places in the computer.

VOLTAGE COMPARATOR

Figure 6-14 shows the op amp connected as a *voltage comparator*. The term comparator is used in various applications. A logic circuit called a comparator will be studied later in this chapter. In other applications, circuits called comparators simply compare frequencies.

The voltage comparator of Figure 6-14 consists of a dc voltage applied to one terminal of the op amp and a varying voltage applied to the other terminal. It is presumed that this op amp cannot be latched. The operation is quite simple. When the input signal at B becomes more positive than the voltage

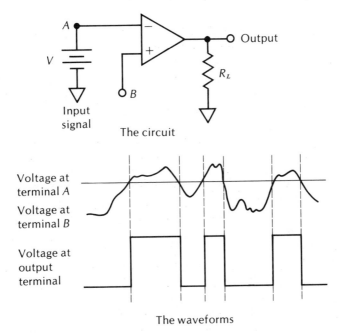

Figure 6-14.   Voltage comparator circuit.

at A, the output of the comparator is high. When the input signal at B is lower than the input signal at A, the output signal is low.

This application can be used in control circuits. For example, it can turn on a heater when the temperature goes below a certain value, and turn it off when the temperature goes above that value. In that case, a thermal transducer (such as a thermistor) can be used to sense the temperature and produce a voltage (through its circuitry) which acts like the voltage at terminal B. The output voltage would be delivered to a control circuit for the heater.

While it is possible to use an ordinary op amp for a comparator, manufacturers produce a special integrated circuit op amp for such applications. It is different from the standard op amp in that it has a much faster *slewing rate*. The slewing rate of an op amp is simply the maximum rate that the output voltage can be changed by changing the input voltage. Ideally, if the input voltage changes instantly from one value to another, then the output voltage should also change instantly. Comparators have an enormously high slewing rate, and this makes them ideal for switching applications.

GYRATOR

Two important advantages of integrated circuits are high reliability and low cost. Another advantage in many applications is their very small space requirement. However, little would be accomplished by using a large number of IC's in a circuit to miniaturize it, if large and bulky inductors were also used in the same system. In addition to their size, weight, and cost, another problem with inductors is that they can produce stray electromagnetic fields which can produce interference.

To get around the problems of using an inductor, the op amp can be connected in a configuration called a *gyrator* (Figure 6-15). Basically, a gyrator accomplishes the same thing as the inductor. It shifts the phase of the voltage and current to 90° with the current lagging the voltage.

ACTIVE FILTERS

A filter circuit permits one frequency, or range of frequencies, to pass but rejects all others. When a filter circuit is made with inductors and capacitors, it is called a *passive filter*. When it is made with amplifiers, it is called an *active filter*. The op amp can be readily adopted to active filter design. Four examples are shown in Figure 6-16.

The lowpass filter permits all the frequencies below the cutoff frequency (f_c) to pass. This feature is accomplished by using a passive lowpass filter at the inverting input, and at the same time, connecting a feedback circuit that is a highpass filter. The capacitor in the feedback circuit delivers the higher frequencies 180° out of phase with the input signal, and therefore tends to cancel these high frequency signals. The overall result is that only the low frequencies are passed to the output terminal.

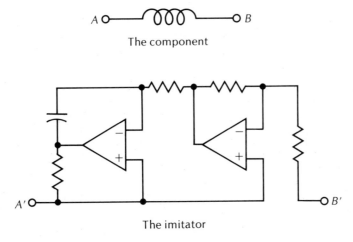

The component

The imitator

Figure 6-15. Op amp gyrator

The highpass filter uses a passive highpass input circuit and a lowpass feedback circuit. High frequencies from the output cannot be fed back to the input because they are grounded by the capacitor in the feedback filter circuit. Again, the result is that the high frequencies pass through the op amp circuit and the low frequencies are attenuated.

A bandpass filter can be accomplished by using a parallel tuned circuit in the feedback network. You will remember that the voltage gain of the op amp is the ratio of the feedback resistance to the input resistance. For a parallel resonant circuit the impedance at resonance is very high. Therefore, the feedback impedance for this circuit is very high at the resonant frequency, so the gain is also high, as shown by the response curve in Figure 6-16.

A notch filter uses a feedback circuit that consists of a passive series-tuned circuit. A series LC circuit has a high impedance at all frequencies except the resonant frequency. Therefore, the impedance of the feedback network in this op amp circuit is high at all frequencies except at resonance. At resonance the feedback impedance is low, so the amplifier response curve has a notch at that point.

The active filters shown in Figure 6-16 are very basic. In practice the feedback networks and the input networks can be much more complex, but the theory of operation is still the same.

OSCILLATORS

An electronic oscillator is simply an amplifier connected with regenerative feedback. An op amp can be readily put to use as an oscillator. There are

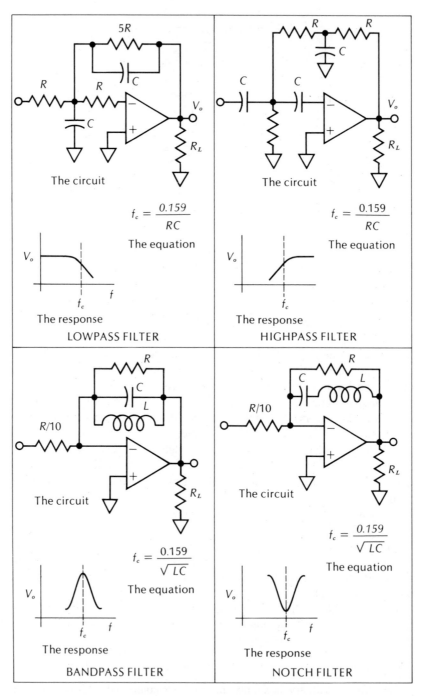

The circuit

$$f_c = \frac{0.159}{RC}$$

The equation

The response

LOWPASS FILTER

The circuit

$$f_c = \frac{0.159}{RC}$$

The equation

The response

HIGHPASS FILTER

The circuit

$$f_c = \frac{0.159}{\sqrt{LC}}$$

The equation

The response

BANDPASS FILTER

The circuit

$$f_c = \frac{0.159}{\sqrt{LC}}$$

The equation

The response

NOTCH FILTER

Figure 6-16. Active filters

two basic kinds of oscillators: those that produce pure sine waves are *sinusoidal oscillators,* and those that produce nonsinusoidal waveforms are *relaxation oscillators.*

In the sinusoidal oscillator of Figure 6-17 the feedback signal consists of two branches. These branches are actually highpass filters and lowpass filters connected in parallel. When resistor R is adjusted so that the basic frequency of the highpass filter is almost the same (but not quite the same) as the frequency of the lowpass filter, then the feedback signal will consist of only one frequency (or very narrow range of frequencies). In other words, only one signal can be amplified.

In the inverting amplifier configuration the output signal is 180° out of phase with the input signal only in the case of a purely resistive feedback network. The feedback network for the sinusoidal oscillator of Figure 6-17 is not purely resistive, and therefore it tends to shift the phase of the feedback signal. Shifting the phase produces regenerative rather than degenerative feedback.

The nonsinusoidal oscillator circuit of Figure 6-18 is accomplished by using a regenerative feedback network consisting of R_2 and R_3. Since the signal is returned directly to the noninverting stage, oscillation will occur. The feedback network comprised of R_1 and C_1 sets the gain of the amplifier at a high value for one particular frequency and thus determines the frequency of oscillation for this square wave generator.

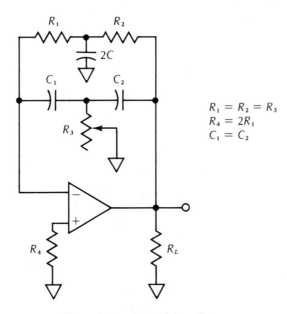

$$R_1 = R_2 = R_3$$
$$R_4 = 2R_1$$
$$C_1 = C_2$$

Figure 6-17. Sinusoidal oscillator

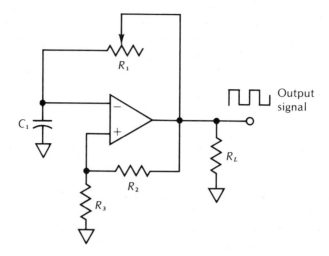

Figure 6-18.   Square wave generator

Basically, the square wave generator is an oscillator that is driven to saturation on both the positive and negative half cycles.

The examples of op amp applications given here do not represent all of the possibilities. They show the versatility of this amplifier and the reason for its great popularity.

Chopper-Stabilized Op Amps

Most differential op amps are direct coupled from the input signal terminals to the output; that is, any change in the dc level at the input amplifier will result in a change in the output voltage. While the two amplifiers making up the differential amplifier are quite evenly matched in an IC, it is never possible to keep them perfectly matched when there are temperature changes. A small change in temperature can produce an unbalance in the differentiating amplifier which causes a change in the dc output level. This condition is referred to as *temperature drift*.

Another problem with the differential amplifier is that very low amplitude ac signals are difficult to amplify due to noise generated by the amplifier itself.

The chopper-stabilized amplifier of Figure 6-19 is a special op amp that has a very low temperature drift. It is capable of producing an output for very very small changes in input voltage—even if these input signal voltages are a varying dc value. If an amplifier has any appreciable temperature drift, it is useless for amplifying small changes in dc input at the terminals. In such a case, one could not tell whether the change in output was due to a change in signal or to drift.

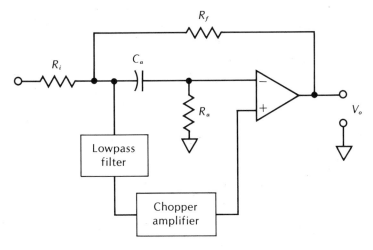

Figure 6-19. Chopper stabilized op amp

In the block diagram of a chopper-stabilized amplifier (Figure 6-19), note that the input signal consists of an inverting input from a highpass filter (C_a and R_a) and an input from a chopper amplifier to the noninverting terminal.

The input to the chopper amplifier is a lowpass filter which eliminates all high frequencies. The resulting dc and low-frequency input to the chopper amplifier is converted into pulses (that is, it is *chopped*), and these pulses are readily amplified by high gain ac amplifiers. The output of the chopped amplifier normally eliminates the pulses through another filter, and delivers only a dc voltage to the noninverting input.

The voltage at the output terminal E_o consists of the amplified version of the low-frequency and high-frequency signals together. Usually the Bode plot for this kind of stabilized amplifier is a small hump in the middle. This is the crossover point between gain due to the high-frequency input and low-frequency input from the chopper amplifier. This small hump is of no consequence to the overall operation.

Chopper-stabilized op amps are used in measuring instruments. They convert small changes in dc voltage from a measuring transducer to an output that is more easily measured.

Current-Differencing Amplifier

There are a few problems inherent with op amps. The fact that they require a positive and negative power supply makes it difficult to combine them in circuits with other electronic components. Their high gain and elaborate circuitry are wasted in some applications. Finally, many op amps cannot operate on the low supply voltages used for some logic gates. In other words, they are not compatible with the logic circuitry.

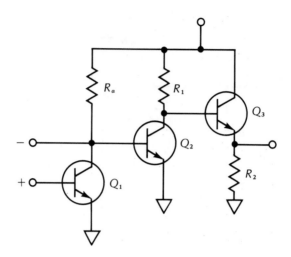

Figure 6-20.   Basic idea behind CDA

The *current-differencing amplifier* (CDA) can be used to advantage in some applications where the op amp is too elaborate, or where its supply requirements are difficult to meet.

Figure 6-20 shows the basic circuitry involved in the CDA. Transistor Q_2 is in a common-emitter configuration. The output signal is direct coupled to

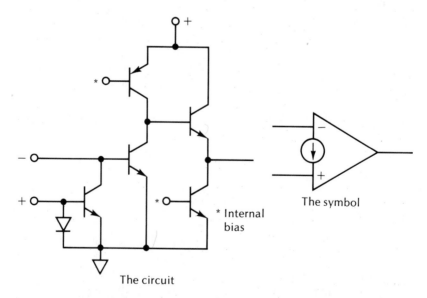

Figure 6-21.   The current-differencing amplifier

emitter follower Q_3. This combination (Q_2 and Q_3) gives a high input impedance, high gain, and low output impedance.

To get a noninverting input, an additional stage (Q_1) is used. Since Q_1 inverts the signal and Q_2 inverts it again, the output signal is in phase with a signal at the + input.

Figure 6-21 shows the actual circuit used in one CDA and its symbol. The circuit has a few basic differences from the one in Figure 6-20. The collector load resistor has been replaced with a *PNP* transistor that serves as a linear resistor. Resistor R_2 is replaced by *NPN* transistor Q_5, which also serves as a linear resistor. Diode D_1 across the emitter-base junction of Q_1 helps to stabilize that amplifier against temperature changes.

The overall result is that the amplifier behaves very much like an op amp, and it can be used in the same wide variety of applications. Since base currents form the input signals rather than voltages, it is a current-operated device. The input acts like a differential amplifier with its – and + inputs: thus the name, *current-differencing amplifier*.

DIGITAL INTEGRATED CIRCUITS

A good place to start the study of digital integrated circuits is by learning the names of the basic logic gates, their symbols, and the truth table that is used to describe their operation.

Each of the basic logic gates can be described with a mathematical equation that is based on Boolean algebra. Boolean algebra is a very useful tool for understanding gates, especially *compound gates* which are made up of a number of basic gates. However, it is possible to understand logic circuitry by not resorting to Boolean algebra if you know how the truth table works.

The AND Gate

Figure 6-22 shows the principle of the AND gate. The circuit is represented by two switches, *A* and *B*, in series with lamp *L* (which can also be called the *load*) and a voltage source. No current will be delivered to the load unless both switches are closed. Therefore, in order to get output *L*, it is necessary to have both inputs *A* and *B*.

This type of circuit can be made with electronic components, and this is true of all the other gates that you are now studying. The only purpose for using switches here is to simplify the meaning of the AND gate.

You will notice that three symbols are used for the AND gate: the MIL symbol, the NEMA symbol, and the ANSI symbol. The three symbols were developed by different standards groups. While the MIL symbol is more popular, many manufacturers prefer to use NEMA symbols, and you should learn both types.

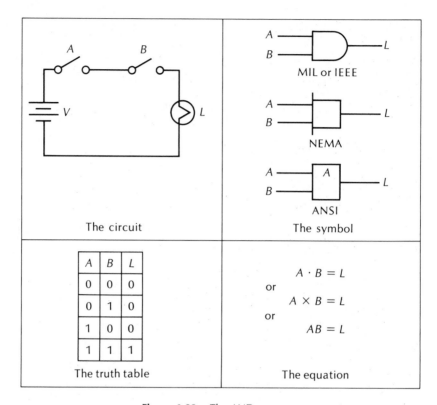

The circuit

The symbol

The truth table

The equation

Figure 6-22. The AND gate

The truth table is a shorthand method of explaining the different conditions that can occur in the circuit. It is based upon the idea that switches and other ON/OFF devices can be represented with the two numbers of the binary system: 0 and 1. The output of a gate has two conditions *(voltage* or *no voltage)* so it can also be represented with the two binary numbers.

We will define the open condition of the switch as being binary 0 and the closed position of the switch as being binary 1. We will use binary 0 to indicate that the lamp is off, and binary 1 to indicate that the lamp is on.

In the truth table the horizontal sections are called *rows,* and the vertical sections are called *columns.* The first row shows that when A and B are both open, the lamp is off. The second row is marked 0 1 0. This means that when A is open and B is closed, the lamp is still open. In other words, you cannot get the lamp to light by using only switch B. In the third row you see that when switch A is closed and switch B is open, the lamp is still off. This means that you cannot light the lamp simply by closing only switch A. The fourth row shows that if both A and B are closed (that is, in a 1 condition), the lamp is on.

It takes a long time to tell about all the possible conditions for the input (switches) and output (lamp) by using the English language. However, with the truth table all of the input and output conditions can be summarized very quickly.

In the Boolean equation, the multiplication sign is used for the word *AND*. It is shown in three forms in Figure 6-22. All three equations should be read, "*A* and *B* equals *L*."

To simplify your work in logic circuitry, you should memorize the symbols, truth tables, and equations for all of the logic gates being presented here.

The OR Gate

The conditions for the OR gate are shown in Figure 6-23. In the English language the word *OR* has two different meanings and this leads to some confusion. Suppose you read the statement, "There will be a light if switch

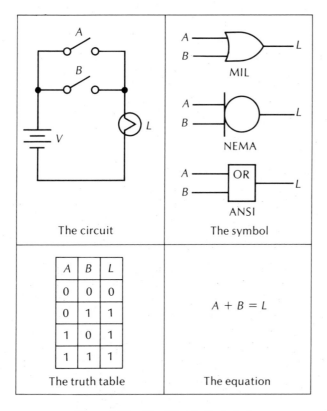

Figure 6-23. The OR gate (inclusive)

A or *B* is closed." Does this mean that there will be a light if switch *A* or *B* (but not both) is closed, or does it mean there will be a light if you close either *A* or *B* or both?

Although the English language does not make this clear, the logic circuitry is very definite about the difference between the two conditions. If you mean that either switch *A* or *B,* or both, can cause the lamp to be on, then you are talking about *inclusive OR* gates. This type of OR gate is shown in Figure 6-23. You can see in this illustration that the switches are in parallel in the circuit. The lamp will be on if you close *A* or *B,* or if you close both switches. As with all other gates, there are three symbols: the MIL, the NEMA, and the ANSI. Any of these symbols can be used to represent the circuit shown in Figure 6-23. The same symbols would be used for any inclusive OR circuit, whether it uses electronic components or relays or any other ON/OFF device.

The first row of the truth table shows that if *A* and *B* are both open, the lamp is off. The second row shows that if you close switch *B* but leave switch *A* open, the lamp will be on. Row 3 shows that if you close switch *A* but do not close switch *B,* the lamp will be on. Row 4 shows that if you close both switches, the lamp will be on. Note that in row 4 both switches can be activated to produce an output. This is what makes the circuit an inclusive OR.

The Boolean symbol for inclusive OR is a + sign. You should read the equation as follows: "*A* or *B* equals *L* ." Do not get in the habit of saying "*A* plus *B* " because this is not what the equation means. It means either *A* or *B* or both will produce an *L.*

The EXCLUSIVE OR Gate

Figure 6-24 shows what is meant by an *EXCLUSIVE OR* circuit. The circuit shows that you can produce a current through the load by closing switch *A* or by closing switch *B.* However, you cannot produce a current through the load if you close both switches. When the switches are in the down position, they are closed.

The symbols for the EXCLUSIVE OR circuit are similar to the ones used for inclusive OR. Study them carefully so that you do not get them confused.

The difference between the truth table for the EXCLUSIVE OR circuit and the truth table for the inclusive OR circuit is in the fourth row. With an EXCLUSIVE OR circuit, closing both switches does not produce a 1 in the *L* column. This means that if you close both switches, there will be no current through the load.

There are two ways to write the equation for EXCLUSIVE OR. To understand the first equation you must know that putting a line over a letter is called an *overbar.* An overbar indicates the word *NOT.* Read the first equa-

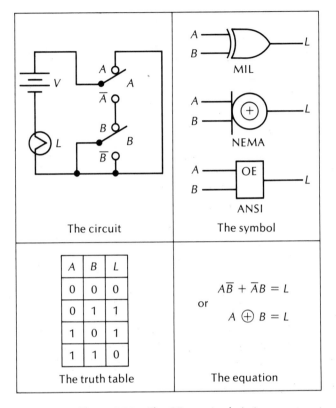

Figure 6-24. The OR gate (exclusive)

tion as follows: "*A* and NOT *B* or NOT *A* and *B* equals *L*." The expression "*A* and NOT *B* ($A\overline{B}$)" means that *A* is not closed and *B* is closed.

Another way to write the equation for EXCLUSIVE OR is to use a plus sign with a circle around it. You will notice this symbol is also used in the NEMA circuit symbol.

The NOT Gate

Figure 6-25 shows the condition for a NOT gate. The word *NOT* is used interchangeably with the word *inverter* in reference to this logic gate. A relay circuit is used to represent the NOT gate. The relay circuit is drawn two different ways so that you will become familiar with both of the methods used in industry. In both circuits the switch is marked with an *A*. Notice that when the switch is open, the relay is in the deenergized position and the lamp *(L)* is connected through the normally closed contact to the supply voltage. Therefore, with the switch open, the lamp is on.

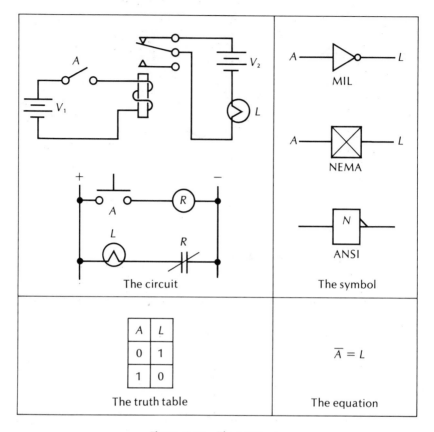

Figure 6-25. The NOT gate

If you close switch *A*, the relay will be energized, thus opening the normally closed contacts. The circuit for the lamp is open, and the lamp is off.

To summarize, the lamp is off when the switch is closed, and it is on when the switch is open. These conditions are shown by the truth table. The first row shows that when the switch is open, the lamp is on. The second row shows that when the switch is closed, the lamp is off. The overbar is used to represent the NOT condition in the equation. This equation should be read as follows: "NOT *A* equals *L*."

The NAND Gate

The word *NAND* in logic gates stands for Not AND. It is actually a negated or inverted AND circuit. The circuit is shown in Figure 6-26. The relay coil is energized when switches *A* and *B* are both closed. When the relay is in the

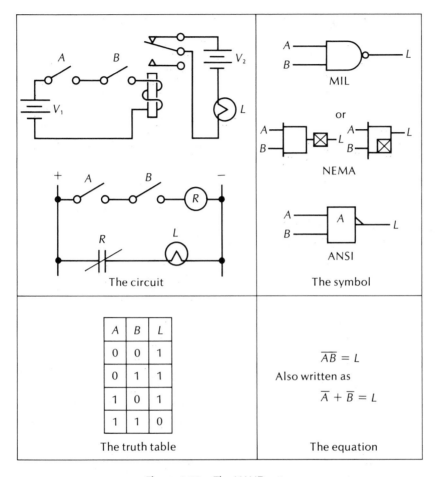

Figure 6-26. The NAND gate

deenergized condition, lamp L is on. Closing both switches causes the relay to be energized, and the lamp will go off.

The symbol for the NAND circuit consists of an AND symbol with an inverted output. In the MIL symbol the small circle at the output represents the inverter. In the NEMA symbol the box with a cross in it represents the inverter.

The truth table shows all of the possible conditions. You will observe that the fourth row shows that the lamp is off when both switches are closed.

The equation for the circuit is $\overline{AB} = L$. You should read this "NOT A and B equals L."

The NOR Gate

The NOR gate is shown in Figure 6-27. It is a negated OR gate. Switches A and B are in parallel. Closing either or both energizes the coil of the relay. The lamp is connected through the normally closed terminals to the battery. The symbols are comprised of an OR symbol plus an inverter.

The truth table shows all of the possible conditions. Note that the lamp is only on if both switches are open, as indicated by the first row.

The equation shows that A OR B with an overbar is equal to L. This should be read "NOT A or B equals L."

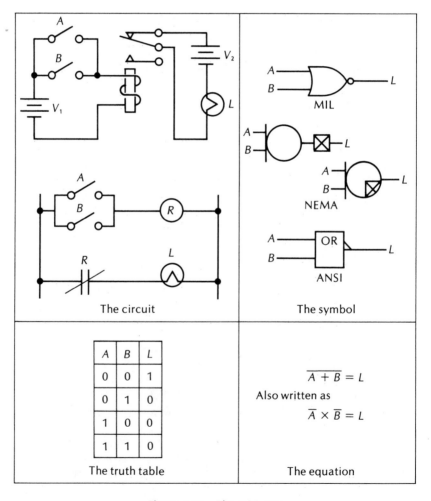

A	B	L
0	0	1
0	1	0
1	0	0
1	1	0

The circuit

The symbol

The truth table

$$\overline{A + B} = L$$

Also written as

$$\overline{A} \times \overline{B} = L$$

The equation

Figure 6-27.   The NOR gate

You should be able to recognize the conditions for all of the basic gates by looking at the truth table alone, by its symbol, or by its equation. The actual circuit is probably the least important of the four things shown in each illustration, because the circuitry may vary widely from one manufacturer to another but the overall result, which you are interested in as a technician, will always be the same. In the schematic diagram for an industrial electronic system, you will not normally find the electronic circuitry. It would be impossible to draw the complete circuitry for a large complex system on one single schematic. Instead, you will see the logic symbols used and their interconnections.

In each of the gates shown in these illustrations there are only two inputs and one output. This is not a necessary condition for the gate. You can have more inputs, but there is usually one output terminal. However, the one output of the gate can be connected to a number of different circuits.

The number of inputs is referred to as the *fan-in,* and the number of outputs is called the *fan-out.* The most important limitation to the fan-in and fan-out is in the amount of input power and output power the circuit can handle. With integrated circuit gates the inputs and outputs will be signal voltages or signal currents rather than switches.

The basic gates are combined into logic systems that can be used for control applications. These logic systems are discussed in the next chapter.

Logic Families

A significant advance in control technology occurred when the logic gates were fabricated as integrated circuits. Throughout the years of early development, manufacturers invented different approaches to produce the logic gates on integrated circuits. This development has given rise to the various *logic families.* A family can be thought of as a system of logic which employs one kind of electronic circuitry for producing a gate. A few of the more important examples, such as diode logic, TTL, and CMOS, will be discussed here.

DIODE LOGIC

Diode logic was the first electronic logic employed in large scale. As a matter of fact, early vacuum-tube computers employed large numbers of diode logic circuits.

Figure 6-28 shows the basic principle of diode logic. The input signals can be in either of two states: 0 volts or 1 volt. Ground is 0 volts and it is represented by binary 0. A positive 1 volt is indicated by the binary 1. This is called *positive logic.* (In a negative logic system binary 0 would represent 0 volts and binary 1 would represent -1 volt.)

In order for a current to flow through the load, either X_1 or X_2 (or both) must be forward biased by a positive voltage at terminals A and B. The truth

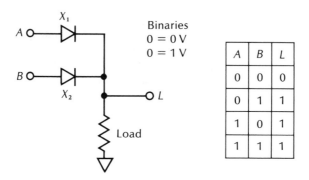

Binaries
0 = 0 V
0 = 1 V

A	B	L
0	0	0
0	1	1
1	0	1
1	1	1

Figure 6-28. Positive logic DL circuit and its truth table

table indicates the conditions of the circuit. Notice in this truth table that when both A and B are low, or zero, there is no output to the load.

Do you recognize the type of logic illustrated in Figure 6-28?

You should know immediately from the truth table that this is an OR gate. In each of the logic families discussed in this chapter, the electronic components can be connected to produce any of the six gates described earlier. However, we will show only one simplified gate for each family.

RTL LOGIC

One of the first integrated circuit transistor logic systems was resistor-transistor logic, or RTL. It is illustrated in Figure 6-29. The input signals are to the base of the transistors through coupling resistors R_1 and R_2. The output signal is taken from across a common load resistor R_3. In RTL circuitry, binary 0 is 0 volts, and binary 1 is 3.6 volts. The resistors in the base circuits of the transistors limit the base current when a binary 1 is connected to A or B.

Look at the truth table and decide what kind of a logic gate is represented here before you read further.

Note that the truth table shows that there is an output (3.6 volts) only when A and B are in the 0 condition. If there is an input to either A or B, the transistor is saturated and the related transistor becomes saturated and grounds the output. In other words, the gate is a NOR.

RTL circuits are obsolescent as far as integrated circuitry is concerned, but they are still being used in many other applications in which the logic circuit is made with discrete components. Also, there are still a lot of older systems in use that have RTL circuitry.

An important characteristic of RTL is that the power supply voltage is about 3.6 volts. One popular numbering system for IC RTL's was the 700, 800, and 900 series. Thus, if you see a digital integrated circuit with a number in the 700's, 800's, or 900's, there is a good chance that it is an RTL

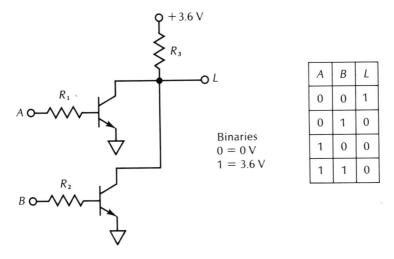

Figure 6-29. An RTL circuit and its truth table

circuit. The difference between the three is in the temperature range over which the integrated circuit can operate.

RCTL LOGIC

One reason for the demise of the RTL circuit is the fact that it is relatively slow compared to other types of logic which were developed later. However, the RTL circuit can be made to operate much more quickly by placing a capacitor across the input resistors (R_1 and R_2 in Figure 6-29). When you do this, you have a type of logic family called RCTL (resistance-capacitance-transistor logic). RCTL circuits are faster than RTL circuits; otherwise they have many of the same basic characteristics.

DTL LOGIC

A much faster transistor integrated circuit logic is shown in Figure 6-30. This is called DTL (diode-transistor logic).

The circuitry shown in Figure 6-30 represents a particular logic gate. Before you read further, study the truth table and decide what type of gate this is.

When the voltage at A is 0 V (in the 0 condition), diode X_1 conducts through resistor R_1. The base voltage of Q_1 is reduced to about 0.7 V, and Q_1 is cut off. This makes the output (L) voltage +5 V.

When A is in the high (+5 V) condition, as represented by binary 1, diode X_1 is cut off. Base bias flows, and Q_1 conducts to saturation, thus grounding the output (L).

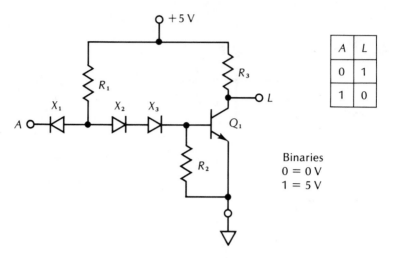

A	L
0	1
1	0

Binaries
0 = 0 V
1 = 5 V

Figure 6-30.   A DTL circuit and its truth table

The logic gate in Figure 6-30 is a NOT gate.

TTL LOGIC

The diode-transistor logic circuit of Figure 6-30 is the basis for the TTL design of Figure 6-31. The two emitter-base junctions of Q_1 are like PN junction diodes. They are forward biased by A or B or both. TTL stands for transistor-transistor logic.

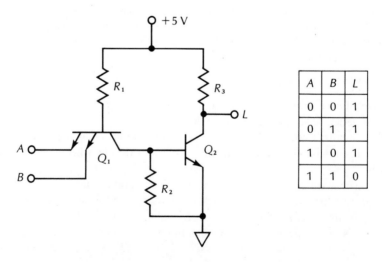

A	B	L
0	0	1
0	1	1
1	0	1
1	1	0

Figure 6-31.   A TTL circuit and its truth table

The collector junction of Q_1 is also like a *PN* junction diode. It takes the place of X_2 and X_3 in the DTL circuit of Figure 6-30.

Grounding A or B (or both) removes the forward bias for Q_2 and output L goes to +5 V. Applying +5 V to both emitters at A and B permits Q_2 to saturate and grounds output L.

The truth table is shown in Figure 6-31. What type of logic does it indicate? The truth table indicates that the circuit is a NAND gate. The TTL logic family (or T²L logic as it is often called) is sometimes referred to as *NAND logic*. In the next chapter you will learn how one type of gate—such as NAND—can be used to make all of the other basic gates. The basic NAND gate used with TTL logic is slightly more complex than the one shown in Figure 6-31, but the basic principle of operation is the same.

The most popular TTL family is the 54/74 series or the 5400/7400 series. Thus, if you see a logic gate with the number 5409 or 7409, you are safe to assume that this is a TTL logic circuit. That immediately identifies the power supply voltage as being +5 V.

ECL LOGIC

The fastest logic family has been emitter-coupled logic (ECL). Figure 6-32 shows the principle of operation.

The other types of logic families that have been discussed are called *saturation logic*. Their operation depends upon the operation of diodes or

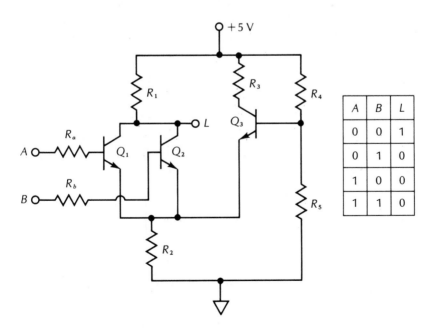

Figure 6-32. An ECL circuit and its truth table

transistors at either of two levels: cutoff and saturation. It takes a finite amount of time to switch a transistor from cutoff to saturation since the semiconductor material must be filled with charge carriers for saturation to occur. If the transistor is saturated, it takes time to sweep out all of the charge carriers to get the transistor into the cutoff condition.

The high speed of ECL is because the transistors in the gates are not saturated or cut off. Instead, they operate between two levels of conduction.

When A and B are both in the low or 0 condition, both Q_1 and Q_2 are cut off. Transistor Q_3 conducts because it is forward biased with a voltage divider at its base. With Q_1 and Q_2 cut off, output L is at +5 V.

When A is placed at +5 V, or a 1 condition, transistor Q_1 conducts. The increased drop across R_2 cuts Q_3 off, and all of the current flows through Q_1. This produces a drop across R_1, and output L goes to a lower voltage, or 0 condition. The same thing happens when B is connected to a +5 V input, or when both are connected to +5 V.

I²L OR MTL LOGIC

The letters I²L represent the words *integration-injection logic*. This family is also called *merged transistor logic*. The basic components for this family are shown in Figure 6-33. A *PNP* transistor (Q_1) is used to obtain base current for Q_2. This transistor replaces the resistor normally used for base current. That is one of the important features of I²L—there are no passive components. This feature is important because passive components require more room than transistors, and they require extra steps for fabrication.

Another important feature of the I²L technology is the transistor with multiple gates. This fabrication also makes possible a higher density of parts.

Manufacturers can use their bipolar IC fabricating plants to make I²L gates. These gates have a low power requirement and medium speed.

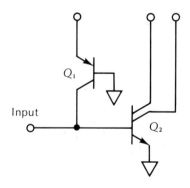

Figure 6-33. Components in I²L logic

CMOS LOGIC

The logic circuits that use transistors discussed up to now have employed bipolars. You will almost always find *NPN* transistors on any integrated circuit package because of the way they are fabricated. The disadvantage of a bipolar transistor compared to a JFET or MOSFET is that the base or input circuit to the transistor draws current and thus dissipates power. Furthermore, since current flows in the base circuit, its impedance must be low in comparison to the input circuit of a vacuum tube or field-effect transistor.

For these reasons it was a logical step to develop logic families using MOSFETs. Two kinds of MOSFETs are available: *enhancement* and *depletion*. The enhancement MOSFET was the most logical one to use because it does not have an idling current in the absence of an input signal. This keeps the power dissipation low. An integrated logic circuit that uses a P-channel enhancement-type MOSFET is referred to as *PMOS*, and one that uses N-channel enhancement MOSFET's is referred to as *NMOS*. Both types of systems have been used in special applications. The most significant MOS technology is the CMOS gate. The letters CMOS stand for *complementary metal-oxide-semiconductor* field-effect transistors. A CMOS logic gate employs both N-channel and P-channel enhancement MOSFET's in the same circuit.

An example of a CMOS gate is shown in Figure 6-34. By now you should recognize the truth table for this circuit as being one for an inverter, or NOT gate. In this circuit Q_2 is actually the amplifier or gate device and Q_1 serves simply as a load resistor for Q_2. An input to the common source amplifier (Q_2) will be inverted at L, so this is a NOT circuit.

An advantage of the simplified circuit shown in Figure 6-34 is that no passive components (resistors or capacitors) are necessary for its operation. Also, there is no idling current in the absence of an input signal.

The original CMOS was developed by RCA and it was referred to as *COS/MOS* (COmplementary Symmetry Metal-Oxide-Semiconductor FET). After other companies started making these devices, they were referred to as *CMOS*, a term which encompasses all of the ones being made.

In the CMOS family, then, different manufacturers will refer to their products according to their own particular nomenclature. Again, RCA is COS/MOS, Motorola is McMos, etc.

The most important of these families are those CMOS devices which have a 4000 number, sometimes referred to as the *4000 series* of CMOS. If you see a CMOS circuit with a number in the 4000's, you can be sure that it is a complementary MOS device.

The families discussed here do not represent the complete range available in industry today. These are only representative to show you how the gates can be made in different families, and how these families are referred to in circuit description.

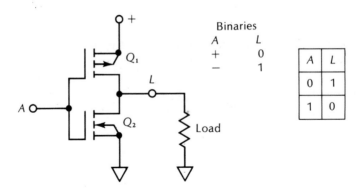

Figure 6-34. A CMOS gate and its truth table

It is interesting to compare some of the important characteristics of a few families, and this is done in Table 6-1.

Comparison of Logic Families

Table 6-1 compares some of the important characteristics of logic families. The designer of a control system must consider all of these factors when deciding which to use in his design.

Additional factors, not shown in Table 6-1, must also be considered. If speed is not an important factor, the designer would choose a slower logic family that costs less. If his company is committed to using a certain logic family, then that is the one he will use regardless of the tradeoffs.

In the next chapter it will be shown that all of the basic gates can be made by using either NOR or NAND gates. A company that makes logic IC's will use one of these basic gates. The choice is related to the ease of fabricating for a particular family.

Table 6-1 shows one column for the amount of power required for each gate and another column for the propagation delay—that is, the amount of time that it takes a signal to pass through the gate. Some designers claim that the product of these columns, called the *speed-power product,* is more valuable for evaluating a family. The reasoning here is that speed and power are tradeoffs and the product makes it possible to better evaluate this tradeoff when comparing logic families.

The fan-out characteristic tells the designer how many outputs can be driven by a gate in the family.

As the technology develops, there are improvements in speed, power requirements, cost, etc. The information in Table 6-1 shows the proper relationship between the gates, but specific values may require updating from time to time.

TABLE 6.1.   Comparison of logic families (values are typical in all cases)

Family	Type of Logic (Basic Gate)	Power per Gate (W)	Propagation Delay (ns)	Fanout (Leads per (Gate)
RTL (supply +3.6 V)	NOR	12×10^{-3}	50 (max	5
DTL (supply +5 V)	NAND	8 to 12×10^{-3}	30 (max)	8
TTL (supply +5 V)	NAND	12 to 22×10^{-3}	6 to 12	10
ECL (supply −5.2 V)	OR-NOR	40 to 55×10^{-3}	1 to 4	25
CMOS (supply +3 to +18 V)	NOR or NAND	50×10^{-9}	50	over 50

In the RTL circuit the power supply voltage is normally 3.6 V plus or minus 10 percent. There have been RTL families that employ 3 V power supplies, but 3.6 V is more common. For the propagation delay it takes 50 ns maximum for the signal to pass through the gate. This is a relatively long time and it was responsible for the use of DTL in place of RTL. Power dissipation for the gate is only 12 mW. This is a small power dissipation, but large compared to the amount of power dissipated for the CMOS.

The DTL and TTL circuits both use power supplies of +5 V, but the TTL (or T^2L as it is sometimes called) gate has a somewhat wider range of tolerances. Note that the propagation delay for the TTL circuit is only about 10 ns per gate, which is a very fast time compared to RTL. Power dissipation for TTL was somewhat higher than that for RTL.

The CMOS gate has a decided advantage as far as this particular list of characteristics is concerned. In the first place you will note that the power supply voltage can be any value from 3 to 18 V. This range is very important in terms of circuit design. The RTL, ECL, and TTL devices must be operated from regulated power supplies to keep their input voltages within the ranges specified. However, the CMOS device has such a wide range of operating voltages that it is not necessary to use a regulated supply. (This is also true for the I^2L family of logic.)

Manufacturers of CMOS claim that because this type of logic can use any power supply voltage, it is directly *compatible* with the other types of logic. In other words, they claim that you can put a CMOS circuit in the same logic system as a TTL circuit. Unfortunately, there is more to compatibility than the power supply alone. The propagation delay must also be considered. If you put a CMOS gate in the same system as a TTL, it is possible that at some point down the line, when two signals are supposed to come together at the same time, they would arrive at different times because of the

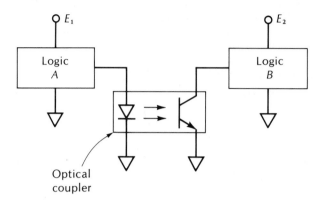

Figure 6-35. A method of coupling unlike logic gates

different types of devices. For this reason, the idea of compatibility is disregarded by some designers.

Figure 6-35 shows one method of interfacing two different logic families. The optical coupler isolates the two families that operate from different power supplies (E_1 and E_2).

MAJOR POINTS

1. One of the more popular linear IC's is the op amp.
2. The name operational amplifier comes from the fact that it was first used in analog computers to perform arithmetic operations such as addition, subtraction, multiplication, and division.
3. Some of the IC op amp features that make it popular are high open-loop gain, differential input, high reliability, low-power requirement, linear rolloff, high-input impedance, and low-output impedance.
4. Depending upon the type of op amp and the application, one or more of the following circuits may be used: antilatchup, offset balancing, and frequency compensation.
5. Because of its very high gain, the output of the op amp is almost entirely dependent on the feedback circuitry.
6. When used as an inverting amplifier, the voltage gain (A_v) equals the ratio of the feedback resistor (R_f) to the input resistor (R_i).
7. The input impedance of the inverting op amp is numerically equal to the input resistance.
8. Differential op amps are not useful for amplifying very low values of ac and very small changes in low voltage dc. The temperature drift can be greater than the signal being amplified.

9. Chopper-stabilized op amps are used because of their very low drift.
10. Current-differencing amplifiers (CDA) are simpler in circuitry and can be operated from a single positive supply. Also, they can be operated with a low dc voltage.
11. Six basic logic gates are used as building blocks for industrial electronic systems: AND, OR, EXCLUSIVE OR, NOT, NAND, and NOR.
12. Some companies use MIL symbols on schematic drawings of their equipment. Other companies use NEMA symbols. It is important to learn both types.
13. A truth table tells all of the possible conditions that can occur in a logic gate.
14. Boolean algebra equations can also be used to tell all of the possible conditions in a logic gate.
15. Basic symbols for the Boolean equations are $\times$ for AND and $+$ for OR. An overbar is used for inverter.
16. All of the basic logic gates can be created with switches and relays. In fact, relay logic has been used in industrial control systems.
17. Different manufacturers have devised their own basic IC logic gates, thus creating a number of logic families.
18. The fastest logic is ECL. TTL is a popular type. CMOS requires very little power drain.
19. No two logic gates are truly compatible unless they have the same propagation delay, power supply requirements, and operating temperature range. CMOS is sometimes called a *compatible* logic family because of its ability to operate with a wide range of power supply voltages.
20. Interfacing between unlike logic families can sometimes be accomplished with an optical coupler.

PROGRAMMED REVIEW

(Instructions for using this programmed section are given in Chapter 1.)

The important concepts of this chapter are reviewed here. If you have understood the material, you will progress easily through this section. Do not skip this material because some additional theory is presented.

---

1. An oscillator is an amplifier circuit that employs

    A. Degenerative feedback. (Go to block 9.)
    B. Regenerative feedback. (Go to block 17.)

---

2. Your answer to the question in block 30 is wrong. Read the question again, then go to block 7.

---

3. The correct answer to the question in block 17 is A. As shown in Figure 6-8 the value of R_a is calculated by the equation:

$$R_a = \frac{R_i \times R_f}{R_i + R_f}$$

Substituting 33 kΩ for R_i and 47 kΩ for R_f:

$$R_a = \frac{33\ k\Omega \times 47\ k\Omega}{33\ k\Omega + 47\ k\Omega}$$

$$= 19.3875\ k\Omega$$

$$\simeq 19.4\ k\Omega$$

Here is your next question.
What does this equation mean? $A + B = L$

A. A and B equals L. (Go to block 13.)
B. A or B equals L. (Go to block 21.

4. The correct answer to the question in block 31 is B. The resonant frequency is found by the equation

$$f_r = \frac{0.159}{\sqrt{LC}}$$

Substituting $L = 0.01$ and $C = 0.01 \times 10^{-6}$

$$f_r = \frac{0.159}{\sqrt{1 \times 10^{-10}}}$$

$$= 0.159 \times 10^5$$

$$= 15,900\ Hz$$

The resistor across the parallel-tuned circuit serves two purposes. First, it sets the gain of the amplifier at resonance. The impedance of the parallel L-C circuit is very high at resonance. This very high impedance in parallel with R results in a total circuit impedance that is, for all practical purposes, equal to R.

The feedback resistor also serves as a swamping resistor. It makes the L-C circuit tune more broadly. This means that the response curve will not be as sharp as for a pure (no resistance) L-C circuit.

Here is your next question.
What is the dB gain of the op amp shown here?

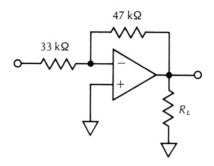

A. 3. (Go to block 16.)
B. 1.5. (Go to block 34.)

---

5. The correct answer to the question in block 28 is B. The first step is to calculate the dB gain of the amplifier.

$$dB\ gain = 20\ log\ \frac{R_f}{R_i} = 20\ log\ \frac{100\ k\Omega}{10\ k\Omega}$$

$$= 20 \times 1 = 20\ dB$$

The Bode plot for the 741 op amp is given in Figure 6-3. It shows that an op amp dB gain of 20 will result in a bandwidth of about 10^5 Hz.

Here is your next question.
To simulate an inductor, an op amp can be connected as

A. A ferrite bead. (Go to block 35.)
B. A gyrator. (Go to block 11.)

---

6. Your answer to the question in block 20 is wrong. Read the question again, then go to block 33.

---

7. The correct answer to the question in block 30 is B. The input signal goes to the noninverting input of the op amp. As shown in Figure 6-10 the gain of this type of amplifier is

$$A_v = 1 + \frac{R_f}{R_i}$$

Substituting $R_f = 1.5$ kΩ and $R_i = 1$ kΩ

$$A_v = 1 + \frac{1.5 \text{ k}\Omega}{1 \ \Omega}$$

$$= 2.5$$

Here is your next question.
An advantage of the chopper-stabilized amplifier is that it

A. Has low drift. (Go to block 14.)
B. Has a broad frequency response. (Go to block 22.)

---

8. Your answer to the question in block 17 is wrong. Read the question again, then go to block 3.

---

9. Your answer to the question in block 1 is wrong. Read the question again, then go to block 17.

---

10. Your answer to the question in block 31 is wrong. Read the question again, then go to block 4.

---

11. The correct answer to the question in block 5 is B. A ferrite bead is a small piece of doughnut-shaped ferrite material that surrounds a wire. It simulates an inductor. A gyrator is an op amp circuit used to simulate an inductor.

Here is your next question.
Compared to CMOS, an advantage of T²L is its

A. Low value of propagation delay. (Go to block 15.)
B. Low power consumption. (Go to block 19.)

---

12. The correct answer to the question in block 33 is B. The relay contact in series with the lamp is normally open. Both *A* and *B* must be closed to energize the relay and close the contacts. Mathematically: $AB = L$.

Here is your next question.
When an op amp saturates, it means that the output voltage does not change, even though the input signal voltage may be changing. This condition is known as

A. Common-mode rejection. (Go to block 29.)
B. Latchup (Go to block 23.)

---

13. Your answer to the question in block 3 is wrong. Read the question again, then go to block 21.

14. The correct answer to the question in block 7 is A. A small amount of dc drift can be tolerated in amplifiers that are used primarily for ac signals. If the amplifier is used for low-voltage dc applications, then the drift becomes very important. It is conceivable that drift could produce a larger change in output voltage than is produced by the low-load signal being amplified. That is why the low-drift feature of the chopper-stabilized op amp is so important.

An example of an application is in the instrumentation field. A very small change in the dc output from a transducer can be amplified by a low-drift chopper-stabilized op amp. The amplified change in the output of the op amp is easily measured.

Here is your next question.
Which of the following should be used with a constant-current device?

A. An RTL NAND circuit. (Go to block 25.)
B. A differential amplifier. (Go to block 28.)

15. The correct answer to the question in block 11 is A. TTI^{2}L logic gates have lower power consumption than CMOS. However, the TTI^{2}L logic gates are much faster; that is, they have a lower propagation delay.

Here is your next question.
The correct truth table for the logic symbol shown here is

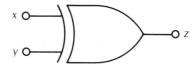

A.

x	y	z
0	0	0
0	1	1
1	0	1
1	1	1

(Go to block 24.)

B.

x	y	z
0	0	0
0	1	1
1	0	1
1	1	0

(Go to block 31.)

16. The correct answer to the question in block 4 is A. This is calculated as follows:

$$\text{dB gain} = 20 \log \frac{R_f}{R_i}$$

Substituting $R_f = 47 \text{ k}\Omega$ and $R_i = 33 \text{ k}\Omega$

$$\text{dB gain} = 20 \log \frac{47 \text{ k}\Omega}{33 \text{ k}\Omega}$$

$$= 20 \log 1.424$$

$$= 20 \times 0.1536$$

$$\approx 3 \text{ dB}$$

The resistors used in the problem are preferred values for carbon composition types.

Here is your next question.
A *logic comparator* has an output whenever the two inputs are the same. Which of these truth tables is for a comparator?

A.

A	B	L
0	0	1
0	1	0
1	0	0
1	1	1

(Go to block 20.)

B.

A	B	L
0	0	0
0	1	1
1	0	1
1	1	0

(Go to block 26.)

17. The correct answer to the question in block 1 is B. Regenerative feedback can be accomplished in op amp oscillator circuits in either of two ways:

When the feedback path is from the op amp output to the noninverting input, it will be regenerative.

When the feedback path is from the op amp output to the inverting input, a phase shifting network is needed. This is the type of feedback used in the sine wave oscillator circuit of Figure 6-17.

Here is your next question.
For minimum offset, the value of R_a in the op amp circuit shown here should be

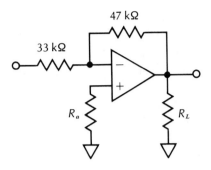

A. About 19.4 kΩ. (Go to block 3.)
B. About 27 kΩ. (Go to block 8.)

---

18. Your answer to the question in block 33 is wrong. Read the question again, then go to block 12.

---

19. Your answer to the question in block 5 is wrong. Read the question again, then go to block 15.

---

20. The correct answer to the question in block 16 is A. The equation for a logic comparator is

$$\overline{AB} + AB = L$$

It means that when there is not an A or B, there is L; or when there is A and B, there is L.

Here is your next question.
Which of the following is the correct truth table for an AND circuit?

A.

A	B	L
1	1	1
1	0	0
0	1	0
0	0	0

(Go to block 33.)

B.

A	B	L
1	1	0
1	0	0
0	1	0
0	0	1

(Go to block 6.)

---

21. The correct answer to the question in block 3 is B. As shown in Figure 6-23, the symbol + is used to mean OR. Therefore, $A + B = L$ means A or B equals L. Remember that this is inclusive OR. Unless otherwise stated, an OR in logic always refers to inclusive OR.

Here is your next question.
One of the specifications for an op amp is slewing rate. This is

A. The maximum rate of change of input level. (Go to block 27.)
B. The maximum rate of change of output level. (Go to block 30.)

---

22. Your answer to the question in block 7 is wrong. Read the question again, then go to block 14.

---

23. The correct answer to the question in block 12 is B. The question accurately describes latchup.

Common-mode rejection refers to the ability of an op amp to reject common-mode input signals. Suppose, for example, that a sine wave signal is applied to the inverting and noninverting terminal at the same time. With no feedback, the gain for the two signals should be the same. That is, two output signals would be 180° out of phase, and they should cancel. In other words, the amplifier should reject these common-mode signals.

Here is your next question.
Write a logic mathematical equation for each of the following gates:

Gate	Equation
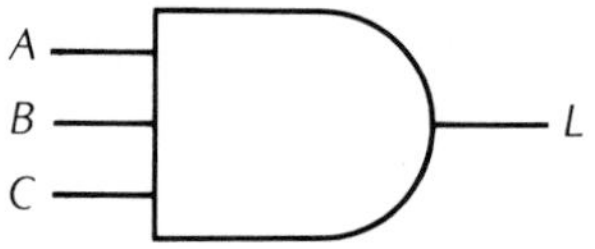 A B C → L (AND gate)	a
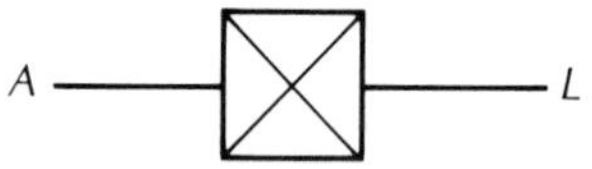 A → L (buffer/inverter box with X)	b
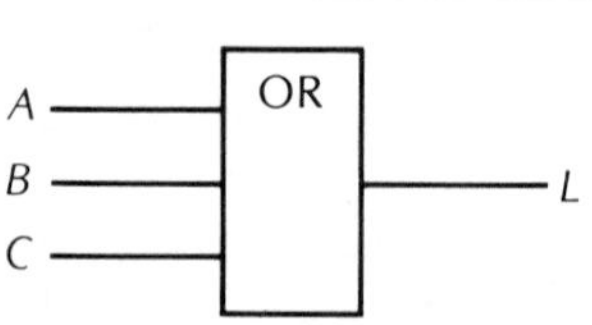 A B C → L (OR gate)	c

(Go to block 36.)

---

24. Your answer to the question in block 15 is wrong. Read the question again, then go to block 31.

25. Your answer to the question in block 14 is wrong. Read the question again, then go to block 28.

26. Your answer to the question in block 16 is wrong. Read the question again, then go to block 20.

27. Your answer to the question in block 21 is wrong. Read the question again, then go to block 30.

28. The correct answer to the question in block 14 is B. (The theory of the differential amplifier was discussed more completely in a previous chapter.)

Here is your next question.
Assume the following circuit uses a 741 op amp. What is the approximate bandwidth of the amplifier?

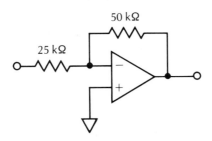

A. 10,000 Hz. (Go to block 32.)
B. 100,000 Hz. (Go to block 5.)

29. Your answer to the question in block 12 is wrong. Read the question again, then go to block 23.

30. The correct answer to the question in block 21 is B. Suppose a pulse is applied to the input of an op amp. The output signal should rise instantly when the leading edge of the pulse arrives at the input. When the trailing edge of the pulse arrives at the input, the output should drop instantly to a lower value. Since no op amp is perfect, the output voltage will not change instantly. Instead, it takes a finite amount of time to change. The rate of change at the output is called the *slewing rate*.

Here is your next question.
The gain of the following op amp is approximately

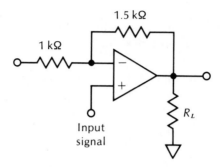

1.5 kΩ

1 kΩ

R_L

Input
signal

A. 1.5. (Go to block 2.)
B. 2.5. (Go to block 7.)

---

31. The correct answer to the question in block 15 is B. The symbol shown is for an EXCLUSIVE OR gate. It has no output if both inputs are off, or if both inputs are on.

Here is your next question.
The bandpass filter shown here will have a resonant frequency of about

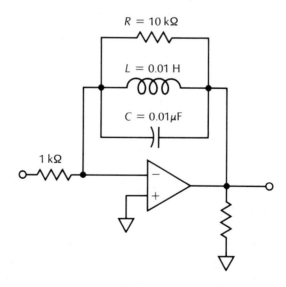

$R = 10 \text{ k}\Omega$

$L = 0.01 \text{ H}$

$C = 0.01 \mu\text{F}$

1 kΩ

A. 159 kHz. (Go to block 10.)
B. 15.9 kHz. (Go to block 4.)

32. Your answer to the question in block 28 is wrong. Read the question again, then go to block 5.

33. The correct answer to the question in block 20 is A. The order of the truth table is reversed from the one described in Figure 6-22. You must learn to identify the circuit from what the truth table tells about it—NOT from the order of the binaries in the load column.

Here is your next question.
Which of the following equations is correct for the relay circuit shown here?

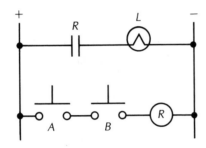

A. $A + B = L$. (Go to block 18.)
B. $AB = L$. (Go to block 12.)

34. Your answer to the question in block 4 is wrong. Read the question again, then go to block 16.

35. Your answer to the question in block 5 is wrong. Read the question again, then go to block 11.

36. You should be able to write these equations from memory.

A. $A \times B \times C = L$.
B. $\overline{A} = L$.
C. $A + B + C = L$.

You have now completed the programmed section.

SELF-TEST WITH ANSWERS

(Answers to this test are given at the end of the chapter.)

1. Another name for the frequency response curve of an amplifier is _____ .

2. The gain of an op amp when no feedback is used is called the _____ .

3. A 4.7 kΩ resistor is used as the feedback resistor in an op amp circuit. The input resistance is 2.7 kΩ. What is the voltage gain of the amplifier?

4. An input signal to a certain op amp drives it into saturation. After the signal is removed, the op amp remains saturated. This condition is known as _____ .

5. An op amp known for its low drift is the _____ .

6. In order to make an op amp oscillator, feedback to the inverting input must be (a) regenerative or (b) degenerative.

7. An op amp circuit that has no voltage gain on a noninverted output is called a _____ .

8. A filter circuit that uses an amplifier is called _____ .

9. Some op amps use an external circuit to obtain a desired rolloff. This circuit is called _____ .

10. A linear integrated circuit containing individual components that are not connected to give a particular output is called _____ .

11. An integrated circuit with an output that is not necessarily related to the input is called _____ .

12. A certain gate requires all of the inputs to be in a 1 condition in order for the output to be in a 1 condition. What type of gate is this?

13. Write the Boolean equation for the following statement: A and B and NOT C equals L.

14. Draw the MIL and NEMA symbols for an EXCLUSIVE OR gate.

15. Write a truth table for a NAND gate.

16. A logic comparator has an output (L) in the 1 condition when the inputs (A and B) are identical. Make a truth table for a logic comparator.

17. Which of the logic gates is fastest—that is, has the lowest propagation delay?

18. In comparing TTL and CMOS logic gates, which does not require a regulated power supply?

19. Represent the following statement with a truth table: The output is in a 1 condition provided none of the inputs (A, B, and C) are in a 1 condition. Be sure to include all of the possible conditions for the input.

20. Write an equation for the output of Figure 6-38.

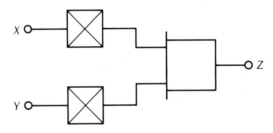

Figure 6-38.

Answers to Self-Test

1. Bode plot
2. Open-loop gain
3. 1.74
4. Latchup
5. Chopper-stabilized type
6. (a) Regenerative
7. Follower
8. An active filter
9. Frequency compensation
10. An array
11. A digital circuit
12. AND
13. $AB\overline{C} = L$
14. See Figure 6-24.
15. See Figure 6-26.
16.

A	B	L
0	0	1
0	1	0
1	0	0
1	1	1

17. ECL
18. CMOS
19.

A	B	C	L
0	0	0	1
0	0	1	0
0	1	0	0
0	1	1	0
1	0	0	0
1	0	1	0
1	1	0	0
1	1	1	0

20. $\overline{xy} = z$

CHAPTER 7

COMPOUND GATES

In the previous chapter you studied basic gates, such as AND, OR, and NOT, that make up all of the digital systems used in integrated circuit logic. These basic gates are used in other applications as well, so it is important to memorize their symbols and truth tables.

Usually, basic gates are used in combinations to make a family of logic such as TTL or CMOS. A designer can choose all of the basic circuitry required for a digital system by using integrated circuits from one family. TTL is a good example. The TTL IC logic chips have numbers in the 5400 or 7400 range. As an example, 7400 is a *quad NAND,* that is, four NAND's within one single integrated circuit package. Using only the IC packages in the 5400 or 7400 series, a designer can make a complete system for performing industrial operations.

Although it is normally not the job of the technician to design systems, he or she should have a basic familiarity of how logic gates are combined. That is one of the subjects of this chapter. There are several different methods by which this subject can be studied. For example, a whole system of mathematics, called *Boolean algebra,* has been devised that enables you to determine the output of combined gates mathematically.

Another way to study the subject of combined logic gates is to use truth tables. This method does not require math, but does require a knowledge of the truth tables for the basic gates. The truth table method will be used in this chapter. It is very useful when no more than six or seven gates are being combined. Beyond that, it is necessary to resort to Boolean algebra because of the complexity of the truth tables involved. To summarize, truth tables are not the best method for designing logic systems, but it is the best way to start the study of gate combinations.

Next to the basic gates the most important devices in digital systems are the *two state,* or *bistable,* devices. A good example is the *flip-flop.* Flip-flops are so important that they are sometimes considered to be one of the basic devices of the logic system. They will be studied in this chapter and their use in a simple counting circuit will be explained.

Just as there are different basic gates, there are also different basic types of flip-flops. Each has advantages and disadvantages. A knowledge of these will be helpful when analyzing digital systems.

THEORY

Making Truth Tables for Combined Gates

Figure 7-1 shows three AND gates combined to produce a single output. The illustration actually shows how the output can be obtained in a step-by-step procedure. The first illustration shows only the AND gates. The second illustration shows the outputs of AND gate 1 and AND gate 2, written as $(A \times B)$ and $(C \times D)$. These outputs are actually the inputs for AND gate 3. In other words, the two inputs $(A \times B)$ and $(C \times D)$ will be ANDed in gate 3. The output of AND gate 3 is $(A \times B) \times (C \times D)$ as shown in the illustration.

Figure 7-2 shows how a truth table can be made to show all of the outputs for all possible combinations of inputs. Inputs A, B, C, and D can be any

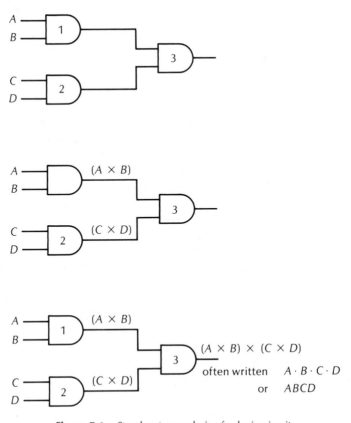

Figure 7-1. Step-by-step analysis of a logic circuit

A	B	C	D	A × B	C × D	ABCD
0	0	0	0	0	0	0
0	0	0	1	0	0	0
0	0	1	0	0	0	0
0	0	1	1	0	1	0
0	1	0	0	0	0	0
0	1	0	1	0	0	0
0	1	1	0	0	0	0
0	1	1	1	0	1	0
1	0	0	0	0	0	0
1	0	0	1	0	0	0
1	0	1	0	0	0	0
1	0	1	1	0	1	0
1	1	0	0	1	0	0
1	1	0	1	1	0	0
1	1	1	0	1	0	0
1	1	1	1	1	1	1

Note binary count
in these columns

Figure 7-2.   Truth table for circuit of Figure 7-1

combination of 1's and 0's. To be sure to get all of the possible combinations, the best technique is to start by making all of the inputs 0, and then in each row under *ABCD*, enter the numbers obtained by binary counting until all of the inputs are at 1.

Table 7-1 shows the binary numbers that are equivalent to decimal numbers 0 through 32. Note that the four right-hand columns are the same as the first four columns in the truth table of Figure 7-2. (Remember that columns are vertical in tables, and rows are horizontal.)

The binary numbers in Table 7-1 should be studied carefully. It is important for a technician to be able to make a binary count if he is to make an effective truth table. Although the table stops at decimal 32, it is an easy matter to learn the counting format so that it could be extended to numbers

TABLE 7.1   COUNTING WITH BINARY NUMBERS

Decimal	Binary	Decimal	Binary
0	000000	17	010001
1	000001	18	010010
2	000010	19	010011
3	000011	20	010100
4	000100	21	010101
5	000101	22	010110
6	000110	23	010111
7	000111	24	011000
8	001000	25	011001
9	001001	26	011010
10	001010	27	011011
11	001011	28	011100
12	001100	29	011101
13	001101	30	011110
14	001110	31	011111
15	001111	32	100000
16	010000		

above 32. Until the format has been learned, you can refer to Table 7-1 when writing truth tables.

As a quick check to make sure that you have included all of the possibilities, it is useful to know that the number of possibilities is equal to 2 raised a power equal to the number of inputs. Table 7-2 lists powers of 2 from 2^0 to 2^{20}. For example, the number of inputs in Figure 7-1 is four *(A, B, C,* and *D)*, so the total number of possibilities of combinations of 0's and 1's is 2^4 or 16.

To summarize, in order to list all possible input combinations of 0 and 1, write the binary count in the columns for the input terminals. If there are *n* inputs, there should be 2^n rows in the truth table.

TABLE 7.2   POWERS OF 2

Number (n)	Exponent (2^n)	Number (n)	Exponent (2^n)
0	1	10	1,024
1	2	11	2,048
2	4	12	4,096
3	8	13	8,192
4	16	14	16,384
5	32	15	32,768
6	64	16	65,536
7	128	17	131,072
8	256	18	262,144
9	512	19	524,288
		20	1,048,576

After the inputs have been put in the table, the next step is to write the outputs of each gate. For Figure 7-1 it is seen that A and B are ANDed; that is, they are combined in an AND gate. The output $(A \times B)$ will be present only when the inputs are both at the 1 condition. Looking at the first two columns in Figure 7-2 it is seen that for the first 12 rows there is at least one 0 input for A or B, so the output will be 0. In the last four rows, both inputs of A and B are 1's, so column $A \times B$ will have 1's in the last four rows.

As shown in Figure 7-1, C and D are also ANDed in AND gate 2. Again, referring to Figure 7-2, column $C \times D$ will have a 1 only when both inputs (both C and D) are at 1. Study the table and note that there is a 1 in the $C \times D$ column every time that both C and D in the third and fourth columns are 1.

To get the output of AND gate 3, the two inputs $(A \times B$ and $C \times D)$ are ANDed. In other words, it is only necessary to AND the two columns in the truth table headed by $A \times B$ and $C \times D$. As with all other AND situations, there can only be an output when both of these columns have a 1. That only occurs in the very last row.

The overall truth table shows that there is only an output for the circuit in Figure 7-1 when all of the inputs are at digit 1. The complete circuit of Figure 7-1 could be replaced with a single AND gate that has four inputs, $ABCD$, as shown in Figure 7-3.

Simplifying Combined Logic Gates

Figure 7-4 shows two different circuits that have the same output. One circuit requires three gates (two AND's and an OR), but the other circuit requires only two gates (one OR and one AND). With a little experience it is possible to write the truth table for a system, and then analyze it to determine if the system could be simplified. The procedure will be explained with reference to Figure 7-4.

Start first with the circuit that has two AND's and one OR. At gate 1 the inputs A and B are ANDed, and in gate 2 the inputs A and C are ANDed. The outputs of these two gates are inputs for the OR, so the output for gate 3 can be written simply as $AB + AC$.

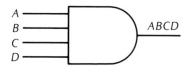

Figure 7-3. A four-input AND gate

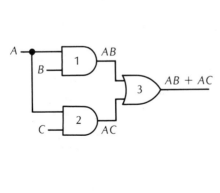

A	B	C	AB	AC	AB + AC
0	0	0	0	0	0
0	0	1	0	0	0
0	1	0	0	0	0
0	1	1	0	0	0
1	0	0	0	0	0
1	0	1	0	1	1
1	1	0	1	0	1
1	1	1	1	1	1

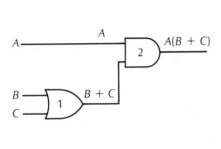

A	B	C	B + C	A(B + C)
0	0	0	0	0
0	0	1	1	0
0	1	0	1	0
0	1	1	1	0
1	0	0	0	0
1	0	1	1	1
1	1	0	1	1
1	1	1	1	1

Figure 7-4. Gate circuits with the same output

The truth table is written by first making columns for all the possible inputs *(A, B,* and *C)*. List all the combinations of 1 and 0 for these three columns. This is done by entering the digital count from 000 to 111. Since there are three inputs, the number of possible combinations (or rows) should be 2^3, which is equal to 8. Note that there are eight rows in the truth table.

The output of gate 1 is obtained by ANDing columns *A* and *B*. As with all AND circuits, it can only be a 1 when both *A* and *B* are at 1, as shown in the truth table column headed *AB*. The output of gate 2 is obtained by ANDing *A* and *C*. Again, there is an output only when *A* and *C* are both at 1. In this case, that occurs in the sixth and eighth rows.

The output of the system is obtained by ORing the *AB* and *AC* columns. For an OR gate there is an output of 1 when there is an input of 1 in either or both of these columns. Note, then, that the output is *AB + AC*. For the complete system there is a 1 in the last three rows. If you study the truth table carefully you will see that there is never a 1 in the $AB + AC$ column unless the A input is at 1. At the same time, one or both of the columns in B or C must also be at 1 in order to get an output. This leads to the conclusion that the output is A and B or C.

The truth table for the circuit in Figure 7-4 that has one OR and one AND will be studied next. As before, there are three inputs, A, B, and C, so the first three columns list all possibilities when all of the numbers in the digital count are listed. The output of the OR circuit is $B + C$. This column is obtained by ORing the second and third columns $(B$ and $C)$. There will be a 1 in the $B + C$ column whenever B or C is at 1, or when both are at 1.

The output of the complete circuit is obtained by ANDing the first column A with the fourth column $B + C$. This means that there will be an output 1 only when both the A column and the $B + C$ column are at 1. This only occurs in the last three rows. Note that the output columns of the truth tables are identical.

Readers with a math background will recognize that when output $AB + AC$ is factored, the result is $A(B + C)$. This is written as follows:

$$AB + AC = A(B + C)$$

So, the results that were obtained by using truth tables could also be obtained with mathematics. Some basic math rules of Boolean algebra, however, must be learned before the math results are consistent.

Logic circuits like the ones shown in Figure 7-4 are related to practical problems encountered in industrial electronics. Suppose, for example, a drill press in a shop can be operated either manually or with automatic control equipment. After the drill press has been in operation a few weeks, it becomes apparent that careless operation is resulting in holes being drilled in the work platform and something has to be done to prevent this.

The designer reasons the problem out as follows: "I want a safety system that will permit the drill press motor to run if operated manually OR by the automatic control equipment, but only under the condition that there is material to be drilled on the worktable."

"Let L be the load, that is, the motor being operated. Since the motor can operate with either of two conditions (manual or automatic), I will use an OR gate to feed the motor (first step in Figure 7-5). But the manual condition can only occur if there is work on the table, so one input to the OR will be manual AND work on the table (second step in Figure 7-5). Likewise, there can be an automatic operation only if there is work on the table (automatic AND work on the table). (This is the third step in Figure 7-5.)" Letters can also be assigned to represent the conditions:

A = work on the drill press table
B = manual switch ON
C = automatic control ON
L = drill press motor runs

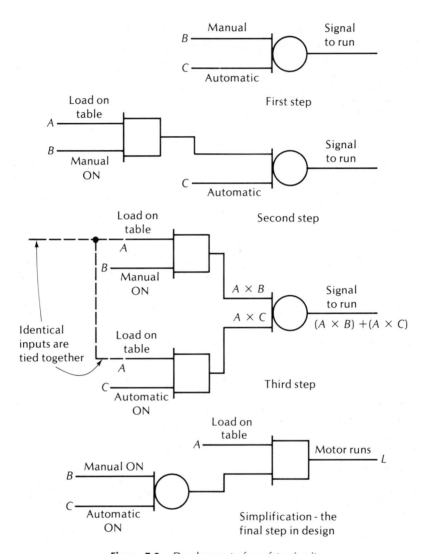

Figure 7-5. Development of a safety circuit

The basic circuit is now as shown in Figure 7-5 with two AND's and one OR. If the designer knows his basics, he will immediately realize that the circuit could be simplified to one AND and one OR as shown in the last step of Figure 7-5.

Another designer, working the same problem, might arrive at the simplest possible circuit directly. It depends upon how the person interprets the information and sets up the problem. But, in either case, the same circuit will result.

From this simple example you can see that the basic rules of logic circuitry that you are now learning can be applied to practical problems. More important than being able to design the system is that you understand how the inputs are combined to make an output. If, for example, the machine operator complains that the system works on manual, but not on automatic, your knowledge of the system tells you the following:

- The motor is OK.
- There is an input from the "Work on Table."
- The manual switch is OK.

Therefore, the trouble must be that no input signal is being received from the "automatic control."

In Chapter 10 you will learn a method of quickly determining if an AND gate is working properly, so you can troubleshoot this system rapidly— *provided you know the basic theory of logic gates, and how to determine output of combined logic gates!*

It is now apparent why it is important to understand gate combinations. Engineers are proficient in writing gate combinations and can often simplify a complex circuit by analyzing truth tables or by employing Boolean algebra. Reducing the gates decreases the complexity of the system and therefore decreases its cost. It is usually not the job of technicians to design circuits. However, it is their job to know what outputs should be present for a given combination of logic gates, so they can test that combination and determine if it is working properly. The method of truth tables just described makes it possible for them to do this.

Glitches in Combined Logic Circuits

It is a common practice to use square wave or pulse signals as inputs to a logic system. These signals have only two steady voltage levels, and the two levels can be used to represent digital 1 and digital 0. Figure 7-6 shows how the signals are used. Keep in mind the fact that the 1 and 0 levels do NOT represent voltage values. Instead, they are two voltages that represent digits 1 and 0.

When several signals are combined in logic gates the output signal can be obtained by drawing all of the inputs in rows with special attention paid to vertically aligning the timing of the circuit, as illustrated in Figure 7-7. Here an AND circuit has three input signals A, B, and C. The desired output is represented as L. The signals for A, B, and C are shown below the AND gate. Remember that each signal represents the two possible logic states (1 and 0). Since it is an AND circuit, there can be an output only when all the inputs are at logic 1. However, with the three signals shown, this condition

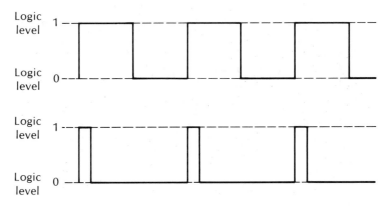

Figure 7-6. Binary digits represented by square waves and pulses

never occurs because at least one of the signals is at 0 for all times. There-fore, the output of this three-input AND should always be zero.

A circuit such as this might be used in a safety condition where it is desired to show that all input signals are at proper amplitudes and phase. If it is desired to have a 1 output when all of the input signals are present, it is only necessary to add an inverter to the output. The result is a NAND circuit.

Now consider the same circuit, but with different signals, as shown in Figure 7-8. Here the first two signals, *A* and *B,* are the same as in Figure 7-7, but signal *C* is arriving a little bit late. Since there are now brief

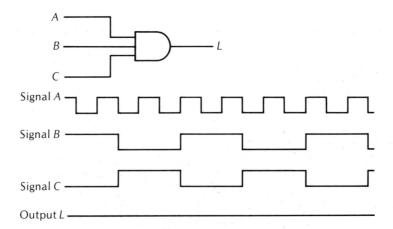

Figure 7-7.   Zero output of an AND gate

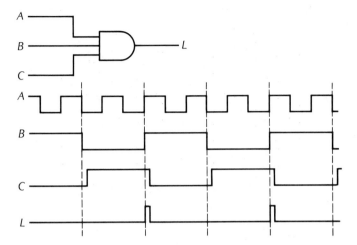

Figure 7-8. Glitches due to late arrival of signal C

periods of time when all the inputs are at logic 1, there will be an output at *L* which is represented by a series of very short duration pulses that indicate that the inputs are not properly timed. The pulses are called *glitches*. Glitches are short duration signals that occur where there are not supposed to be signals. They can cause a breakdown in a digital system. When technicians are testing circuits like this, they must always be aware of the possibility of glitches and their causes.

Figure 7-9 shows a common cause of glitches. Two input signals *A* and *B* are in phase and are to be compared in an AND circuit. It is desirable for the output of this AND to be logic 0 when the input signals are in phase. To accomplish this, signal *B* is inverted with an *inverter* (NOT) gate so that the two inputs to the AND will not be at logic 1 or 0 at the same time.

At least, this is the way the circuit is supposed to work. The problem is that the inverter, like all logic gates, has a *propagation delay*. In other words, it takes a small amount of time for the signal to get through the NOT

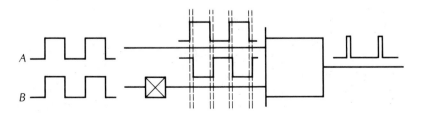

Figure 7-9. Glitches caused by propagation delay

gate. The output signal of the NOT is not exactly 180° out of phase with the signal at A. The slight delay makes it possible for both inputs to be at logic 1 for a very short period of time on alternate half cycles. The overall result will be glitches in the output as shown in Figure 7-9.

To summarize, glitches can occur when signals are routed through a logic system in such a way that the propagation delays cause the signals to get out of synchronization.

NAND and NOR Logic Systems

It is customary for integrated circuit logic manufacturers to make only one type of logic gate—either NAND or NOR. All of the other basic gates can be made from either of these. TTL logic, for example, is usually made of NAND gates, which CMOS is usually made of NOR's.

To understand how this works, consider first the basic NAND gates of Figure 7-10. It is sometimes useful to be able to draw the NAND with separate AND and NOT symbols, as shown in the illustration, in order to simplify the truth table for the system.

A few basic rules that will be helpful in combining NAND's or NOR's to obtain the basic gates are illustrated in Figure 7-11. The truth tables that verify the relationships are given with the tables. The results show that when the inputs of either AND or OR gates are connected, then the output will always be the same as the input. Also, a double negative, or NOT-NOT combination, results in an output identical with the input. These relationships will be useful in obtaining the basic gates from NAND or NOR logic.

MAKING NOT GATES FROM NAND GATES

The simplest basic gate is the NOT or inverter, and the method of obtaining it from NAND logic is shown in Figure 7-12. The gate drawings show that

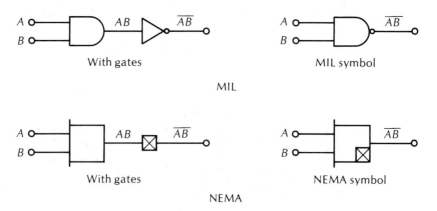

Figure 7-10. NAND used for making basic gates

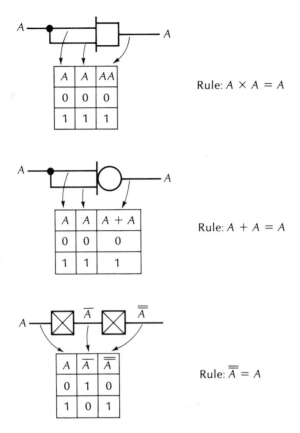

Rule: $A \times A = A$

Rule: $A + A = A$

Rule: $\overline{\overline{A}} = A$

Figure 7-11. Some basic combinations

the inputs of the NAND are connected to obtain the NOT. When this is done, the output is always the reverse of the input. It might seem at first that this is a difficult way to obtain a NOT circuit. Since a conventional amplifier will produce an output that is 180° out of phase with the input, it can be used as a NOT circuit. However, manufacturers assure us that because of the integrated circuit manufacturing process, it is simpler to make one single gate—in this case, NAND—and derive all other gates from it.

They do not show the NOT as a special case of NAND. Instead, they use a simple NOT symbol. For example, the hex (six) NOT integrated circuit package in the TTL family shown in Figure 7-13 uses NOT symbols and carries a number of 7406 plus the pinouts for TTL integrated circuit logic. Note that the manufacturer indicates that they are NOT circuits by using the basic MIL symbol.

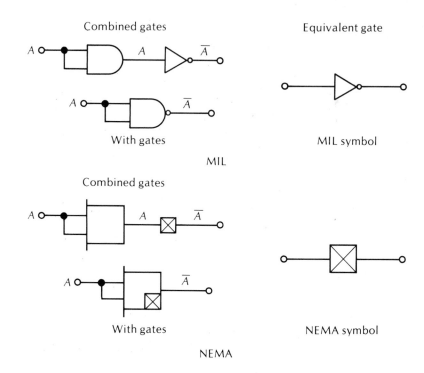

Figure 7-12. Getting a NOT from NAND logic

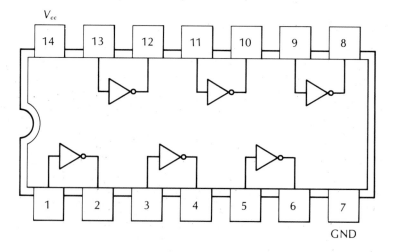

Figure 7-13. A TTL hex NOT

MAKING AND GATES FROM NAND GATES

Figure 7-14 shows how an AND gate may be obtained with NAND's. Note that the output of the first NAND is $\overline{AB}$ (NOT A and B). The output of the second NAND is NOT $\overline{AB}$. As shown in Figure 7-11, the double negative is actually a positive. Truth tables can be used to show that the output of the two NAND's in Figure 7-14 are actually AND's of the input. See Figure 7-15. First, the two inputs are listed with all possibilities by making a binary count from 00 to 11. The next step is to AND the A and B. This is done in the third column. This column has a 1 only when both inputs are at 1. The AND column is inverted to get the NAND column $\overline{AB}$. For the NAND gate connected as a NOT, the output is A and B. It is obtained by inverting $\overline{AB}$ to get $\overline{\overline{AB}}$, as shown in the fourth column in Figure 7-15. The fifth column of Figure 7-15 represents the output of an AND gate.

MAKING OR GATES FROM NAND GATES

Figure 7-16 shows how an OR can be obtained by using AND's. The truth table for this circuitry is shown in Figure 7-17. Again, the A and B inputs are listed with all possibilities. NAND gates 1 and 2 are connected as NOT's so the A and B must be inverted in the truth table. This is done in columns 3 and 4. The $\overline{A}$ and $\overline{B}$ columns are next ANDed because a NAND is really an AND followed by a NOT. Note that column 5 is obtained by ANDing NOT A and NOT B.

The AND gate is followed by a NOT, so the column marked $\overline{A}\ \overline{B}$ must be inverted again. Column 6 gives NOT $\overline{A}\ \overline{B}$, which is the output of the circuit in Figure 7-16.

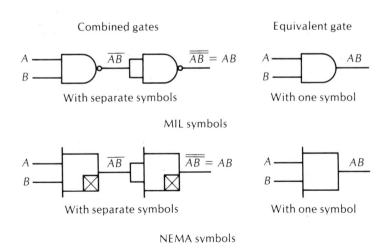

Figure **7-14.** Getting an AND from NAND logic

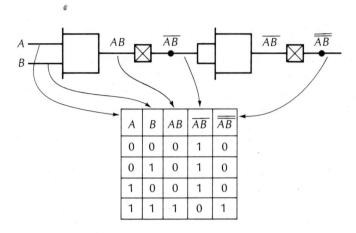

Figure 7-15. Getting an AND from NANDS

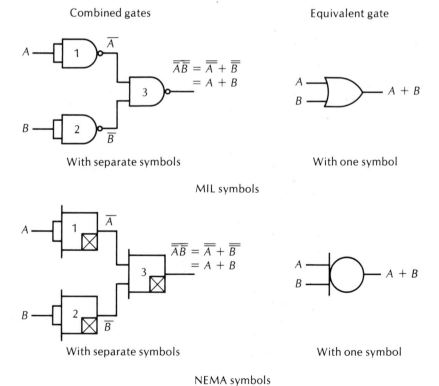

Figure 7-16. Getting an OR from NAND logic

A	B	$\overline{A}$	$\overline{B}$	$\overline{A}\overline{B}$	$\overline{\overline{A}\overline{B}}$	A + B
0	0	1	1	1	0	0
0	1	1	0	0	1	1
1	0	0	1	0	1	1
1	1	0	0	0	1	1

Figure 7-17. Truth table for OR circuit
made of NAND gates

In column 7 the conditions for *A* OR *B* are listed. This column is obtained
by taking the OR of *A* and *B* in the first two columns. Note that the seventh
column is identical to the sixth column; therefore, it follows that $\overline{\overline{A}\,\overline{B}} = A +$
B. A simple rule that is sometimes referred to as *DeMorgan's theorem* can
now be introduced. *Whenever there is an overbar for two gates, it is OK to
break the bar and change the sign.* Going back to Figure 7-17, the sixth
column has an overbar for NOT *A* and NOT *B* ($\overline{\overline{A}\,\overline{B}}$). Breaking this bar gives
NOT NOT *A* or NOT NOT *B* ($\overline{\overline{A}} + \overline{\overline{B}}$). As explained before, a double
overbar is really a positive. Therefore, $\overline{\overline{A}}$ or $\overline{\overline{B}}$ is the same thing as *A* or *B*.

MAKING NOR GATES FROM NAND GATES

To obtain a NOR from NAND logic, it is only necessary to obtain first the
OR as shown in Figure 7-16, and then follow it with an inverter. The
connection is shown in Figure 7-18.

The truth table for the gates of Figure 7-18 is shown in Figure 7-19. The
first two columns represent all the possibilities for the inputs *A* and *B*.
Columns 3 and 4 represent the outputs of NOT gates 1 and 2. These outputs
are NANDed in two steps in columns 5 and 6. The first step is to AND the
inputs $\overline{A}$ and $\overline{B}$; the second step is to invert them. As shown in the previous
example, this results in *A* + *B* at the output of gate 3. It is now only
necessary to invert this result to make $\overline{A + B}$, which is another way of
saying NOR. This output is shown in column 7.

MAKING EXCLUSIVE OR AND LOGIC COMPARATOR FROM NAND GATES

All of the basic gates, except the EXCLUSIVE OR and the COMPAR-
ATOR, have now been obtained with NAND's. Figure 7-20 shows how to
obtain both the COMPARATOR and EXCLUSIVE OR with NAND logic.
The truth table for these systems is shown in Figure 7-21. If you have
followed the previous examples carefully, it should be obvious that the outputs of gates 1 and 2 will be NOT *A* and NOT *B*. These outputs are given in
columns 3 and 4. The output of gate 3 is the NAND of *A* and *B*, which is

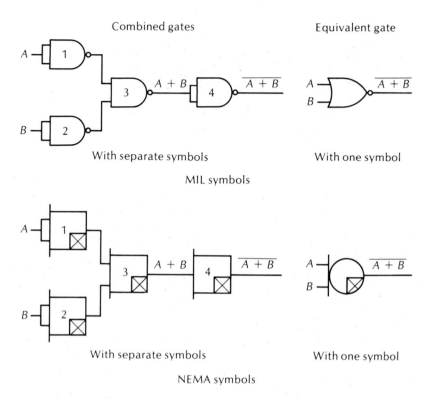

Figure 7-18. Getting a NOR from NAND logic

A	B	$\overline{A}$	$\overline{B}$	$\overline{A} \cdot \overline{B}$	$\overline{\overline{A} \cdot \overline{B}}$ $(A + B)$	$\overline{A + B}$
0	0	1	1	1	0	1
0	1	1	0	0	1	0
1	0	0	1	0	1	0
1	1	0	0	0	1	0

Figure 7-19. Truth table for NOR circuit made of NAND gates

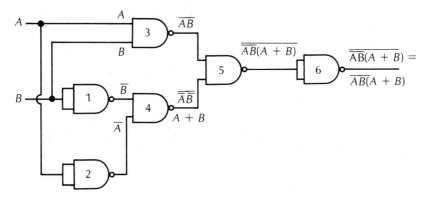

Figure 7-20. Using NAND gates to make a logic comparator and an exclusive OR

obtained in two steps in Figure 7-21. The first step is to get AB which is given in column 5, then invert this column to obtain $\overline{AB}$ in column 6. Column 6 then represents the output of gate 3.

The output of gates 1 and 2 are NANDed by gate 4. As shown in the illustration, this gives an $\overline{\overline{A}\,\overline{B}}$ which has been shown to be the same as OR. (Remember, break the bar and change the sign.) Now the input to gate 5 is $\overline{AB}$ ANDed with $A + B$. Column 8 shows this result. It is immediately obvious that column 8 represents a logic comparator because there is a 1 only when A and B (columns 1 and 2) are identical. To obtain a comparator, then, it is necessary to use only gates 1 through 5 in Figure 7-20.

	Output of gate 1		Output of gate 3				
		Output of gate 2		Output of gate 4			Exclusive
						Comparator	OR
A	B	$\overline{A}$	$\overline{B}$	AB	$\overline{AB}$	A + B	$\overline{AB(A + B)}$	$\overline{\overline{AB}(A + B)}$
0	0	1	1	0	1	0	1	0
0	1	1	0	0	1	1	0	1
1	0	0	1	0	1	1	0	1
1	1	0	0	1	0	1	1	0

Figure 7-21. Truth table for circuit of Figure 7-20

An additional inverter (gate 6 of Figure 7-20) will invert all the digits in column 8 in Figure 7-21. The result is obviously the EXCLUSIVE OR of the first two columns in that truth table.

In the drawing of the gates (Figure 7-20) note that the double overbar at the output of gate 6 represents a double NOT, so the overbars can be removed as shown.

In this section the methods of using NAND's to obtain all of the basic gates have been shown. This is how TTL manufacturers make their logic. It is just as easy to make all of the basic gates using only NOR's, but this exercise will be left to the reader.

Flip-Flops

Next to the six basic gates, the most often used device is the *flip-flop*. Normally, flip-flops are made on integrated circuit chips by using either NAND's or NOR's, depending on the family of logic.

Figure 7-22 shows the most basic kind of flip-flop available. It is called *S-R* for *set* and *reset,* and it is sometimes referred to as a *latch*.

Integrated circuit flip-flops have only two possible outputs, marked Q and $\overline{Q}$. Either Q is at 1 and $\overline{Q}$ is at 0, or Q is at 0 and $\overline{Q}$ is at 1. When $Q = 1$ and $\overline{Q} = 0$, the flip-flop is said to be *high;* when $Q = 0$ and $\overline{Q} = 1$, the flip-flop is said to be *low*.

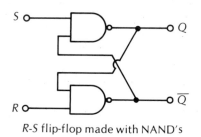

R-S flip-flop made with NAND's

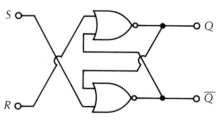

R-S flip-flop made with NOR's

Figure 7-22.   Two basic flip-flops made with gates

It should be understood that the output terminals marked Q and $\overline{Q}$ in Figure 7-22 are sometimes marked 1 and 0.

The best way to understand the action of the R-S flip-flop is to follow the step-by-step operation described in Figure 7-23. Normally, the inputs to the NAND flip-flop are held at the 1 condition. It is possible to switch either of the inputs to 0 momentarily, but because the switches are spring loaded, they will not remain in that position. Instead, they will return to the 1 condition. Normally, square waves are used instead of mechanical switches and batteries for the inputs to the flip-flop.

OPERATION OF THE FLIP-FLOP

For the first condition of the flip-flop in Figure 7-23, it is considered to be *low*. In other words, $\overline{Q}$ is at 1 and Q is at 0. When the flip-flop is in the low condition, it is sometimes said to be *reset*. If switch R is operated momentarily, it will cause the two inputs to gate B to be 0. However, since this is a NAND, the output will not change. It follows, then, that the condition of this flip-flop cannot be changed by switching R when the flip-flop is in the low condition.

If switch S is operated momentarily when in the first condition, one of the inputs to gate A will be switched to 0. This switching will cause the flip-flop to switch to the second condition in Figure 7-23.

The numbers in parentheses represent the previous state—the inputs and outputs of the flip-flop in the first condition. When switch S is switched momentarily to the 0 position, one of the inputs to NAND A goes to 0, and the output of that NAND goes to 1. The 1 at the output of Q is also delivered to the input of gate B. Gate B now has two 1 inputs, and its output must go to 0 at $\overline{Q}$. The 0 at $\overline{Q}$ is fed back to the input at gate A, so now gate A has two 0's when S is at 0, or a 0 and 1 when S is at 1. In either case, the output of gate A is 1. The condition for the flip-flop shown in the second condition is now high. If S is switched to 0 again, it will not change the output from 1. The third condition is now in effect.

In order to get the flip-flop into the low condition again, it is necessary to flip switch R momentarily. This is an important consideration regarding flip-flops. One switch puts it in the high condition, and the other switch puts it in a low condition. When the inputs are both at 1, the flip-flop is in a stable condition and will not change.

Figure 7-24 shows a truth table for the different conditions possible for the S-R flip-flop of Figure 7-23. In the first row both S and R are in the 1 condition, and the flip-flop is in the high condition as indicated by 1 and 0 in the Q and $\overline{Q}$ columns. In the second row R is switched to 0 momentarily, which changes the output of the flip-flop to a low condition. When R is returned to 1 again, as shown in the third row, the flip-flop remains in the low condition. This is the stable condition for the S-R flip-flop made of NAND's.

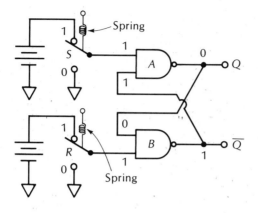

First condition:
Flip-flop is in low
condition

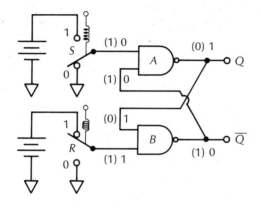

Second condition:
Flip-flop is
switched to high
condition

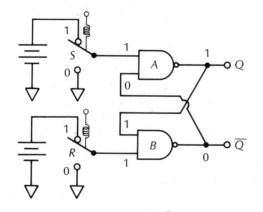

Third condition:
Flip-flop stays
in high condition

Figure 7-23. An R-S flip-flop

S	R	Q	$\overline{Q}$
1	1	1	0
1	0	0	1
1	1	0	1
1	0	0	1
1	1	0	1
0	1	1	0
1	1	1	0
0	1	1	0
1	1	1	0
0	0	Not allowed	

Figure 7-24. Conditions for operation of NAND flip-flop

In the fourth row R is flipped to 0 again, but it has no effect on the output of the flip-flop. Remember that in order to change the condition, it is now necessary to operate the other switch. Row 5 shows both S and R in the 1 condition again and the flip-flop is still in the low condition.

In row 6, S is momentarily switched to 0, thus returning the flip-flop to the high condition, with Q at 1 and $\overline{Q}$ at 0. In row 7, S and R are both at 1 and the flip-flop remains in the high condition.

In row 8, S is flipped again momentarily but it has no control over the output of the flip-flop. In order to switch the flip-flop to low again, it is necessary to switch R to 0. Row 9 indicates that S and R are in the 1 condition again and the flip-flop is still high.

An important characteristic of this flip-flop is that *it must not be possible to switch both S and R to the 0 condition at the same time.* For example, if the inputs are square waves, it must not ever be possible for both waves to go to 0 at the same time. This is sometimes considered to be an important disadvantage of the *S-R* flip-flop.

The flip-flops made with NOR's in Figure 7-22 are designed in such a way that S and R must normally be held at 0 unless it is being switched. Except for this difference, the operation of the NOR flip-flop is similar to that of the NAND type just described.

Figure 7-25 shows the symbol normally used for flip-flops. The high and low conditions are both shown.

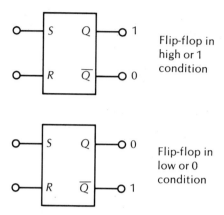

Figure 7-25. Symbol for an R-S flip-flop

TOGGLE FLIP-FLOPS

For the toggle flip-flop of Figure 7-26 it is only necessary to change the input at the T terminal in order to change the output. This is different from the S-R flip-flop in which one switch and then the other had to be operated in order to change the output condition. The toggle flip-flop, which is usually called a T *flip-flop*, can be made by assembling an S-R flip-flop out of NAND's, then ANDing gates 1, 2, 3, and 4 as shown in Figure 7-26. Sometimes these toggle flip-flops are available in integrated circuit packages, and the symbol for the IC toggle flip-flop is shown in Figure 7-26.

An important feature of toggle flip-flops made with NAND's is that they normally toggle on the trailing edge of a square wave input. Thus, if you apply a square wave to a T terminal, the output will change every time the square wave switches from 1 to 0. This is a general rule, but it is not a universal rule. Flip-flops can be made in which the toggling occurs on the rising edge rather than the trailing edge. They are usually made with NOR's.

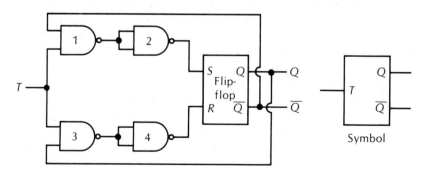

Figure 7-26. Toggle flip-flop

It is necessary to refer to the manufacturer's literature to determine when the switching should actually take place.

Figure 7-27 shows the waveforms for a toggle flip-flop under two different conditions. For one condition the input is a series of pulses. The other is an input square wave to the T terminal. In both cases the flip-flop actually changes its condition on the trailing edge of the input signal. Since this change occurs only once for each cycle, it is clear from Figure 7-27 that the output of the Q or $\overline{Q}$ terminals will have a frequency that is equal to one-half that of the input frequency to the T terminal. This important fact should be remembered because flip-flops are often used in circuits called *divide by 2*. In such circuits the output is equal to the input frequency divided by 2. By using a number of flip-flops, it is possible to divide a frequency by 4, or 8, or any number.

CLOCKED FLIP-FLOPS

Figure 7-28 shows another special case, called a *clocked flip-flop*. Its important feature is that the condition of the flip-flop cannot be changed by switching S or R alone. The clock must be in the positive condition or a 1 before the input to S or R can change the output condition. This type of flip-flop is important because it cannot be triggered accidentally by noise pulses at the S or R terminals. It will change conditions only when you are ready for it to change, and you indicate this by applying a 1 to the clock terminal.

CLEAR TERMINALS

The CL terminal should not be confused with the CLR terminal which is often shown on flip-flop symbols. This is a *clear terminal*. An application of

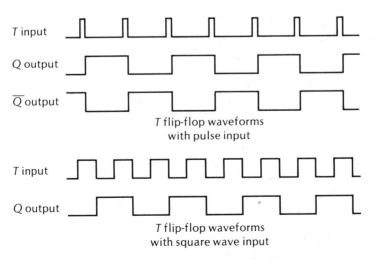

T flip-flop waveforms
with pulse input

T flip-flop waveforms
with square wave input

Figure 7-27. Waveforms for a *T* flip-flop

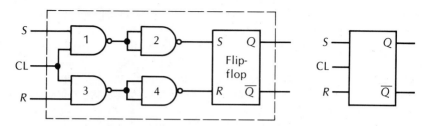

Figure 7-28. A clock flip-flop

a 1 input to the clear will always switch the flip-flop to the high condition. This feature is useful when many flip-flops are used together, and you want them all to return to the high condition simultaneously.

MASTER-SLAVE *J-K* FLIP-FLOPS

It has already been pointed out that a basic *S-R* flip-flop has the disadvantage of having a disallowed condition. When the two inputs of the NAND flip-flop are simultaneously switched to the low condition, the condition of the output cannot be determined. Usually one of the flip-flops will switch first, but there is no way of telling which. The overall result is that it will go either high or low depending on which flip-flop predominates.

To get around this disadvantage, a *master-slave J-K* flip-flop is used. Figure 7-29 shows an example made of NAND gates. It is actually two

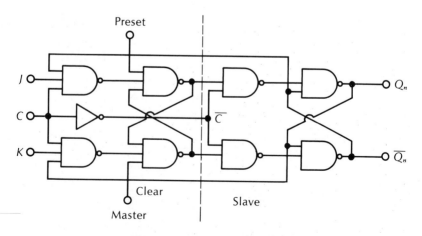

Figure 7-29. A master-slave flip-flop

flip-flops connected in series. The second flip-flop, called the *slave,* will not switch until the condition of the first one (the *master*) is stabilized. No one knows the reason for calling the circuit in Figure 7-29 a *J-K* flip-flop. (It is possible that the designers of the original *J-K* flip-flop simply hunted around to find letters that were not used extensively in electronics.) To switch the master it is necessary to change either *J* or *K* and at the same time apply a clock positive 1 to the *CL* terminal. The inverter causes this clock signal to put a 0 at the slave $\overline{CL}$ terminal, so the slave cannot change conditions because its clock is at 0.

The appropriate input pulse is applied to *J* or *K* and the master is switched. When the input clock is returned to 0, the clock at the slave goes to 1, and it changes its condition according to the inputs from the *Q* and $\overline{Q}$ terminals of the master.

J-K flip-flops are used extensively in logic systems because of their reliability and positive switching conditions. A *J-K* flip-flop made with NAND logic can be toggled by connecting the *J* and *K* terminals to a 1 and then applying a toggle signal to the *CL* terminal.

Ripple Counters

One important application of flip-flops is in counting circuits. Toggle switch flip-flops can be used to make counters.

Figure 7-30 shows toggle flip-flops connected in an *upcounter* circuit. Light-emitting diodes are connected to the *Q* outputs of the flip-flops. They represent the three binary positions for the counter. The purpose of this counter is to display counts in binary form.

For the starting condition, all the flip-flops are at the 0 or low condition and all the lights are out, as illustrated by the unshaded circles. These are NAND flip-flops that toggle on the trailing edge of an input pulse.

When the first input pulse is applied, the trailing edge of the pulse causes flip-flop 1 to switch to the high condition. As a result, the 2^0 LED is switched to ON. The other lights will be off at this time because flip-flop 2 and flip-flop 3 do not change their condition. The reason is that *Q* of FF_1 went from 0 to 1, and it is only possible to trigger these flip-flops when the signal goes from 1 to 0.

At the end of the second pulse, flip-flop 1 will switch to the low condition. Since Q_1 switches from 1 to 0, it causes flip-flop 2 to go to the high condition. Note that at the end of the second pulse the second light is on. The other two are off. This represents the binary 010, which is the same as decimal 2. (Refer to Table 7-1.)

At the end of the third pulse, flip-flop 1 switches again to the high condition. Now both lights in the right column are on.

At the end of the fourth pulse, Q_1 switches to low, causing Q_2 to switch to low also. Since Q_2 is switching to a low condition, it causes the toggle of

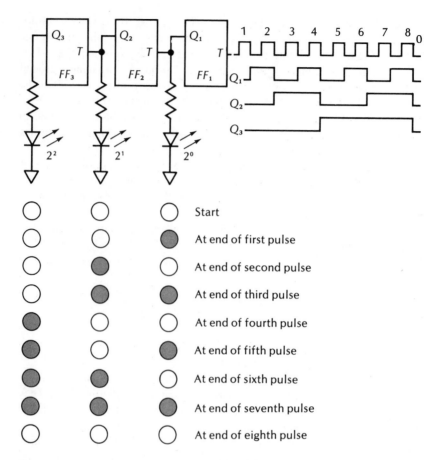

Figure 7-30. An upcounter made with *T* flip-flops (presumed to be TTL NAND flip-flops)

flip-flop 3 to switch that unit to a high condition. Therefore, at the end of the fourth pulse the count is 100. Binary 100 is the same as decimal 4.

At the end of the fifth pulse, flip-flop 1 again switches to high. Flip-flop 2 is not affected. Flip-flop 3 will still be in the high condition and the decimal equivalent of 5 is registered.

At the end of the sixth pulse the first flip-flop switches to low, causing flip-flop 2 to switch to high. Now the count is 110—equivalent to decimal 6.

The next pulse switches the first flip-flop to high, and all of the flip-flops are in the high condition. All of the LED's are ON, representing binary number 111. Binary 111 is the same as decimal number 7.

The eighth pulse switches the first flip-flop to low, which in turn switches the second flip-flop to low. Flip-flop 2, in turn, switches the third flip-flop to low. Thus, at the end of the eighth pulse the binary 000 is represented, which is the starting condition. This circuit will continue to count as long as it is connected.

To make a *downcounter* the flip-flops are connected as shown in Figure 7-31. In this case the initial condition is that all of the flip-flops are in the high condition and all of the lights are on. At the end of the first pulse the first flip-flop is switched off. The toggles are connected to the $\overline{Q}$ outputs, so the second flip-flops are not affected.

At the end of the second pulse the second flip-flop switches to high again. This switches the $\overline{Q}$ terminal from 1 to 0 and turns off the second flip-flop. The rest of the operation is obvious from the waveforms illustrated in Figure 7-31.

The counters shown in Figures 7-30 and 7-31 are called *ripple counters*. The reason is that the input pulse changes the first flip-flop, which changes the second flip-flop, which changes the third one, etc. Therefore, the count ripples along the flip-flops until all of them are set in the required condition.

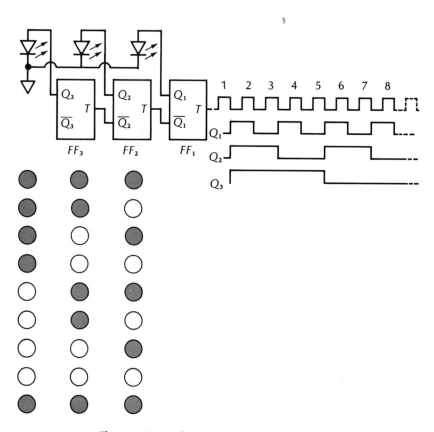

Figure 7-31. A downcounter made with *T* flip-flops

Counting circuits in which all of the flip-flops change at once are called *synchronous counters*.

It is easy for electronic systems to count in binaries, as shown in these illustrations, because it only requires devices that have two stable states: ON and OFF. To count in the decimal system it would be necessary to have components that have 10 stable states. Such devices are not readily available. However, it is a simple matter to convert the binary numbers or binary outputs into decimal numbers by using what is known as a *binary-to-decimal* converter. This is a single-chip integrated circuit designed specifically for this purpose.

From the circuit of Figure 7-30 it is obvious that a counter that uses three flip-flops will count to decimal 7, then start over. Suppose it is desired to count only to decimal 5, then start over. This is represented by binary number 101, and Figure 7-32 shows the conditions of the flip-flops. The first and third flip-flops are high, and the second one is low at the Q terminals. The outputs of the $\overline{Q}$ terminals will be 010. The 0's are passed through inverters to change them to 1's, and the three 1's are delivered to a NAND gate. The NAND output will be 0, which is sent to the *CLR* terminal to reset the Q's of all of the flip-flops to 0's.

When a technician is troubleshooting a counter, or any digital circuit, he measures input and output pulses. Knowing what conditions should exist for various stages, he can determine if the circuit is working properly. This is essentially the same procedure used in other types of circuits. He knows what the signals should be, then he measures them. Finally, he interprets the measurements as to whether the circuit is functioning properly.

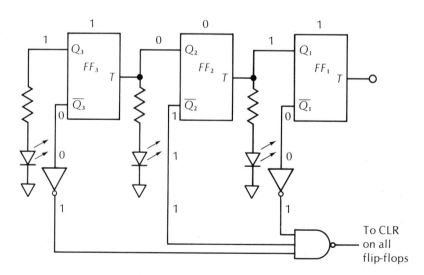

Figure 7-32. Counter designed to count to 101 (decimal 5). The highest number displayed by this counter is 4.

Chapter 11 will describe some of the special test equipment used with digital circuits. Chapter 10 will show how the digital devices and combined logic are used to control machinery and processes.

MAJOR POINTS

1. To be sure that all the possible input conditions have been included in a combination of logic gates, simply list all the digital counts. For example, if there are two inputs—A and B—then all the possible combinations of 1 and 0 inputs are:

A	B
0	0
0	1
1	0
1	1

2. The number of possible input combinations of 1's and 0's in a logic system is 2 raised to a power equal to the number of inputs. For example, with two inputs there are $2^2 = 4$ possible combinations. (See 1 above.)

3. Technicians do not normally design circuits, but they must be able to tell what output can be expected for a combination of gates. They can do this with truth tables.

4. Glitches are undesired output signals in a digital system. A likely cause of glitches is unequal delays caused by routing signals through different gates.

5. When gates are used in combinations, the output can be determined by the use of a truth table and by applying some basic rules.

6. An important rule is that two overbars represent a double negative, which is the same as a positive. For example, $\overline{\overline{A}} = A$.

7. Manufacturers usually make the basic gates from NAND or NOR logic, rather than making six basic gates.

8. Another important rule is sometimes called *DeMorgan's theorem*. It can be simplified to a basic procedure called "break the bar and change the sign." Thus,

$$\overline{AB} = \overline{A} + \overline{B} \quad \text{and} \quad \overline{A + B} = \overline{A}\,\overline{B}$$

This basic rule is useful to technicians when troubleshooting circuits. It is necessary to know what the output of a circuit *should* be in order to know if it is operating properly.

9. Flip-flops are important devices used in logic circuits. They have two stable conditions: *high* and *low* (sometimes referred to as 1 and 0).

10. Flip-flops can be used as upcounters and downcounters.

11. Counters can be programmed to stop at any number, or to return to zero at any number.

12. In ripple counters each flip-flop is controlled by the signal from the previous flip-flop. With synchronous counters, the flip-flops all change at the same time.

PROGRAMMED REVIEW

(Instructions for using this programmed section are given in Chapter 1.)

The important concepts of this chapter are reviewed here. If you have understood the material, you will progress easily through this section. Do not skip this material because some additional theory is presented.

---

1. The output for the logic circuit in the figure below is

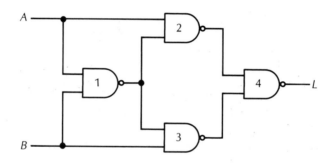

A. $A = B$. (Go to block 9.)
B. $A \oplus B$. (Go to block 17.)

---

2. The correct answer to the question in block 9 is B. The combination of NOT and OR makes NOR, and all basic gates can be made with NOR logic. The combination of AND and OR cannot be used to make all of the basic gates. This restriction becomes apparent when you try to make a NOT out of AND and OR gates.

Here is your next question.
The output of the AND circuit in the figure below will be

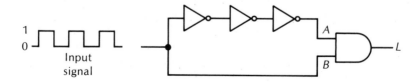

A. Logic 1 at all times. (Go to block 14.)
B. A series of short-duration pulses. (Go to block 16.)

---

3. The correct answer to the question in block 16 is B. The basic rules
    of logic circuits, such as the one given in the question of block 16,
    can always be obtained with truth tables. Input A can have either
    of two values: 0 or 1.

$$
\begin{array}{ccc}
A & 0 & L \\
0 & 0 & 0 \\
1 & 0 & 0
\end{array}
$$
(ANDing first
two columns)

This means that A × 0 = 0.

Here is your next question.
Which of the symbol pairs in the figure below represent a NAND gate?

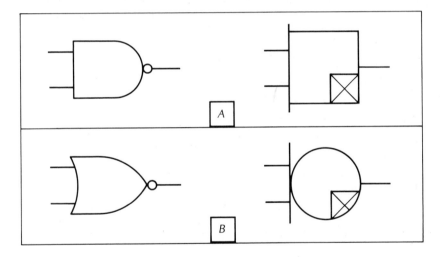

A. The ones shown in part A. (Go to block 23.)
B. The ones shown in part B. (Go to block 10.)

---

4. The correct answer to the question in block 23 is B. Although there
    may be a rare exception, it is almost never possible to directly
    substitute CMOS NAND for TTL NAND. The pinouts and the
    propagation delay for the two types of logic are usually different.
    Also, propagation delays are not the same. It is best not to mix
    logic types in a circuit.

    Here is your next question.
    How many input signal combinations are possible with a four-input
    NAND gate?

A. 16. (Go to block 18.)
B. 8. (Go to block 15.)

---

5. The correct answer to the question in block 25 is B. Generally, flip-flops made with NAND's toggle on the trailing edge. However, this is not a universal rule, so you should consult the manufacturer's literature in each case.

    Here is your next question.
    What decimal count will produce a 0 output at L in the circuit below?

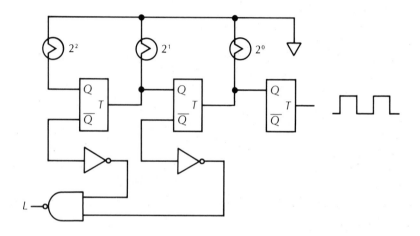

    A. 6 and 7. (Go to block 13.)
    B. 0 and 1. (Go to block 20.)

---

6. Your answer to the question in block 25 is wrong. Read the question again, then go to block 5.

---

7. The correct answer to the question in block 13 is B. The binary representation of decimal number 12 is 1100. Four flip-flops are needed to count to this number.

    Here is your next question.
    The counter in the figure below will clear when it is at decimal

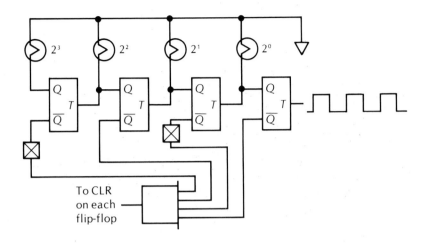

A. 16. (Go to block 27.)
B. 10. (Go to block 21.)

---

8. Your answer to the question in block 16 is wrong. Read the question again, then go to block 3.

---

9. The correct answer to the question in block 1 is A. The expression $A = B$ means that the circuit is a comparator. In other words, there is an output when A and B are both zero, and there will be an output when A and B are both 1.

The output of the circuit can be obtained by the truth table method.

Inputs		Output of NAND 1 (NAND of columns 1 and 2)	Output of NAND 2 (NAND of columns 1 and 3)	Output of NAND 3 (NAND of columns 2 and 3)	Output of NAND 4 (NAND of columns 4 and 5)
A	*B*				
0	0	1	1	1	0
0	1	1	1	0	1
1	0	1	0	1	1
1	1	0	1	1	0

Summarizing the first two columns with the last column,

A	B	Output of NAND 4
0	0	1
0	1	0
1	0	0
1	1	1

This is the truth table for a comparator.

Here is your next question.
All of the basic gates can be obtained from

A. Combinations of AND and OR gates. (Go to block 22.)
B. Combinations of NOT and OR gates. (Go to block 2.)

---

10. Your answer to the question in block 3 is wrong. Read the question again, then go to block 23.

---

11. Your answer to the question in block 23 is wrong. Read the question again, then go to block 4.

---

12. The correct answer to the question in block 18 is B. The rule is to break the overbar and change the sign.

    Here is your next question.
    A certain flip-flop is in the low condition. This means that

    A. Q is at 1 and $\overline{Q}$ is at 0. (Go to block 19.)
    B. Q is at 0 and $\overline{Q}$ is at 1. (Go to block 25.)

---

13. The correct answer to the question in block 5 is A. Three binary digits are displayed. Whenever the 2^2 and 2^1 flip-flops are in the high condition, the first two digits are 1's, and the $\overline{Q}$ outputs are 0's. The 0's are inverted, and two 1's are delivered to the NAND circuit to produce a 1 output at L.

    It does not matter what condition the third flip-flop (2^0) is in, so a 0 output occurs from the NAND for binary numbers 110 and 111. These binary numbers correspond to decimal numbers 6 and 7.

    Here is your next question.
    How many flip-flops are needed to make a ripple counter that can count to decimal 12?

    A. 3. (Go to block 26.)
    B. 4. (Go to block 7.)

14. Your answer to the question in block 2 is wrong. Read the question again, then go to block 16.

15. Your answer to the question in block 4 is wrong. Read the question again, then go to block 18.

16. The correct answer to the question in block 2 is B. The three NOT circuits invert the input signal. Theoretically it is 180° out of phase with the other AND input, and if this were true the output would be 0 at all times. This follows because whenever A goes to 1, B goes to 0; and, whenever A goes to 0, B goes to 1.

The actual output is a series of pulses because the inverters have a propagation delay, meaning that the two signals are not exactly 180° out of phase. For a brief period both inputs are at logic 1, and during those brief periods the output is able to go to logic 1.

The waveforms are shown below.

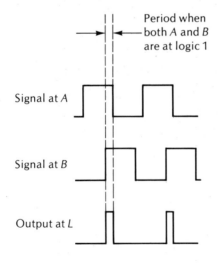

Here is your next question.
Which of the following logic statements is correct?

A. $A \times 0 = A$. (Go to block 8.)
B. $A \times 0 = 0$. (Go to block 3.)

17. Your answer to the question in block 1 is wrong. Read the question again, then go to block 9.

18. The correct answer to the question in block 4 is A. The number of possibilities is equal to 2 raised to the number of inputs, $2^4 = 16$.

Here is your next question.
Which of the following equations is valid?

A. $\overline{AB} = \overline{A}\,\overline{B}$. (Go to block 24.)
B. $\overline{AB} = \overline{A} + \overline{B}$. (Go to block 12.)

19. Your answer to the question in block 12 is wrong. Read the question again, then go to block 25.

20. Your answer to the question in block 5 is wrong. Read the question again, then go to block 13.

21. The correct answer to the question in block 7 is B. Decimal 10 is represented by binary number 1010. All inputs to the AND gate must be at logic 1 in order to get a CLEAR output. The inputs to the AND gate will be 1's when the 2^3 and 2^1 flip-flops are high, and when the 2^2 and 2^0 flip-flops are low. Note the inverters that convert the low (0) outputs from 2^3 and 2^1 to 1's.

Here is your next question.
Use a truth table analysis to show that the circuit below is an EXCLUSIVE OR.

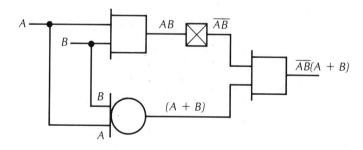

(Go to block 28.)

22. Your answer to the question in block 9 is wrong. Read the question again, then go to block 2.

23. The correct answer to the question in block 3 is A. The symbols are shown in Chapter 6.

Here is your next question.
The NAND gates in TTL logic operate from a +5 V regulated supply. The NAND gates of CMOS logic can also be operated from a +5 V regulated supply. Can a CMOS NAND gate be substituted for a TTL NAND gate in a circuit?

A. Yes. (Go to block 11.)
B. No. (Go to block 4.)

---

24.   Your answer to the question in block 18 is wrong. Read the question again, then go to block 12.

---

25.   The correct answer to the question in block 12 is B. In the low condition the $\overline{Q}$ output is at logic 1.

Here is your next question.
Normally, a T flip-flop made with NAND logic can be expected to trigger on the

A. Leading edge of the toggle input pulse. (Go to block 6.)
B. Trailing edge of the toggle input pulse. (Go to block 5.)

---

26.   Your answer to the question in block 13 is wrong. Read the question again, then go to block 7.

---

27.   Your answer to the question in block 7 is wrong. Read the question again, then go to block 21.

---

28.   Inputs

A	B	AB	$\overline{AB}$	$A + B$	$\overline{AB}(A + B)$
0	0	0	1	0	0
0	1	0	1	1	1
1	0	0	1	1	1
1	1	1	0	1	0

Summarizing the inputs and output:

A	B	L
0	0	0
0	1	1
1	0	1
1	1	0

This is the truth table for an EXCLUSIVE OR circuit.

You have now completed the programmed section.

SELF-TEST WITH ANSWERS

(Answers to this test are given at the end of the chapter.)

1. A disadvantage of the basic latch is that it can be put into an undetermined condition by switching both inputs at the same time. To avoid this, the flip-flop can be designed to have (a) a master and slave, (b) an input and output, (c) a him and her, (d) none of the above.

2. Toggle flip-flops can be connected to make a ripple counter. How many flip-flops are needed to produce a decimal equivalent of 8?

3. The binary display of a ripple counter is shown below:

    ∅∅∅0

    This represents a decimal count of _____ .

4. Either of two basic logic gates can be used to make all of the other gates. The two are _____ and _____ .

5. When making circuits from basic gates, different families of logic should not be used. For example, in the circuit of Figure 7-16, gates 1 and 2 must be from the same logic family. One reason is that power supply requirements may be different for different families. Another reason is the differences in (a) appearance, (b) weight, (c) cost, (d) propagation delay.

6. You can make a comparator by inverting the output of _____ .

7. A certain J-K flip-flop is in the low condition. To get it into the high condition, the next input signal must be delivered to (a) the J terminal, (b) the K terminal. (Disregard clock terminal input.)

8. When writing the outputs of a certain circuit the following expression is obtained: $\overline{AB}$. According to DeMorgan's theorem, this is the same as (a) $\overline{AB}$, (b) $\overline{A} + \overline{B}$, (c) $A + B$, (d) $\overline{A} + \overline{B}$.

9. A NAND-type J-K flip-flop can be made to toggle by (a) connecting J and K to 0, and applying the toggle signal to the clock terminal; (b) connecting J and K to 1, and applying the toggle signal to the clock terminal.

10. An undesired output from a circuit of combined logic gates is called a _____ .

Answers to Self-Test

1. (a)
2. 4
3. 14
4. NAND and NOR
5. (d)
6. An EXCLUSIVE OR
7. (a)
8. (d)
9. (b)
10. Glitch

CHAPTER 8

FLIP-FLOP CIRCUITS, MICROPROCESSORS, AND MICROCOMPUTERS

You have studied one example of a flip-flop circuit—the ripple counter—in Chapter 7. A disadvantage of ripple counters (also called *asynchronous counters*) is that they are slow compared to synchronous counters. The word "slow" is relative, since these devices can easily make a million counts per second.

Synchronous counters will be discussed in this chapter. Although they have the advantage of speed, they have the disadvantage that they produce a greater demand on the power supply. In practice both types—ripple and synchronous—are used. The distinguishing feature of synchronous counters is that the clock signal is delivered to all of the toggle terminals at the same time. (With ripple counters the clock signal goes to only the first flip-flop toggle terminal.)

A microprocessor is an integrated circuit that is used as the heart of a computer. It has applications in industrial electronics besides its use in computers.

All microprocessors have the same basic sections. The difference between microprocessors made by different manufacturers is primarily in the way information—called *data*—is moved in between the sections.

Microprocessors are used in control circuits. It is important to understand how they work if you are going to be able to understand complete systems.

THEORY

Comparison of CMOS and TTL Flip-Flops

Although the basic action of flip-flops is the same regardless of which logic family they are connected to, there are some differences in the way they are connected in circuits.

Figure 8-1 shows a TTL flip-flop connected so that it will toggle when a square wave signal is delivered to the clock terminal. There are some important things to note in this circuit:

- The power supply for TTL circuits is regulated at 5 V.
- The J, K, CLR (clear), and set terminals are all wired to logic 1, which is the $+5$ V supply voltage.

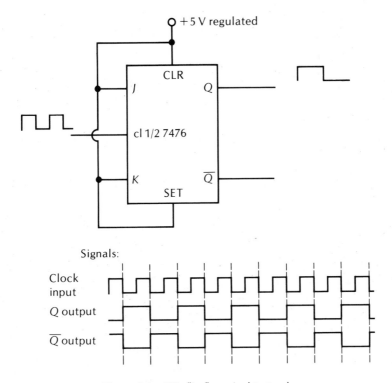

Figure 8-1. TTL flip-flop wired to toggle

- The input signal is to the clock terminals. When the flip-flop is wired this way, the clock terminal corresponds to the toggle input described in the previous chapter.
- The circuit toggles only when the clock goes from a high (logic 1) to a low (logic 0). This is shown by the dashed line between the signals.
- The output signal frequency at Q and at $\overline{Q}$ is one-half the input frequency. For this reason it is called a *divide-by-2* circuit. Note that the $\overline{Q}$ output is the inverse of the Q output.

The TTL 7476 is actually a dual flip-flop, that is, two flip-flops in a single package.

If the Q output of Figure 8-1 is connected to the toggle of a second flip-flop, the output signal of the combination will have a frequency that is one-fourth the input frequency. This is shown in Figure 8-2. The circuit is shown with the input to the flip-flop so that the LED's will be able to display a binary count.

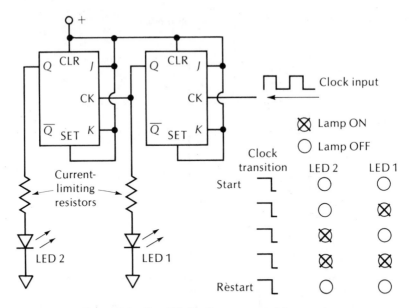

Figure 8-2. Two TTL flip-flops connected for counting

The sequence for the lamp operation is given here:

Start	00	binary number for decimal 0	
	01	binary number for decimal 1	
	10	binary number for decimal 2	
	11	binary number for decimal 3	
Restart	00	binary number for decimal 0	
	01	etc.	

Additional flip-flops can be added to the circuit of Figure 8-2 to get a higher count.

Figure 8-3 shows a binary counter made of CMOS. This circuit is different from the TTL circuit in a number of ways:

- The flip-flops toggle on the *leading* edge of the clock pulse, whereas TTL flip-flops toggle on the *trailing* edge.
- The set and clear terminals must be *grounded* (at logic 0) to get a toggle action. Compare this with the TTL toggled flip-flops in which the SET and CLR terminals must be at logic level 1 to get toggle action.
- The clock signal to the second flip-flop comes from the $\overline{Q}$ terminal of the previous flip-flop, but in the TTL circuit it comes from the Q terminal.

The counters of Figures 8-2 and 8-3 can both be made to count down instead of up by connecting the LED's to the outputs of the $\overline{Q}$ terminals rather than to the Q terminals.

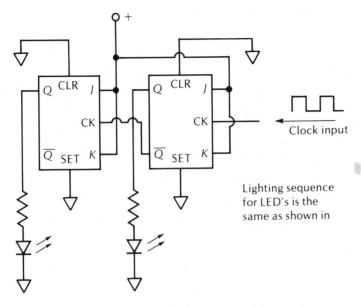

Figure 8-3. Two CMOS flip-flops connected for counting

For the remainder of this chapter the circuits discussed will employ TTL rather than CMOS integrated circuits. In all cases there is a CMOS equivalent. However, keep in mind that there are basic differences in the way the families of circuits are connected to achieve a given result.

SYNCHRONOUS COUNTERS

In the ripple counters of Figures 8-2 and 8-3 each flip-flop is toggled by a signal from the preceding stage. The count is made by signals rippling through the stages.

The J and K terminals must both be connected to logic 1 to get the toggle action. If the J or K terminal, or both terminals, are switched to logic 0, the counting stops. This fact is useful in making a counter count up or down to a certain value and then stop.

The TTL synchronous counter of Figure 8-4 has the clock signal delivered to all of the flip-flops, rather than to only the first as in the case of the ripple counter. Therefore, a clock transistion from logic 1 to logic 0 makes it possible to switch all flip-flops at the same time. However, only the ones that advance the count will actually switch.

Assuming that all of the flip-flops are low ($Q = 0$ and $\overline{Q} = 1$) the first negative-going clock transition will switch FF_1 to high. Note that FF_2 cannot switch to high because its J and K terminals are at logic 0 when the

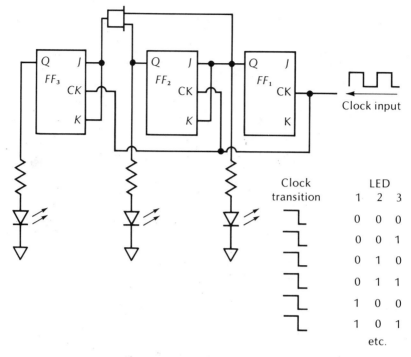

Figure 8-4. Synchronous TTL counter

clock transition appears on the *CL* terminal. (Remember, *FF₁* is low when the transition appears, so the logic 0 on *FF₁* is connected to *J* and *K* of *FF₂*.)

FF₃ cannot switch until both *FF₁* and *FF₂* are high, so that two logic 1's are delivered to the AND gate. The output of the AND gate will make the *J* and *K* of *FF₃* high and allow it to change states each time *FF₁* and *FF₂* go high.

Part of the count is shown in Figure 8-4. Appendix I shows the binary count for higher numbers.

Since all flip-flops can change at one time (when the count is all 1's), the demand on the power supply can be high when a large number of flip-flops are used in a synchronous counter. This is the major difficulty of synchronous counters. However, since the flip-flops all change at once, it is not necessary to wait for the count to ripple through the counter.

DIGITAL DISPLAYS

The binary numbering system is not a convenient way to display the output of a counter. Few people can read binary numbers. Furthermore, it takes a large number of digits to display a number. For example, five binary digits are needed to display decimal number 16. (Decimal 16 is displayed as binary 10000.)

A decoder can be used to convert binary numbers to a coded output for operating a seven-segment display. An example of this type of circuit is shown in Figure 8-5.

The synchronous counter comprises four flip-flops in a single integrated circuit package. The input clock signal is a square wave with a frequency of 1 Hz. This low frequency is used so that the changes in display can be observed. A much higher frequency—up to 20 MHz—is used in practical applications.

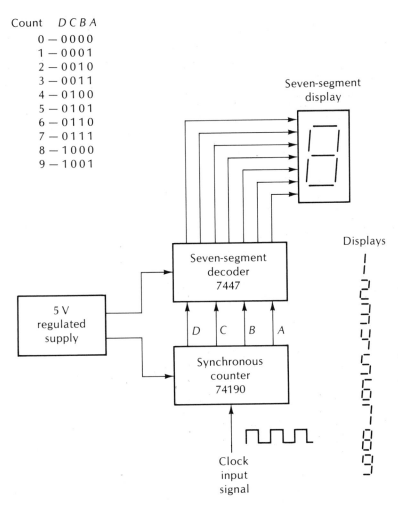

Count   D C B A
0 — 0 0 0 0
1 — 0 0 0 1
2 — 0 0 1 0
3 — 0 0 1 1
4 — 0 1 0 0
5 — 0 1 0 1
6 — 0 1 1 0
7 — 0 1 1 1
8 — 1 0 0 0
9 — 1 0 0 1

Seven-segment display

Displays

Seven-segment decoder 7447

5 V regulated supply

D C B A

Synchronous counter 74190

Clock input signal

Figure 8-5. Circuit for decimal display

The outputs of the four flip-flops in the counter are their signals, representing a binary count from 0 to 9. After 9 is reached the circuit is reset to all zeros, then starts counting again.

The decoder converts the binary output to signals that operate the various parts of the display. For example, a binary input signal of 1000 will cause all output lines to the display to go to logic 1, and decimal number 8 is displayed.

Although only one digit is displayed in Figure 8-5, a number of circuits can be connected together to get any number of digits in the display.

HOW COUNTERS ARE USED IN INDUSTRIAL SYSTEMS

Counters are basic circuits used in many different types of industrial electronic systems. Two examples will be given to demonstrate their application.

The circuit of Figure 8-6 counts boxes on a conveyor belt. Each time a box passes between the light source and the light-to-voltage transducer, there is a negative-going pulse. The pulse is amplified and squared so it can be used as the clock signal for the counter.

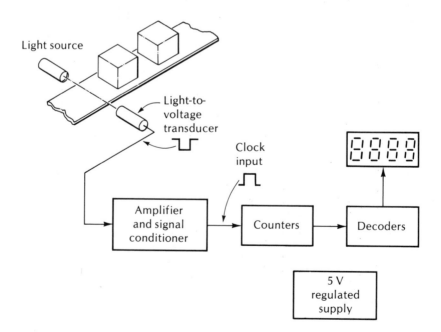

Figure 8-6. Circuit used for counting boxes on a conveyor belt

Each clock pulse causes the counters to increase their binary output by one digit. The decoder converts the output for operating the seven-segment display.

Four counters and four decoders would normally be used to get the four decimal digits as shown in the circuit of Figure 8-6. However, a single decoder can be used if it is multiplexed. This simply means that the decoder output is switched from digit to digit. The switching is so rapid that all of the digits appear to be ON at all times.

Figure 8-7 shows how a counter can be used to determine the speed of a motor in revolutions per second. A magnet, moving past the pickup coil, induces a voltage in that coil. Although an amplifier is not shown, the induced voltage would have to be amplified and shaped in order to be suitable as a clock pulse for the counter.

The 1-second pulse delivered to the *enable* terminal allows the counter to count for that period of time, then stops the count. The displayed number will be the number of pulses (and revolutions) per second.

The *latch* feature in the decoder holds the count on the display while the next count is being made.

MICROPROCESSORS AND MICROCOMPUTERS

A microprocessor is an integrated circuit that performs some of the important tasks in a microcomputer. For example, it performs the arithmetic operations and memory functions. Figure 8-8 shows the basic parts of a

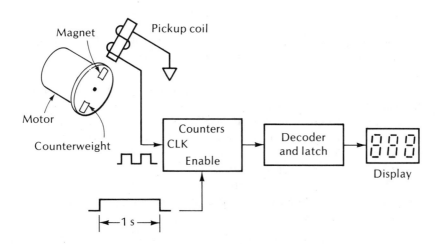

Figure 8-7. Circuit for measuring revolutions per second

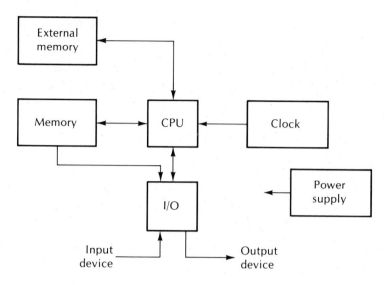

Figure 8-8. Internal and external parts of a microcomputer

microcomputer. The CPU (central processing unit) is a microprocessor in this system.

The input device delivers the instructions to the computer in a coded form. A set of instructions is called the *program*. The code consists of *bits* (*bi*nary digi*ts*) which are combined to make a *word*. If 8 bits are used for each step in the program, then each 8-bit combination is called a *word*.

Typewriters, teletype machines, combinations of switches, and keyboards are all examples of input devices. In industrial electronic systems, punched tapes or magnetic tapes may be coded to instruct machinery. The tape reader then serves as an input device to the system. Appendix I shows a punched-tape code.

The coded input signals go to an I/O (input/output) section that delivers the signal to the proper section.

The CPU performs math or logic manipulations on the input signal. For example, if two numbers are to be multiplied, the multiplication is per-formed in the CPU.

The internal and external memory sections are used to store data in the form of binary numbers. The location of the data in the memory is called the *address*. Several different kinds of memory are used in microprocessor and computer systems.

1. A *random-access memory* (RAM) can be used to store data. The data stored can be retrieved by the CPU, and the stored data can be changed by signals from the CPU.

2. A *read-only memory* (ROM) has the data put in by the manufacturer and this data can be retrieved by the CPU. However, the CPU cannot change the data. As an example, the manufacturer may store the value of π in the ROM. This value can be retrieved and used for solving problems, such as finding the area or circumference of a circle. However, the value cannot be changed.

3. A *programmable read-only memory* (PROM) is like the ROM in the sense that the data stored in it can be retrieved, but not changed, by the CPU. However, the data can be changed with special equipment made for that purpose; it employs ultraviolet lights to change the data.

The terms *volatile* and *nonvolatile* are also used in reference to memory. If the memory is lost when the power is turned off, it is a volatile type. If the memory remains and is usable when the power is turned off and on, then it is a nonvolatile type.

The clock in the system delivers one or more pulse waveforms to time all of the operations and to move the data into and out of the system. It also moves the data within the system.

The output device displays the result of the computer operation. Typewriters and teletype machines may be used for this purpose. Cathode-ray tubes are also popular display devices. In large computers special high-speed printers are favored. And, if the computer is used to operate machinery, then the machine could be called the *output device*.

There are so many uses for microprocessors and microcomputers in industry that it would be impossible to list them in a chapter. However, a few examples will give you an idea of the range of possibilities.

- Machine control
- Information storage and retrieval
- Inventory control
- Automatic testing
- Control of fluids
- Intrusion alarms
- Generation of characters and drawings on picture tubes
- Speed control
- Automatic weighing
- Paint mixing
- Automatic welding and riveting

MAJOR POINTS

1. Counters are synchronous if the clock signal is applied to all of the flip-flops in the counter simultaneously.
2. Asynchronous counters are slower than synchronous counters, but do not put as much demand on the power supply.

3. A microprocessor is an integrated circuit used as the CPU of a microcomputer.
4. Flip-flops in different families have some basic differences in their operation, so their connections into counting circuits are different.
5. A *J-K* flip-flop can be connected so that it will toggle, so it can be used in counting circuits.
6. Toggled flip-flops divide the clock signal by 2.
7. A counting circuit will count up or down depending upon where the LED's, or other display devices, are connected.
8. Decoders change the binary outputs of four flip-flops in a counting circuit to signals that operate a seven-segment display.
9. A single chip containing four flip-flops can be used for counting.
10. A counter can be used to determine the number of events per second (such as revolutions per second) by enabling it for 1-second periods.
11. A program is a set of instructions for a computer.
12. A word is a combination of binary digits, or bits.
13. A microcomputer uses a microprocessor, I/O section, memories, and clock signals.
14. Random-access memories can be reprogrammed. The CPU can put information into the RAM. It can also take information out of a RAM.
15. Read-only memories cannot be programmed.
16. PROMS are read-only memories that can be programmed with special equipment.
17. Volatile memories are erased when the power supply is shut off.

PROGRAMMED REVIEW

(Instructions for using this programmed section are given in Chapter 1.)

The important concepts of this chapter are reviewed here. If you have understood the material, you will progress easily through this section. Do not skip this material because some additional material is presented.

---

1. The value of π is stored permanently in a memory used in a calculator. The memory is

A. Volatile. (Go to block 9.)
B. Nonvolatile. (Go to block 17.)

---

2. Your answer to the question in block 13 is wrong. Read the question again, then go to block 20.

---

3. The correct answer to the question in block 17 is A. The highest number is obtained when all flip-flops are high, so the output is

111. That is equal to decimal 7. (Binary numbers are given in Appendix I.)

Although the highest number of the counter is 7, it actually makes 8 counts. Remember that 000 is a count. A basic rule is that flip-flops can make 2^N counts, where N is the number of flip-flops; and the highest number they can display is 2^N-1.

Here is your next question.
To stop a *J-K* flip-flop from toggling, a signal is sent to

A. The *J* and *K* terminals. (Go to block 8.)
B. The clock terminal. (Go to block 5.)

---

4. Your answer to the question in block 18 is wrong. Read the question again, then go to block 13.

---

5. Your answer to the question in block 3 is wrong. Read the question again, then go to block 8.

---

6. Your answer to the question in block 20 is wrong. Read the question again, then go to block 15.

---

7. Your answer to the question in block 15 is wrong. Read the question again, then go to block 10.

---

8. The correct answer to the question in block 3 is A. The polarity of the stopping signal to the *J* and *K* terminals depends upon which family it is in.

Here is your next question.
Which of the following statements is correct?

A. A bit is a combination of words. (Go to block 11.)
B. A word is a combination of bits. (Go to block 18.)

---

9. Your answer to the question in block 1 is wrong. Read the question again, then go to block 17.

---

10. The correct answer to the question in block 15 is B. Draw the segments on a separate sheet of paper and darken sections *A, F, G, C,* and *D.* You will see the number 5 as it looks on a seven-segment display.

Here is your next question.
Which of the following terminals on a decoder is used to hold a display?

A. Strobe. (Go to block 12.)
B. Latch. (Go to block 19.)

---

11. Your answer to the question in block 8 is wrong. Read the question again, then go to block 18.

---

12. Your answer to the question in block 10 is wrong. Read the question again, then go to block 19.

---

13. The correct answer to the question in block 18 is B. There is no difference in cost between ripple and synchronous counters.

Here is your next question.
Which of the following is used for both flip-flop counters and microprocessors?

A. Clock. (Go to block 20.)
B. CPU. (Go to block 2.)

---

14. Your answer to the question in block 17 is wrong. Read the question again, then go to block 3.

---

15. The correct answer to the question in block 20 is A. The CPU is instructed to get information out of memory when a certain coded binary is sent to the CPU.

Here is your next question.
The segments of a seven-segment display are usually lettered as shown here, but lowercase letters are often used. To display the number 5, which sections should be emitting light?

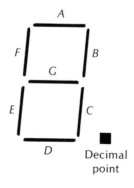

Decimal
point

A. *B* and *E*. (Go to block 7.)
B. *A, F, G, C*, and *D*. (Go to block 10.)

---

16. Your answer to the question in block 19 is wrong. Read the question again, then go to block 21.

---

17. The correct answer to the question in block 1 is B. You would not want the value of π to be erased every time you turned the calculator off.

Here is your next question.
When three flip-flops are used to make a counter, the highest decimal number that could be obtained is

A. 7. (Go to block 3.)
B. 8. (Go to block 14.)

---

18. The correct answer to the question in block 8 is B. A word is a combination of binary digits, or bits.

Here is your next question.
For some applications a ripple counter may be

A. Too expensive when compared with a synchronous counter. (Go to block 4.)
B. Too slow when compared with a synchronous counter. (Go to block 13.)

---

19. The correct answer to the question in block 10 is B. The strobe terminal is used to turn the display on and off rapidly, and thereby control the display brightness.

Here is your next question.
Is this statement true or false: An electric typewriter can be used as an input and an output device.

A. The statement is true. (Go to block 21.)
B. The statement is false. (Go to block 16.)

---

20. The correct answer to the question in block 13 is A. A microprocessor may require several different clock signals.

Here is your next question.
The location of information in a memory is called the

A. Address. (Go to block 15.)
B. Pad. (Go to block 6.)

21. The correct answer to the question in block 19 is A. In order to use an electric typewriter for an input and/or output device, the keys must be wired specially for this type of operation. The keys must generate a binary-coded signal to operate as an input device, and they must be operated by a binary-coded signal in order to be useful as an output device.

You have now completed the programmed section.

SELF-TEST WITH ANSWERS

(Answers to this test are given at the end of the chapter.)

1. When a flip-flop is connected into a counting circuit, the block signal is delivered to the _____ .

2. With four flip-flops in a counting circuit, the maximum number of counts possible is _____ .

3. What is the highest count for a counting circuit comprised of four flip-flops? _____ .

4. A memory that retains data, even though the power is removed, is said to be _____ .

5. A memory that is programmed by the manufacturer, and cannot be reprogrammed, is called _____ .

6. A word is a combination of _____ .

7. If a clock signal goes to all of the flip-flops at the same time, it is a _____ counter.

8. A disadvantage of ripple counters, when compared to synchronous counters, is that _____ .

9. To stop a flip-flop, a signal is delivered to the _____ terminals.

10. To hold a display on a seven-segment readout, a signal is delivered to the _____ terminal of the decoder.

Answers to Self-Test

1. Clock or toggle
2. 16
3. 15
4. Nonvolatile
5. ROM
6. Bits
7. Synchronous
8. They may be too slow for some applications.
9. J and/or K
10. Latch

CHAPTER 9

POWER SUPPLIES

One section that industrial electronic systems have in common with all other electronic systems is the power supply. Dc power is needed for operating the amplifying components, logic and switching circuits, control circuits, etc. *Regulated* supplies for both ac and dc are needed for operating some industrial systems. A power supply is regulated when the output voltage or current is maintained at a constant value despite changes in input voltage or output load resistance.

The study of power supply circuitry can provide a foundation for many of the control circuits used in the other parts of the industrial electronic system. Modern power supplies may employ thyristors, closed-loop feedback circuitry, and voltage and power amplifiers.

Both ac and dc power are needed in industrial electronic systems. There are four possible configurations for changing existing power into another form or another value. They are:

1. Input dc, output ac
2. Input dc, output dc
3. Input ac, output dc
4. Input ac, output ac

Each configuration will be discussed in this chapter, but the main emphasis will be on the dc power supply. This type of supply changes ac power from the commercial power line into dc voltages and currents for operating electronic systems.

The term *power supply* is somewhat misleading. This circuit does not actually supply power; instead, it converts the power from one form (ac) to another (dc). In this chapter, "power supply" will always refer to one that converts ac to dc.

The power supplies used in industrial electronic systems can be divided into three groups: *unregulated, semiregulated,* and *regulated.* Unregulated supplies change the input power of the system into dc voltages. This type of supply is used in applications where the designer knows exactly what voltage and current is going to be required by the electronic system, and when that voltage and current does not change during normal operation.

Semiregulated power supplies provide some means of holding the output at a constant value of voltage or current, but no automatic circuitry is used to readjust the output when the load changes or when the line voltage changes.

Regulated power supplies have a *sense circuit* for monitoring the output voltage or current. The sensed voltage or current is compared with a known accurate value. If the voltages (or currents) do not match, a *control circuit* is set into operation. The control circuit readjusts the output of the supply to the proper value.

Regulated supplies compensate for changes in load resistance as well as for changes in input line voltage. In the electronic form the regulated supply behaves like an operational amplifier in that both use negative feedback.

Examples of the three types of power supplies are discussed in this chapter.

THEORY
Circuits for Converting Power

Power for operating electronic systems is of two kinds: dc power, which has the same polarity at all times; and ac power, which periodically reverses polarity. In some applications it is necessary to convert from one type of power to the other or change an ac or dc voltage (or current) from one level to another. The four possibilities are shown in Figure 9-1.

INVERTERS

An inverter is a power supply system that changes dc to ac. One method of doing this is to use a dc motor to turn an ac generator. This is illustrated in

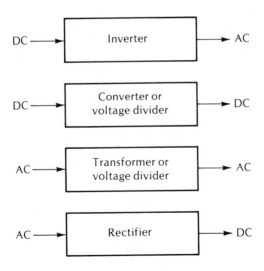

Figure 9-1. Four methods of converting power

Figure 9-2. The motor-generator inverter combination is usually constructed in the same housing.

An electronic inverter is also shown in Figure 9-2. It consists of an oscillator, which is a circuit that changes dc to ac, and a transformer. The electronic type has become more popular because it requires no moving parts, and thus reduces the maintenance requirements.

A disadvantage of inexpensive electronic inverters is that the output waveform is not a pure sine wave, but is more closely related to a square wave. This waveform is not suitable for operating some kinds of electrical equipment. High-priced inverters include waveshaping circuits—usually inductors and capacitors that produce a pure sine wave output voltage.

CONVERTERS

A converter is a power supply system that changes a dc voltage from one value to another. A mechanical converter (also known as a *rotary converter*) is a motor-generator combination. This is shown in Figure 9-3. Although shown as separate units in the drawing, the motor and generator are usually

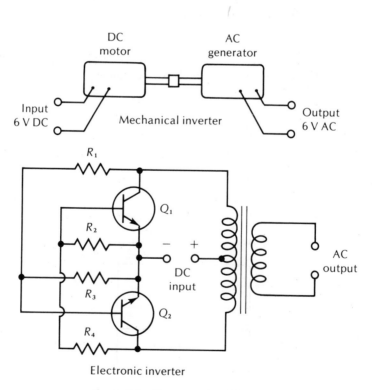

Figure 9-2. Two examples of inverters

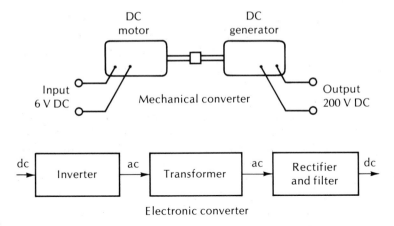

Figure 9-3. Two examples of converters

constructed in the same housing. In this particular application a 6 V motor turns a generator that produces 200 V output.

Rotary converters are still used in some applications, but in electronics they have been largely replaced by inverter-rectifier combinations of the type illustrated in Figure 9-4. (Rectifiers that convert ac to dc are discussed in a later section.)

When the dc voltage is being dropped to a lower value, a voltage divider can be used. An example is shown in Figure 9-4. The values of the resistors and their power ratings must be calculated on the basis of expected load currents flowing through the divider. In the example shown, positive and negative voltages are achieved with a single power supply and a common point in the divider.

TRANSFORMERS FOR AC TO AC CONVERSION

Transformers are used to change an ac voltage from one value to another. Figure 9-5 shows a transformer with both higher and lower voltages on the secondary. The ratio of primary to secondary turns is called the *turns ratio*. Mathematically, the turns ratio is equal to the voltage ratio.

$$\frac{N_p}{N_s} = \frac{V_p}{V_s}$$

In a perfect transformer the input power is equal to the output power, or

$$V_p + I_p = V_s + I_s$$

From this equation it is obvious that if the secondary voltage (V_s) is increased, then the secondary current (I_s) must be decreased in order to keep

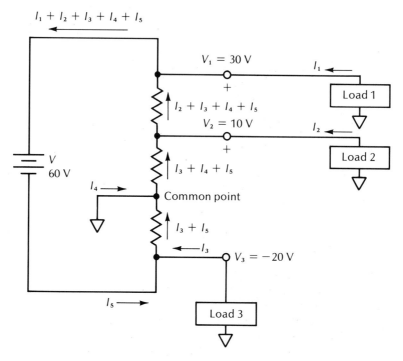

Figure 9-4. Voltage divider for dc supply

the output power equal to the primary power. In other words, when the secondary voltage is stepped up, the secondary current capability is stepped down.

A smaller conductor size can be used for a step-up winding due to the lower current rating, but greater insulation is required. If the secondary voltage is stepped down, the secondary current-carrying capacity can be greater provided the wire used for winding the secondary has a larger diameter.

Transformers are sometimes used to split the phase of the ac power by using the grounded center tap as shown in Figure 9-6. There are two reasons for using this type of transformer. First, some full-wave rectifiers require a center-tapped secondary. This type of rectifier will be discussed further in the next section. Second, some circuits (such as operational amplifiers) require both positive and negative voltages. These voltages can be readily obtained by rectifying and filtering the voltages of the center-tapped secondary.

Transformer Losses

There is no such thing as a perfect transformer with no power loss. Iron-core transformers have *hysteresis loss,* which occurs when the transformer core

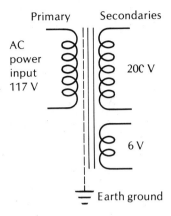

Figure 9-5. Power transformer

becomes permanently magnetized on each half cycle. A certain amount of input power must be used to demagnetize the core on each half cycle before magnetization in the opposite direction can occur.

There is also an *eddy current loss* that results from flux around the primary current cutting through the iron core. This is in accordance with Faraday's law which states that a moving flux cutting across a conductor causes a voltage to be induced in that conductor. Eddy current losses cause the transformer to become hot, which represents a loss in power. Using a *laminated core* (a core made of thin sheets of metal) reduces eddy current losses but has no effect on the amount of hysteresis loss.

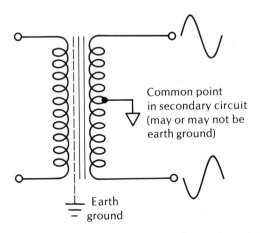

Figure 9-6. Transformer with center-tapped secondary winding

A third loss is *copper loss,* which is due to the resistance of the copper wire used in winding the primary and secondary. Current flowing through this resistance dissipates power.

Although a perfect transformer is not realizable, very high efficiencies have been obtained, and transformer losses are usually low enough to be disregarded in most power supply calculations.

The dashed line between the primary and the secondaries in Figure 9-5 is a *Faraday shield.* It prevents capacitive coupling between the primary and the secondary. These windings are both conductors separated by insulators, and therefore they form a capacitor that will transfer voltages from the primary to the secondary. Capacitive coupling permits coupling of transient voltages from primary to secondary. (Transient voltages consist of voltage spikes and noise that come in on the power line and that are usually caused by nearby motors or electrical appliances.) These spikes, when coupled into the secondary, can be of sufficient amplitude to destroy semiconductor components.

It is not possible to determine the turns ratio of a transformer by measuring the primary and secondary resistances. The reason is that the wire sizes of various windings are not normally the same, so it is not possible to tell whether a lower resistance is due to a lower number of turns or a larger wire size.

Replacement Transformers

If a power transformer should fail, and if the exact replacement is not possible, some basic rules can be applied to determine the substitute. The primary voltage and current ratings should be the same. Some transformers are designed to operate at either 50 or 60 Hz. They can be used in countries other than the United States that have 50 Hz power. A transformer designed to operate on 60 Hz connected into a 50 Hz line will result in excessive primary current and likely cause a transformer burnout. The reason for this is that the primary inductive reactance (X_L) is equal to $2\pi fL$. As the frequency (f) decreases, the reactance decreases and the current increases. Therefore, with a substitute transformer it is necessary to observe the manufacturer's specifications about frequency of operation.

The secondary voltage ratings of the transformer must be duplicated in a replacement. These voltage values are usually readily available from the manufacturer's literature. The secondary current rating should also be observed, but a larger current rating can be substituted with no problems, provided the voltage ratings are the same.

The physical size of the transformer may be a consideration in making the replacement. A word of caution is needed here: Smaller transformers are not necessarily more desirable! The smaller size *could* mean less iron in the magnetic circuit and that, in turn, means greater transformer losses and greater heat during its operation.

Isolation Transformers

Isolation transformers are of special importance to a technician. They protect him when troubleshooting electronic equipment which may, under certain conditions, place him in direct contact with the ac power line.

Figure 9-7 shows how the isolation transformer works. It is a 1:1 transformer, which means that the secondary voltage is equal to the primary voltage. One-half of an ac power line is normally grounded. A person who gets connected between the hot line and ground, such as the person in position *a* in Figure 9-7, can receive a lethal shock. This position actually places him directly across the power line. In the secondary of an isolation transformer there is no earth ground connection. Therefore, the person in position *b* or position *c* is not likely to be electrocuted if he comes in contact with one of the lines. Of course, if he gets across *both* lines he can still receive a lethal shock.

A word of caution is needed here. Because of the capricious nature of electricity, it is *never* desirable to get in contact with an ac line under any conditions, regardless of whether or not it is on an isolated secondary.

Variable transformers that change a line voltage into a range of voltage values from 0 to about 130 V are popular as bench items, but they are *not* normally isolation transformers. Actually, they are autotransformers as shown in Figure 9-8. Therefore, *there is no protection against accidental contact with one side of the power line!*

AC VOLTAGE DIVIDERS

Voltage dividers can be used when the ac voltage is to be stepped down. The two examples shown in Figure 9-9 are similar to the voltage dividers used in stepping down dc and ac voltages.

When capacitors are used as voltage dividers, no power is dissipated. This feature assumes that they are perfect capacitors. However, undesirable phase shifts may occur, so capacitive voltage dividers have limited application.

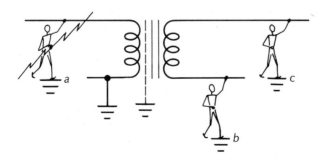

Figure 9-7. Isolation transformer to protect technicians from electrical shock

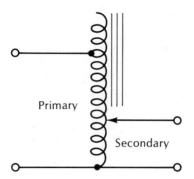

Figure 9-8.   Adjustable power transformer

When two capacitors are connected in series, as shown in Figure 9-9, the larger voltage appears across the smaller capacitor. The voltage ratios are in inverse proportion to the capacitances.

$$\frac{C_1}{C_2} = \frac{V_2}{V_1}$$

This characteristic is true for both ac and dc voltages across the capacitive divider.

Rectifier Diodes

Rectifiers are used in circuits for converting ac to dc. The most popular kinds employ semiconductor diodes which permit current to flow in only one direction.

Two important characteristics for selecting rectifier diodes are (1) the *forward current*, which is the maximum allowable rms current through the

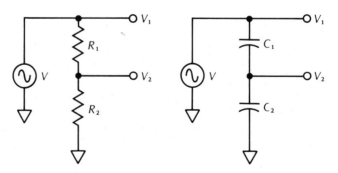

Figure 9-9.   Ac voltage dividers

diode when it is forward biased, and (2) *the peak inverse voltage* (piv), which is the maximum allowable reverse voltage across the diode.

When replacing a diode in a circuit, it is possible to use larger forward-current ratings and larger piv ratings. If a diode with the required piv rating is not available, then it is possible to connect two or more in series as shown in Figure 9-10. The peak inverse voltage in this combination is equal to the sum of the peak inverse voltages of each diode.

The resistors across the diodes in Figure 9-10 are required to assure that the reverse voltage is equally divided. The resistors will have equal values if the diodes have equal inverse voltage ratings. Normally, very high resistance values are needed here to prevent any appreciable reverse current from flowing in the circuit. The capacitors are used to prevent transient voltages from destroying the diodes.

If an increased current rating is needed beyond what is available in the replacement diode selection, it can be obtained by connecting the diodes in parallel as shown in Figure 9-11. The resistors normally have low resistance values. They are connected in series with the diode to assure that each diode will conduct and carry its share of the load.

The theory of the series resistor is easy to understand. A forward starting voltage is needed before a semiconductor diode can conduct. If one of the three parallel diodes in Figure 9-11 has a lower starting voltage than the others, it will begin to conduct sooner. The voltage across the diode will be so low that it will prevent the other two diodes from starting into conduction. The voltage across the cathode resistor assures that the total voltage drop (forward diode voltage and voltage across the resistor) is large enough to

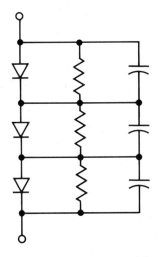

Figure 9-10. Series-connected diodes

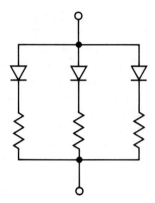

Figure 9-11. Parallel-connected diodes

start all of the diodes. Diodes should not be connected in parallel without the starting resistors, which are usually called *equalizing resistors*.

The discussion on series- and parallel-connected diodes given here is also applicable to other semiconductor components.

Figure 9-12 shows the symbols for the rectifying components used most often in power supply circuits. Directly heated cathode tubes can conduct a large amount of current flow, and they have a large piv rating.

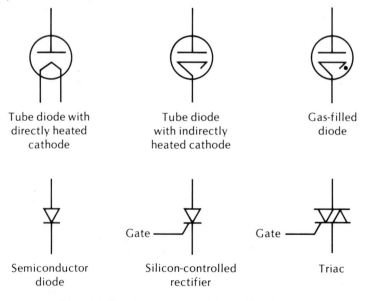

Tube diode with directly heated cathode	Tube diode with indirectly heated cathode	Gas-filled diode
Semiconductor diode	Silicon-controlled rectifier	Triac

Figure 9-12. Components used in rectifier circuits

A disadvantage of the directly heated cathode is that the filament must connect into the ground return circuit of the power supply, making it inconvenient to use this type of rectifying device in some circuits such as the bridge rectifier to be discussed later. A vacuum tube diode with an indirectly heated cathode can have its cathode circuit more readily isolated from the common side of the supply, but its electron emission is usually lower.

The gas-filled diode operates on the principle of avalanching. This principle, illustrated in Figure 9-13, results in much larger current flow in the plate circuit. An electron leaving the cathode strikes a gas molecule and knocks another electron loose. Now two electrons are moving toward the positive anode. They both strike gas molecules and each knocks an electron loose. Now there are four electrons moving toward the positive anode.

The process continues, and many electrons arrive at the anode, making the anode current very high. The atoms that have lost electrons are positively charged. An atom that has lost an electron or gained an extra electron is called an *ion*. The positive ions of gas move toward the cathode where they recombine with electrons and become neutral atoms. The avalanching procedure just described can occur in semiconductor components as well as in gaseous tubes. The overall result is a high current through the diode, and a high available power supply current.

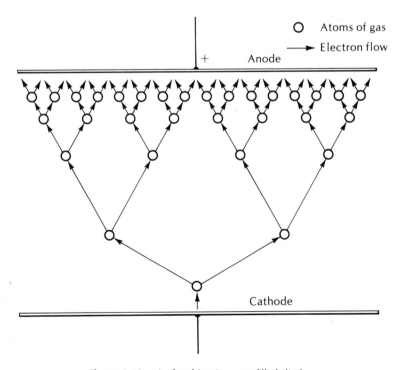

Figure 9-13. Avalanching in a gas-filled diode

Gas-filled diodes are used in high-current rectifying systems, such as battery chargers and heat generators. Special forms of gas-filled diodes, such as *ignitrons,* can produce an enormous amount of current.

There are a number of types of semiconductor diodes, such as the selenium rectifier and the silicon rectifier. Silicon rectifiers are more popular because they can be made in small plastic packages.

The silicon-controlled rectifier (SCR) is a *breakover diode.* It will not conduct until a positive voltage is applied to its gate. SCR's are used in regulated power supplies. The triac is similar to the SCR except that it can conduct equally well in either direction after a positive gate voltage has been applied.

It is important to remember that the leads of semiconductor diodes are identified by marking the cathode side. The cathode may be marked with a dot, or with a positive sign, or a special shaping of the diode package. These methods are shown in Figure 9-14. The bridge rectifier in Figure 9-14 shows typical markings for the ac input (~) and dc output (+ and –).

EVALUATING RECTIFIER DIODES

An ohmmeter can be used to determine if a diode is open or shorted. The test is shown in Figure 9-15. To perform this test correctly it is necessary to know which of the ohmmeter leads is negative and which is positive.

When the negative terminal of the ohmmeter goes to the anode, and the positive lead to the cathode, as shown in Figure 9-15, the diode is reverse biased. This means that the ohmmeter should indicate a very high resistance. When the ohmmeter leads are reversed, also shown in Figure 9-15, the diode is forward biased and the ohmmeter should indicate a low resistance. The terms *low forward resistance* and *high reverse resistance* are relative. The minimum ratio of front-to-back resistance of 100 to 1 is usually considered to be the minimum accepted value. Normally, much higher ratios are obtained.

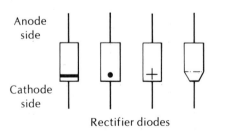

Rectifier diodes

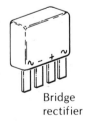

Bridge rectifier

Figure 9-14. Diode markings

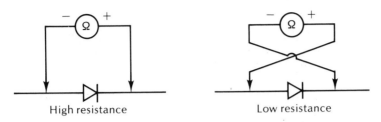

Figure 9-15. Ohmmeter test for diodes

A more accurate test of the diode's condition is obtained with a diode curve tracer of the type shown in Figure 9-16. The circuit in Figure 9-16 plots a characteristic curve of the diode on the screen of the oscilloscope. The circuit is easy to construct and it will check rectifier diodes, zener diodes, bipolar transistors, tunnel diodes, four-layer diodes, and other semiconductor components. Figure 9-17 shows some typical characteristic curves obtained with the diode curve tracer.

Use of a curve tracer requires a qualitative judgment; that is, experience is required to evaluate the curves. In contrast, a quantitative test uses numbers to determine the condition of the diode. With a little experience, a technician can tell more about a diode's condition by using a curve tracer rather than an ohmmeter.

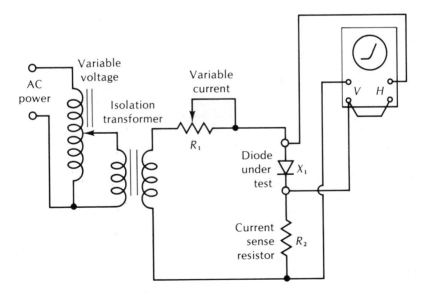

Figure 9-16. Diode curve tracer (actual display may be reversed on screen)

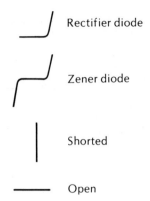

Rectifier diode

Zener diode

Shorted

Open

Figure 9-17. Characteristic curves obtained with a diode curve tracer

Rectifier Circuits

Rectifier circuits are used in power supplies to convert alternating current to direct current. Some of the more popular rectifier circuits will be identified and discussed briefly here. In all of these circuits, a solid arrow is used to represent current flowing in one direction and a dashed arrow is used to represent current flowing in the opposite direction.

Rectifier circuits can be divided into two basic groups: *half wave* and *full wave*. The names refer to the amount of input ac voltage waveform that is converted to dc.

HALF-WAVE RECTIFIERS

Figure 9-18 shows two examples of half-wave rectifiers. The solid arrows in these illustrations and for other illustrations on rectifiers show the electron current path for the half cycle when point *a* is positive with respect to point *b*. The dashed arrows show the electron current path on the half cycle when point *b* is positive with respect to point *a*.

The basic half-wave rectifier produces a pulsating dc current through the load resistor. The waveform for this current is the same as the output voltage waveform shown in the illustration. As indicated by the solid arrows, current flows when *a* is positive with respect to *b* because the rectifier diode is forward biased. No current flows when point *b* is positive with respect to point *a*.

The output voltage waveform across R_L does not look much like dc. As a matter of fact, it is a pulsating type of dc which has very limited applications in electronics. Fortunately, inductors, capacitors, and resistors can be employed in *filter circuits* to smooth this type of waveform into a nearly pure dc output.

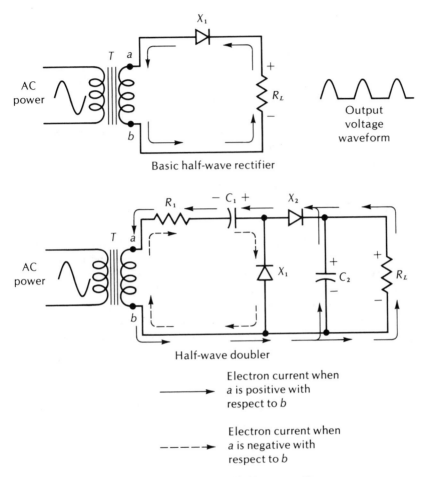

Figure 9-18. Examples of half-wave rectifiers

The half-wave doubler shown in Figure 9-18 gets its name from the fact that the output voltage at R_L is theoretically twice the peak value of voltage from a to b. It is easier to start this discussion with the half cycle in which b is positive with respect to point a. During this half cycle, electron current flows as shown by the dashed arrows and charges capacitor C_1 through rectifier X_1. Resistor R_1 is needed to limit the charging current for capacitor C_1. On the first few cycles of operation this current can be very high, and diode X_1 would have to have a very high current rating to withstand the surge. After C_1 has reached its initial charge, the charging current is reduced. Only enough current to replace any lost charge in C_1 is needed.

It is less expensive to use a *surge limiting resistor* (R_1) than it is to specify a very high forward current rating for the diode. On the next half cycle when point *a* is positive with respect to point *b*, the voltage across the transformer secondary is in series with the voltage across capacitor C_1. In other words, the two voltages are additive as shown in Figure 9-19. The voltage during the half cycle when *a* is positive with respect to *b*, then, is the sum of the voltage across capacitor C_1 plus the voltage across the secondary winding.

During the half cycle when the voltage across C_1 is in series with the voltage across the transformer secondary, electron current flow is shown by the solid arrow. Capacitor C_2 is charged to the sum of the peak voltages V_1 and V_{ab}, making the output voltage across R_L equal to twice the voltage across *ab*.

The circuit is called a *half-wave doubler* because capacitor C_2, which is actually the source for the output voltage across R_L, receives its charge only during one half cycle of ac input. The other half cycle is being used to charge capacitor C_1.

The charge and discharge of capacitor C_2 is an important feature of the half-wave doubler circuit. Figure 9-20 shows the charge and discharge currents. When diode X_2 (of Figure 9-18) conducts, C_2 charges to a voltage equal to $V_1 + V_{ab}$. When X_2 is cut off, C_2 discharges through R_L. The discharge current maintains a positive voltage across R_L.

If C_2 could hold an infinite charge, the voltage across R_L would be pure dc; that is, there would be no fluctuations in the dc voltage. In practice, capacitor C_2 discharges through R_L when X_2 is cut off, and must be re-

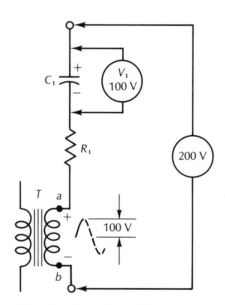

Figure 9-19. Secondary voltage adding to voltage across the capacitor (on positive half cycle)

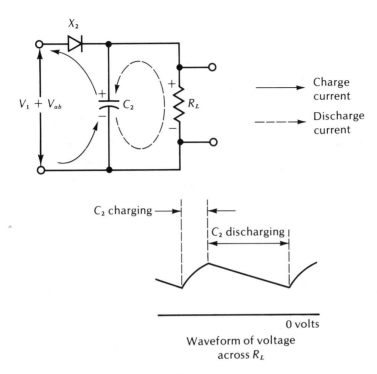

Figure 9-20. Charge and discharge currents for capacitor C_2

charged each time X_2 conducts. The disadvantage of the half-wave doubler is immediately apparent. If there is a large load current, then C_2 will become appreciably discharged on each half cycle, which in turn will produce a large ripple voltage and poor regulation. A typical waveform is shown in Figure 9-20.

FULL-WAVE RECTIFIERS

Figure 9-21 shows the circuit of a full-wave rectifier. Unlike the half-wave rectifier just described, this circuit must have a transformer with a center-tapped secondary in order to operate. The voltages across the secondary at point x and point y are 180° out of phase, so when point x goes positive, point y goes negative.

On one half cycle when x is positive and y is negative, diode X_1 is forward biased. This results in the current flow shown by the solid arrows. Note that the current flows upward through the load resistor R_L. At this time diode X_2 is cut off because of the negative voltage at point y.

On the next half cycle, point x is negative and point y is positive. Diode X_1 is reverse biased due to the negative voltage on its anode. At the same time, diode X_2 is forward biased and current flows in the direction shown by the

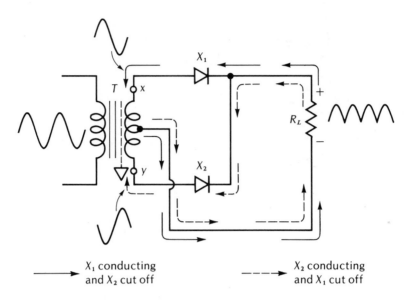

X₁ conducting and X₂ cut off ————→

X₂ conducting and X₁ cut off - - - →

Figure 9-21. Full wave rectifier

dashed arrows. Note that this current also flows upward through resistor R_L.

The center tap of the secondary is considered to be zero volts for both half cycles. It is zero volts because it is halfway between equal positive and negative values.

Since current is flowing through R_L on both half cycles of input power, the output waveform consists of a series of half waves as shown in the illustration. Compare this waveform with the one for the half-wave circuit in Figure 9-18.

An important consideration is the power supply filter ripple frequency. In the half-wave circuit the ripple frequency is equal to the frequency of the power line voltage. Thus, if the rectifier is connected to a 60 Hz power line, the ripple frequency is 60 Hz. In the full-wave rectifier there are two cycles of output for each cycle of input. Therefore, the ripple frequency is twice the power line frequency. A full-wave rectifier connected to a 60 Hz power line has a ripple frequency of 120 Hz. Compared to 60 Hz ripple, it is easier to filter a 120 Hz ripple. Ease of filtering the output waveform is an important advantage of full-wave rectifiers.

Figure 9-22 shows another example of a full-wave rectifier. This circuit is called a *bridge rectifier*. Note that the transformer does not have to be center tapped in this case. In fact, the bridge rectifier circuit can operate without a transformer if the power line voltage is appropriate for the desired use. The

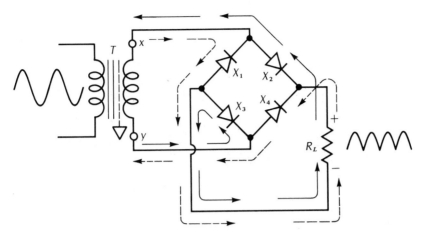

Figure 9-22. Bridge rectifier circuit

only need for the transformer is to step the voltage up or down and to provide isolation from the power line.

On one half cycle, point x is positive with respect to y. During this half cycle the electron current flows away from point y through diode X_3, through the load resistor (R_L) and through diode X_2 to point x.

On the next half cycle, point y is positive with respect to point x. During this half cycle, electron current, shown with dashed arrows, leaves point x, goes through diode X_1, through the load resistor, and through diode X_4 to point y. Current flows through the load resistor during both half cycles of input power, so the output waveform is that of a full-wave rectifier.

At one time the bridge rectifier was considered to be less reliable than the full-wave rectifier type shown in Figure 9-21. The reason was that vacuum tube diodes were used, and they were the most unreliable part in the system. Now that solid-state diodes are used, reliability is no longer a limitation on the use of bridge rectifiers.

An important feature of the bridge rectifier is that it does not require a center-tapped secondary, so any ac voltage can be converted to pulsating dc. This circuit is used in ac measuring instruments that use dc meter movements. The ac current must be converted to a form of dc before it can be measured. Typically, bridge rectifiers are used in such circuits.

Figure 9-23 shows the circuit for a *full-wave voltage doubler*. On one half cycle, point x on the secondary of the transformer will be positive, and point y will be negative. The positive voltage at point x forward biases diode X_1 and reverse biases diode X_2. On the half cycle, then, when x is positive with respect to y, electron current flows as shown by the solid arrows. This current charges capacitor C_1 to the peak value of voltage.

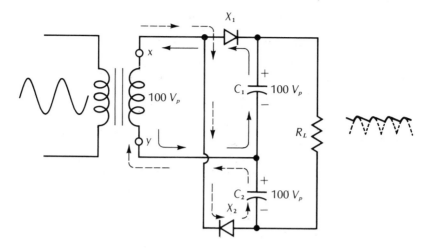

Figure 9-23. Full wave voltage doubler

On the next half cycle, point x is negative with respect to point y. Now, diode X_1 is reverse biased and diode X_2 is forward biased. Electron current flow during this half cycle is shown by the dashed arrows. Current through X_2 charges capacitor C_2 to the peak value of voltage.

The peak value of voltage across C_2 and the peak voltage across C_1 are additive, making the output voltage across R_L twice the value of the peak voltage (V_p) across the secondary. Some current is being delivered to the capacitor circuit on each half cycle and the resulting waveform shows that the ripple frequency is twice that for the half-wave diode. In other words, the ripple frequency is 120 Hz when the full-wave doubler is connected to the 60 Hz power line.

The same number of parts are used for the full-wave bridge as for the half-wave bridge. Each circuit has its own advantage. The half-wave bridge can be constructed with one-half of the secondary of the transformer at point y connected to a common point in the circuit. This is an advantage in certain wiring configurations.

On the contrary, in Figure 9-23 the transformer winding cannot be connected to common at either lead. However, the ripple frequency is higher, so the waveform is easier to filter.

THREE-PHASE RECTIFIERS

As shown in Figure 9-24 three-phase power consists of three individual sinusoidal voltage waveforms separated by 120°, or one-third of a complete cycle. The voltage waveforms in this figure are distinguished by a solid line, a dashed line, and a broken line. These three waveforms will be identified the same way as in the discussion of three-phase rectifier circuits.

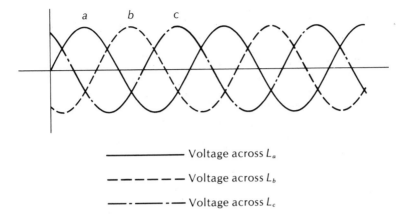

Figure 9-24.   Voltages for three-phase power

Figure 9-25 shows a half-wave three-phase rectifier circuit. Three-phase power is applied to the delta-connected primary, and the output is taken from across the wye-connected secondary. The voltages induced in windings L_a, L_b, and L_c are the voltages illustrated in Figure 9-24. Point d in this circuit is *neutral*, which means it can be considered to be zero volts at all times.

At some moment, point a will be positive with respect to points b and c. This event occurs during the first half cycle of waveform illustrated in Figure

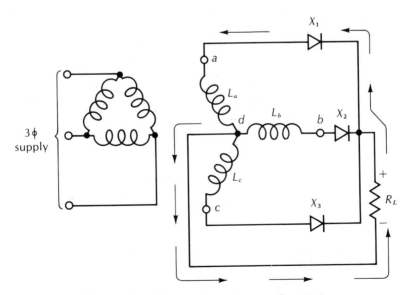

Figure 9-25.   Half-wave, three-phase rectifier circuit

9-24. At the instant that point *a* reaches its peak value, points *b* and *c* are both negative—meaning that diodes X_2 and X_3 are cut off and diode X_1 is forward biased. The electron current path at this instant of time is shown by the solid arrows.

It is difficult to illustrate electron current flow in this type of circuit because diodes X_2 and X_3 are not both cut off during the full half cycle of waveform for *a*. The electron flow illustrated occurs only at the time when the voltages at *b* and *c* are both negative, and the voltage at *a* is positive.

As the voltage at point *a* decreases during the last part of the positive half cycle, the voltage at point *b* begins to go positive and diode X_2 begins to conduct. At the moment it reaches its peak value, the other two diodes are cut off. When point *c* reaches its positive peak voltage, diodes X_1 and X_2 are cut off.

The overall result is that current flows through R_L at all times since the waveforms never reach zero at the same instant. Figure 9-26 illustrates the half waveform of voltage across R_L. The solid dark outline indicates the actual voltage waveform across the load resistor.

Figure 9-27 shows a full-wave three-phase rectifier circuit. Two rectifiers are required for each phase, making a total of six diodes in the circuit. Again, it is very difficult to illustrate electron current flow in these circuits because the voltages are not always positive or zero at the same instant.

Assume that the voltage at point *a* is its maximum positive peak value, forward biasing diode X_1. Refer again to the waveforms illustrated in Figure 9-24, and note that at this instant of time the waveforms at *b* and *c* are both negative. That means that diodes X_4 and X_6 are both forward biased, and the solid arrows show the path of electron flow.

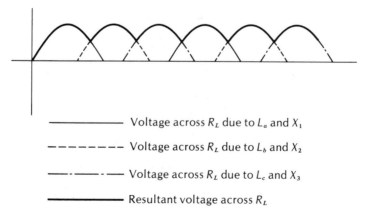

Voltage across R_L due to L_a and X_1

Voltage across R_L due to L_b and X_2

Voltage across R_L due to L_c and X_3

Resultant voltage across R_L

Figure 9-26. Waveforms for the half-wave, three-phase rectifier circuit

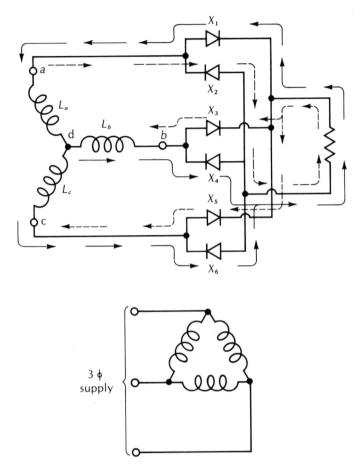

Figure 9-27. A full-wave, three-phase rectifier circuit

When point a is at its maximum negative value of voltage, b and c are both positive and diodes X_2, X_3, and X_5 are forward biased. During this instant the electron current flow is illustrated by the dashed arrows.

It is possible, with a little patience, to trace the conduction for each diode in the circuit. The output waveform is shown in Figure 9-28. The solid line in this illustration shows the actual waveform of voltage across the load resistance. There are six cycles of ripple frequency for each cycle of input waveform, so the ripple frequency is 360 Hz. For many applications, this small amount of ripple is of no consequence, and a filter is not used.

For the half-wave, three-phase circuit of Figure 9-25 there are three cycles of ripple for each cycle of input waveform. So, when connected to a 60 Hz, three-phase power line, the ripple frequency is 180 Hz. The amplitude of the ripple is greater than for the full-wave circuit.

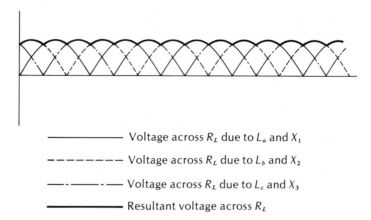

_____ Voltage across R_L due to L_a and X_1

- - - - - - - Voltage across R_L due to L_b and X_2

— · —— · — Voltage across R_L due to L_c and X_3

_____ Resultant voltage across R_L

Figure 9-28.   Waveforms for the full-wave, three-phase rectifier circuit

Power Supply Filters

The output voltage of a rectifier is not a pure dc, but rather, positive or negative pulsations. These pulsations must be filtered if a pure dc voltage is required. The simplest type of filter is a capacitor or inductor (or both) in the output circuit of the power supply. The filter is inserted between the power supply and the load. A lowpass filter is required. This is a filter that is able to pass very low frequencies (down to dc) and reject higher frequencies.

Ideally, the lowpass filter would have a sharp cutoff at some point below the ripple frequency of the supply. An ideal response curve is shown in Figure 9-29. As a general rule, the simpler the filter design the more difficult

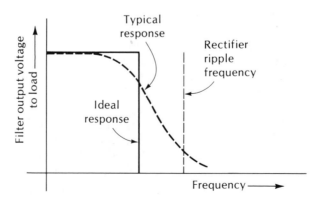

Figure 9-29.   Filter circuit response curves

it is to obtain a steep *rolloff* of the type shown in the ideal characteristic curve. (Rolloff refers to the drop from maximum value to minimum value on the response curve.)

All filters used in power supplies are either *passive* or *active*. Passive filters use nonamplifying components such as inductors, capacitors, and resistors. Active filters use amplifying components such as tubes, transistors, or operational amplifiers.

PASSIVE FILTERS

A *brute force* filter is one that uses very large capacitors and very large inductors as filters. The idea of brute force design is to obtain a response with the sharpest possible rolloff without using a large number of components. Some tradeoffs must be considered. The larger the filter capacitor used, the greater the amount of leakage current that it has. Therefore, it is not possible to use extremely large capacitors without some loss in efficiency and reduction in power supply output. Large filter chokes are costly, bulky, and liable to cause interference with other parts of the circuitry if the flux is not well contained in the magnetic circuit. Furthermore, countervoltages in large inductances can produce destructive currents in power supply components as well as in components in circuits being operated from the power supply voltage.

Figure 9-30 shows a bridge rectifier with a simple capacitive filter of the brute force type. As shown by the solid arrows, capacitor C is charged to the peak value of the positive half cycle from the bridge rectifier. As the positive voltage from the power supply decreases, the capacitor discharges through R_L to maintain the load current constant. The dashed arrow shows the discharge current path. This current prevents the voltage across R_L from dropping to zero volts when the positive pulsations from the rectifier go to zero volts.

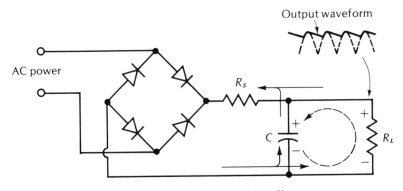

Figure 9-30. A simple capacitive filter

The waveform resulting is shown with a dark heavy outline in Figure 9-30. This waveform approximates a dc. Without the filter, the output voltage would be the half waves shown with dashed lines. There is still a small amount of ripple in the output voltage, but in many applications this can be tolerated.

The series resistor (R_S) in Figure 9-30 protects the diodes of the bridge from the large surge current that occurs when capacitor C receives its initial charge. Thus, R_S is useful only as a *surge limiting resistor* at the time the power supply is first energized. It is normally a small resistance value and does not affect normal power supply operation.

Figure 9-31 shows a brute force inductive filter. The dashed waveform shows the bridge output, and the heavy waveform shows the output with the choke (L) in place. It will be noted that the waveform is less positive than the peaks from the rectifier. This is significant. *The output voltage with a simple choke filter is somewhat less than the output voltage obtained when a simple capacitive filter is used.* Furthermore, a series limiting resistor is not required.

An important consideration for the simple choke filter is the *critical inductance,* which is the smallest value of inductance that can be used for a given power supply current. An inductor operates on the principle of expanding and contracting magnetic fields around a varying current. If the value of current is too small for a given coil, it does not act as an inductor. Another way of saying this is that if the inductor is too small for a given amount of current, the inductive reactance is negligible. Whenever replacing an inductor in a power supply, it is important *never* to use a smaller value of inductance since it is possible to get an inductor that is below the critical inductance value.

In some power supplies the load current may decrease to a very small value for certain periods of time during normal operation. If this is the case,

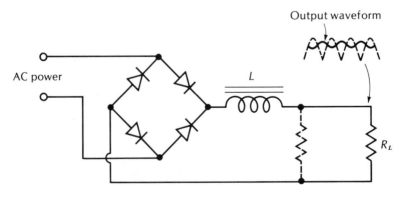

Figure 9-31. A simple inductive filter

then a resistor (R_1) should be connected in parallel with the load to maintain the minimum allowable current through the inductance at all times. The connection of such a bleeder is shown in dashed lines in Figure 9-31.

Combinations of inductors and capacitors are used in power supply filters. Three examples are shown in Figure 9-32. The *LC capacitive input filter* gets its name from the fact that the first component following the rectifier is a capacitor. This capacitor (C_1) charges to the peak value of the rectifier output, and the output voltage from the filter is higher than the output voltage from the choke input filter. The configuration shown for the capacitive input filter is sometimes referred to as a *pi filter* because of its similarity to the Greek letter π.

In many power supplies a resistor is substituted for the choke. This is illustrated in the *RC capacitive input filter* of Figure 9-32. The resistor limits the charging and discharging current of capacitor C_2. In effect the operation of R_1 and C_2 is similar to that of R_L and C in Figure 9-30. Resistors are less

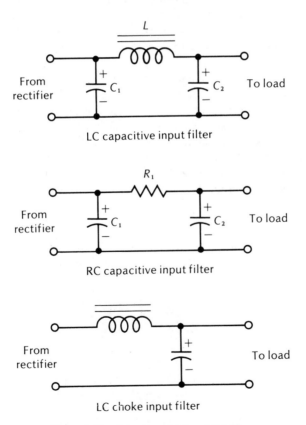

LC capacitive input filter

RC capacitive input filter

LC choke input filter

Figure 9-32. Examples of *LC* and *RC* filters

expensive than chokes, and less bulky. However, they have the disadvantage that they produce a dc voltage drop when the load current flows through them. This means that a lower output voltage is obtainable using a resistor instead of an inductor.

The *LC choke input filter* in Figure 9-32 produces a reasonably pure dc to the load at a somewhat reduced output voltage compared to using a capacitive input filter. As before, a resistor may be substituted for the choke in the *L*-type filter. Choke input filters are *always* used with gaseous rectifier circuits. Gaseous rectifiers are capable of conducting extremely high currents, and the choke is required to protect the rectifier diode from excessive surge currents, especially during warmup.

Very large capacitors are manufactured by rolling the plates into their cylindrical form, as shown in Figure 9-33, thus producing a certain amount of self-inductance. This self-inductance makes it impossible for the capacitor to short circuit high-frequency transients. Transient voltages arriving on the power line pass through the rectifier and appear across the load. In semiconductor circuitry these transients can be destructive. To get around this problem, large capacitance values are often shunted with small capacitances. An example is shown in Figure 9-34. These capacitances offer a short circuit to ground for transient voltages, thereby preventing them from getting to the load.

Power supply filters are sometimes named according to the method of calculating the filter design. *Constant K filters,* for example, are obtained by designing the supply around a value that is a constant referred to as *K*. *M-derived filters* are designed around a value *M* which is obtained from the constant *K*. Butterworth and Tchebysheff (Chebyshev) filters are named for the designers who derived the necessary equations for their operations. Figure 9-35 shows an example of an *M*-derived filter circuit. This is more elaborate than the constant *K* filters of Figure 9-32. It has a parallel *LC*

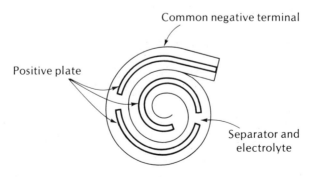

Figure 9-33. Rolling capacitor plates to make them inductors

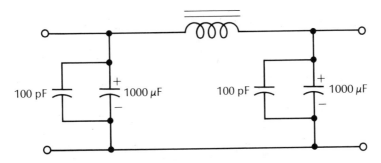

Figure 9-34.   Use of 100 pF transient filters

circuit inserted in series with the load. The impedance of a parallel *LC* circuit is maximum at its resonant frequency. Because of the *LC* circuit, the ripple is greatly impeded, and the rolloff for *M*-derived filters is sharper than for constant *K* filters.

The *ripple factor* or *percent ripple* rating of a power supply is a measure of how well its dc output voltage has been filtered. To calculate the ripple factor, two current values (I_o and I_{rms}) must be determined. The *average value of the load current (I_o)* is the value of load current as measured by a dc meter. The *effective value of the load current (I_{rms})* is the value of load current as measured by a true rms meter. Most ac meters are designed to measure the rms value of pure sine waves only, and they cannot be used for this measurement because the ripple waveform is usually not sinusoidal.

The ripple factor γ can now be determined:

$$\gamma = \left[\left(\frac{I_{rms}}{I_o} \right)^2 - 1 \right]^{\frac{1}{2}}$$

The percent ripple is determined as follows:

$$\text{Percent ripple} = \frac{\text{rms value of ripple voltage}}{\text{average value of power supply voltage}} \times 100$$

The rms value of ripple voltage must be read with a true rms-reading meter, since the ripple waveform is not a pure sine wave. A low value is desirable for both ripple factor and percent ripple. A technician has a great amount of latitude when making the replacement of a defective component in a brute force power supply. For example, if the original filter included a 1000 μF filter capacitor, he could use a 1500 μF capacitor for replacement. It is safe to use a larger value, but it is not desirable to use a smaller capacitance value.

Power supply filters that are designed for specific rolloffs have their component values specified more accurately than for brute force design. In the

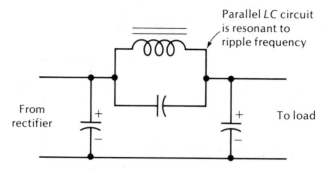

Figure 9-35.   Example of an M-derived filter

latter case, different values should *not* be substituted. To do so would alter the rolloff characteristic and likely result in poor filtering.

ACTIVE FILTERS

Instead of using passive components *(R, L,* and *C)* power supply circuits often employ active filter designs. In the example shown in Figure 9-36, resistors R_1 and R_2 make a voltage divider for obtaining the bias for transistor Q_1. The base voltage of transistor Q_1 is filtered by capacitor C_1. The filtering action of this circuit is equivalent to a brute force design using a capacitor with a value of capacitance multiplied by the beta *(h_{fe})* of the transistor. This means that by using a high gain transistor there is a very large capacitance across R_L.

The most important advantage of the active filter, which is sometimes called an *electronic filter,* is that a very large capacitance is obtained without a high leakage current and typically low breakdown voltage that would occur if an electrolytic capacitor were used directly.

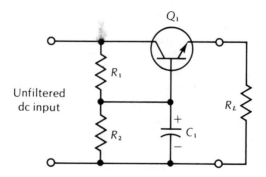

Figure 9-36.   An electronic filter

Regulated Power Supplies

All power supplies have internal resistance. In the case of a battery, internal resistance is due to the resistance of the electrolyte to electron current flow, the resistance of the electrodes, and the terminal resistance.

For a power supply the internal resistance is due to a number of factors such as the forward resistance of the rectifying components, the dc resistance of the power transformer, the resistance of the series filter element (such as the choke), and wire resistances.

The amount of internal resistance has an important effect on the output voltage. To understand why, refer to Figure 9-37. In this illustration R_i represents the internal resistance of the battery. Naturally, this internal resistance cannot be reached. It is distributed throughout the battery, but it is simpler to represent it as a separate resistor in power supply discussions.

When a load current flows through R_i, there is a voltage drop across that resistor. In the illustration, the electron load current is shown with arrows. Note that the current flows through R_i in such a way that it produces a voltage drop that opposes the voltage at V. In this illustration V represents the theoretical battery voltage when there is no internal resistance. The terminal voltage for the power supply is V minus the voltage drop across the internal resistance.

When the load resistance is known, the power supply can be designed by taking the drop across the internal resistance into consideration. However, if the load resistance is variable, then the value of I_L will increase and decrease

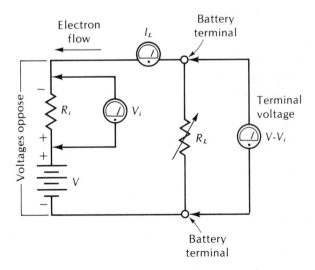

Figure 9-37. Internal resistance of power supplies affects terminal voltage.

as R_L is changed. This causes a fluctuating voltage at the power supply terminal.

It is obvious that the lower the internal resistance of the power supply the better the ability of the power supply to deliver a constant voltage to the output terminals. For a rechargeable battery some control is maintained over the internal resistance by simply keeping the battery in the charged condition. For a battery that cannot be charged (that is, for a *primary battery*), when the internal resistance becomes high enough to cause the terminal voltage to drop below some predetermined value, the battery must be discarded.

Regulation and *percent regulation* of a power supply are measures of how well it maintains its terminal voltage with changes in load current. Mathematically:

$$\text{Regulation} = \frac{\text{no load voltage} - \text{full load voltage}}{\text{full load voltage}}$$

$$\text{Percent regulation} = \frac{\text{no load voltage} - \text{full load voltage}}{\text{full load voltage}} \times 100$$

$$\text{Percent regulation} = \frac{10 - 8.5}{8.5} \times 100$$

$$= 17.6 \text{ percent}$$

For a power supply that has an output voltage of 10 V with no load and 8.5 V when delivering full load current:

Instead of specifying regulation or percent regulation, the designer may give the maximum allowable change in output voltage for a given change in load current.

A low percent regulation is desirable. When a high percent regulation is given with a power supply rating, it has very poor regulation. In other words, a small amount of change in load current will produce a relatively large amount of voltage change at the terminals.

One way to get around the problem of poor regulation is to use a power supply regulator. Figure 9-38 shows two possible types.

The *shunt regulator* is connected directly across the load. As the load resistance changes, the resistance of the shunt regulator also changes so that the total current $I_L + I_{SH}$ is a constant value. By maintaining a constant current through a series resistor (R_S), the terminal voltage of the power supply is held constant.

The *series regulator* produces a voltage drop in series with the voltage drop across the load resistance. The idea of the series regulator is to change the voltage V_S so that the total amount of voltage $V_S + V_L$ equals the input voltage (V). The value of V_L can be maintained constant provided V_S is able

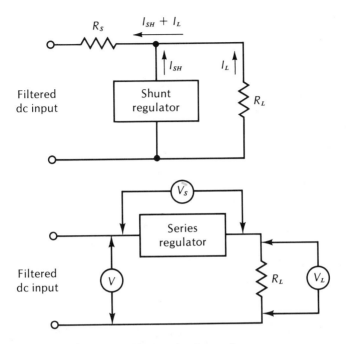

Figure 9-38. Shunt and series regulators

to change to correspond with changes in load current. For example, if the load current increases, the resistance of the series regulator must decrease so that the voltage across the load is not changed.

The shunt regulator is usually a simpler circuit and requires less components, but it is less efficient than the series regulator. Both types are used extensively in electronic circuitry.

SHUNT REGULATOR CIRCUITS

Figure 9-39 shows two examples of shunt regulators. These circuits are identical except for the type of component used to produce the regulation action. The voltage regulator tube (VR) is rapidly being replaced by the zener diode, which is the solid-state equivalent. Both components maintain a fairly constant voltage across them at all times, so that changes in load current or load resistance will not produce changes in output terminal voltage. It is absolutely necessary to have a series resistor R_S in series with a voltage regulator tube or zener diode in order for the components to operate properly.

One disadvantage of the simple shunt regulators of Figure 9-39 is that the components (VR and the zener diode) both generate high-frequency noise

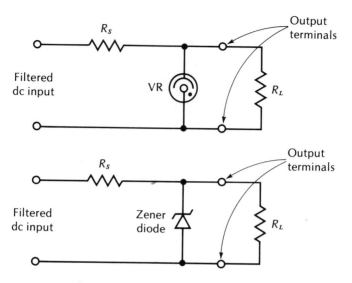

Figure 9-39. VR and zener diode regulators

voltages during their operation. These noise voltages must be filtered in some applications.

Voltage regulator tubes and zener diodes can both be stacked in series to obtain larger regulated voltage values provided the current ratings of the components are identical.

A SERIES REGULATOR CIRCUIT

Figure 9-40 shows a simple series regulator. It consists of a series-pass transistor (Q_1) which obtains its bias from the voltage across the zener diode (X_1). This voltage is filtered by capacitor C. Changes in load resistance will

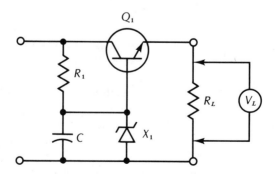

Figure 9-40. Series pass regulator

not seriously affect the output voltage provided the dc input voltage is always sufficient to operate the zener diode circuit.

The voltage (V_L) across the load resistor is equal to the zener diode voltage plus the emitter-base voltage of Q_1. The emitter-base voltage of a transistor is nearly constant under varying current conditions, and the zener voltage is constant. Therefore, the output voltage of the supply is constant.

The circuits of Figures 9-39 and 9-40 are called *open-loop regulators*. Although they maintain a nearly constant output voltage, they have no provision for correcting the output voltage. Open-loop regulators provide good regulation at a low cost. However, for a power supply with a more closely regulated output voltage, it is necessary to use feedback circuitry.

CONSTANT VOLTAGE SUPPLIES

In a closed-loop regulator a portion of the output voltage is sensed and compared with a reference voltage in a *comparator,* or comparison amplifier. If there is any difference between the sensed voltage and the reference voltage, the output control signal is developed which corrects for the difference. Correction is made until the output voltage is equal to the preset value.

Figure 9-41 shows a closed-loop regulator for ac. The input unregulated ac voltage is developed across a motor-driven autotransformer. The dashed line in the illustration indicates a mechanical connection between the motor shaft and the variable arm of the transformer. An autotransformer is shown in the circuit. If it is necessary to isolate the primary and secondary circuits, an adjustable transformer like the one shown in the inset can be used.

Regardless of which transformer is used, the operation is the same. The output voltage is sensed by the sense circuit and compared with a reference. If these two voltages are the same, no dc control voltage is sent to the motor control circuit, and the arm of the transformer adjustment is not disturbed. When there is a difference between the output voltage and the reference, a dc voltage is delivered to the motor control circuit. The motor will adjust the transformer in such a way as to correct for the error.

When the output voltage is too high, the motor turns in one direction to reduce the secondary-to-primary turns ratio in the transformer; and when the output is too low, the motor turns in the opposite direction to increase the turns ratio. Figure 9-42 shows a block diagram of a typical closed-loop regulated dc power supply.

In the circuit of Figure 9-42 a feedback loop controls the resistance of a series control element. Regulators of this type are sometimes called *analog* to indicate that the output is continuously controlled. The alternative to analog closed-loop systems is a *digital* system that periodically senses the output voltage and makes corrections. In the system of Figure 9-42 the ac power is rectified and filtered in a conventional power supply. A filtered dc is delivered to the load resistance through the series control element.

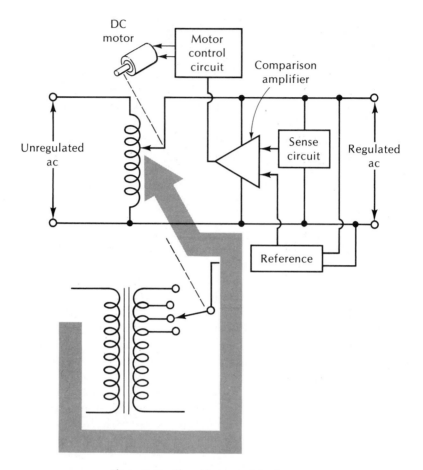

Figure 9-41. Closed-loop regulator for ac

There is a certain amount of voltage drop across the control element. This dc voltage drop is determined by the input voltage from the feedback loop. The output voltage is continually monitored by the sense circuit and compared with a voltage from the reference circuit. If there is any difference between these two voltages, then a modified control voltage is delivered to the series control element.

For example, suppose the voltage across the load resistor (R_L) is too high. The sense circuit senses the high voltage and delivers it to the comparison amplifier. The dc control voltage to the series control element increases the voltage drop across that component. By increasing the voltage across the control element, the voltage across the load resistance is decreased. Therefore, control is effective.

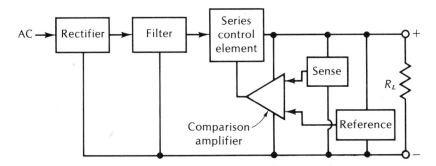

Figure 9-42. Closed-loop regulator for dc

If the output voltage across R_L is too low, the sense circuit will deliver a control voltage to decrease the voltage across the series control element, thereby increasing the voltage across the load resistance.

Figure 9-43 shows a closed-loop shunt regulator. The regulated dc output voltage across R_L is the same as the voltage across the shunt regulator (Q_1). This voltage, in turn, is determined by the amount of current in the base circuit, which is controlled by a comparison amplifier (A_1). The difference between the sense voltage and the fixed reference voltage is what sets the amount of conduction of the comparison amplifier. The shunt regulator is usually a simpler design compared to the series regulator, but it is not as efficient.

Regulated supplies like the ones shown in Figures 9-41, 9-42, and 9-43 have an output voltage that depends upon a reference voltage. Zener diodes are often used for obtaining the reference voltage. Although the voltage across a zener diode is nearly constant, in practice it is not a perfect reference. Changes in power supply line voltage and the load voltage can produce slight changes in the zener (or VR) reference voltage. If the reference voltage changes, then the output voltage is no longer exactly correct.

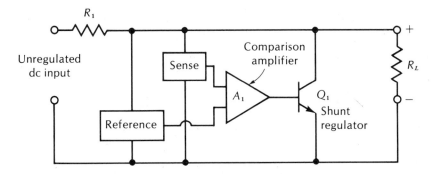

Figure 9-43. Closed-loop shunt regulator

To get around this problem, a separate regulated power supply may be used for the reference. Figure 9-44 shows a constant-voltage supply. In this circuit the reference regulator obtains a dc voltage from a full-wave rectifier and capacitive filter. The regulated voltage supplies a dc to the reference zener diode and produces a much more stable dc reference voltage across variable resistor R_P and series resistor R_R. The comparison amplifier senses the output voltage of the dc power supply being regulated and compares it with this reference. Any error that occurs is used to control the base current in the series regulator.

CONSTANT-CURRENT SUPPLIES

The closed-loop regulated supplies discussed so far regulate the output voltage. They are called constant-voltage power supplies.

It is also possible to have a constant-current supply. An example is shown in block diagram form in Figure 9-45. In order to regulate the current it is necessary to insert a low value of resistance in series with the load to sense the load current. That is the purpose of R_M in Figure 9-45. Any change in load current will produce a change in dc voltage across R_M, and thus change

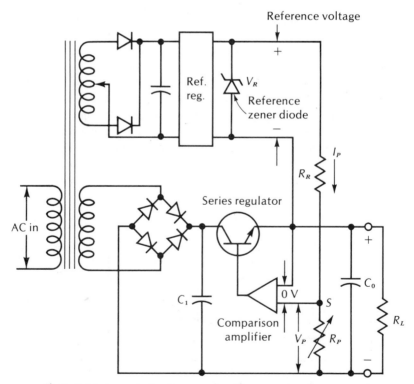

Figure 9-44. A constant-voltage supply with a separate reference supply

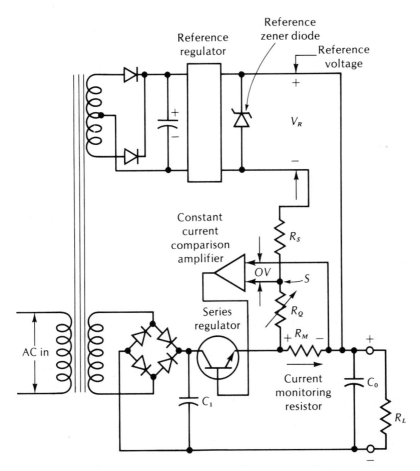

Figure 9-45. A constant-current supply with a separate reference supply

the dc voltage into the comparison amplifier. The output of the comparison amplifier controls the base current of the series regulator. A separate reference supply is used in this illustration, and the zener diode produces a fixed dc voltage across resistor R_Q to provide the reference.

CONSTANT-VOLTAGE/CONSTANT-CURRENT SUPPLIES

Figure 9-46 shows the characteristic curve of a constant-voltage supply and a constant-current supply. It is not physically possible to have a constant voltage and constant current power supply both in the same unit and operating at the same time. The so-called constant voltage/constant current supply shown in Figure 9-47 actually operates as a constant-voltage supply up to a certain value of current, and then once that value of current is reached, the current becomes a constant regardless of the voltage.

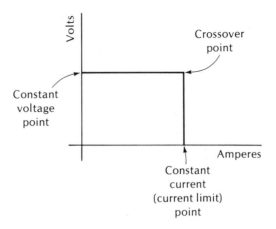

Figure 9-46. Characteristic curve for a constant-voltage/
constant-current power supply

As might be expected the constant voltage/constant current power supply is a combination of the two previously discussed in Figures 9-45 and 9-46.

Many of the more elaborate power supplies have a *current limit* adjustment. It permits the supply to be set for a maximum allowable current value. As long as the load current is below the current limit setting, the supply will function properly. The current limit feature may be thought of as being an adjustable fuse. Current limiters sense the load current the same way as a constant current supply. If the current exceeds a certain predetermined value, the current limiter delivers a voltage to the series-pass transistor to override any dc voltage on the base that is already present. This permits the regulator to function in the normal manner up to the point where the current reaches the limiting value.

SCR REGULATORS

The closed-loop analog regulators discussed so far must conduct current at all times. Power dissipated in the series-pass transistor or in the shunt regulator is power lost to the output terminals of the supply.

The SCR regulator of Figure 9-48 is more efficient because it regulates by preventing the undesired power from being generated. As in the other supplies, the sense and reference voltage are compared in a comparator. The dc voltage from the comparator determines the firing time for the SCR's.

Figure 9-49 shows some typical waveforms in the SCR regulator circuit. The ac voltage delivered to the bridge is assumed to be a pure sine wave. The high output waveform occurs when the SCR's are set to fire at the instant the rectifier output voltage reaches its peak value. The low output waveform occurs when the SCR's are set to fire at a later time. The dark

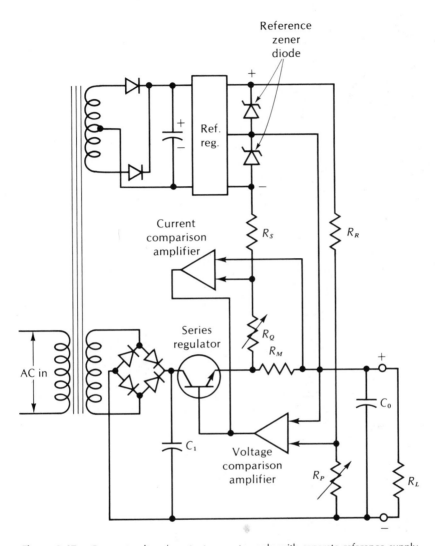

Figure 9-47. Constant-voltage/constant-current supply with separate reference supply

outline shows the actual output waveform. The output power is proportional to the shaded area under the output curves.

The waveforms in Figure 9-49 are more difficult to filter compared to the output waveform of a four-diode bridge. However, this disadvantage is offset by the greater efficiency.

SWITCHING REGULATORS

Figure 9-50 shows the basic idea behind the circuitry for a switching regulator. A conventional rectifier and filter delivers a dc voltage to a switch.

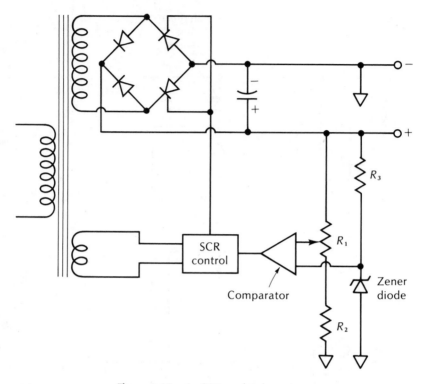

Figure 9-48. An SCR-regulated supply

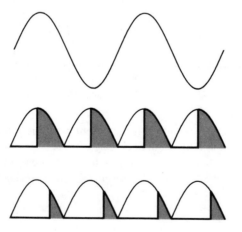

Figure 9-49. Waveforms for the SCR-regulated supply

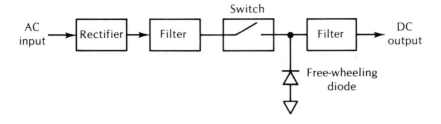

Figure 9-50. Simplified circuit for controlling dc voltage

Although it is shown as a mechanical switch, in practice an electronic switch is used.

The waveforms in Figure 9-51 show how the output voltage can be controlled by rapidly opening and closing the switch. When the ON time is equal to the OFF time, the average value of the output voltage is 50 V. When the ON time greatly exceeds the OFF time, the output voltage approaches the 100 V value. When the switch is closed 75 percent of the time, the output voltage is 75 V.

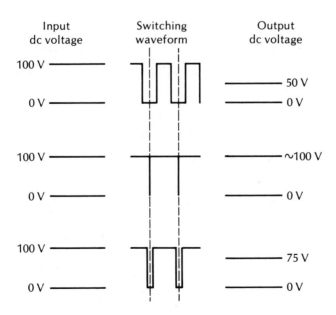

Figure 9-51. Comparison of switching waveforms with dc voltages

The pulse waveform delivered to the filter circuit in Figure 9-50 presents a special problem. At the instant the switch turns the voltage off, a countervoltage is developed across the filter inductor. This countervoltage is sufficiently large to damage semiconductor components. The *free-wheeling diode* in the circuit grounds the countervoltage to prevent such damage. This diode should always be checked when the switching supply is being serviced. If it shows any sign of leakage, it should be replaced before expensive damage results.

Figure 9-52 shows the block diagram of an actual switching regulator. The reference is a zener diode that works through series resistor R_Z. The sense circuit is a series voltage divider of the types shown in other circuits. The 100 kHz oscillator is voltage controlled by a dc voltage from the comparator. The oscillator turns the switching transistor on and off, producing waveforms like the ones in Figure 9-51. The ON time vs. the OFF time is controlled by the dc voltage delivered to the oscillator.

Dc power to the switching transistor is filtered by C_1, and X_1 is a freewheeling diode. The 100 kHz signal from the transistor can be filtered with lower values of C and L compared with the values required for filtering a 60

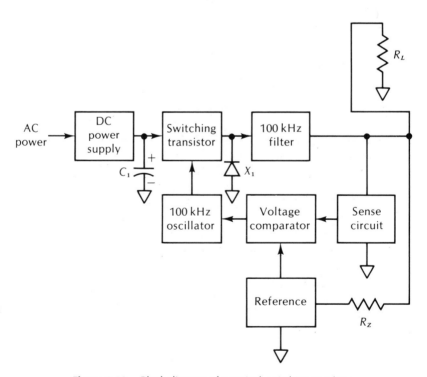

Figure 9-52. Block diagram of a typical switching regulator

Hz ripple. The result is that the switching regulator can be made in a more compact package.

REGULATOR CIRCUITRY

The discussion of regulating systems in this chapter has been on a block diagram basis. The actual circuitry, for the most part, is conventional. For example, series-pass transistors are forms of power amplifiers. This includes power transistors in parallel to get a higher current rating, and darlington connections to get a higher value of beta.

Differential amplifiers are used extensively as comparison amplifiers in which the sense and reference voltages are compared. Differential op amps are especially suitable because of their high dc gain.

Zener diodes are used to obtain voltage references, and voltage dividers are used to obtain sense voltages. These circuits are clearly seen in Figure 9-48.

Some regulated power supplies have two stages of regulation. One stage is the preregulator, which is a form of open-loop regulator. The other stage is the main regulator, which is a closed-loop system.

Figure 9-53 shows two examples of transformer preregulators. In both cases the theory of operation is based on the idea that the output voltage at the secondary is limited in amplitude.

In the ferroresonant transformer a capacitor (C) charges on each half cycle. The transformer is designed in such a way that at the instant the flux approaches its maximum value the capacitor discharges. The discharge current produces a flux that adds to that already existing, and the result is that the transformer is driven into saturation. This action occurs on each half cycle. The output waveform is limited in amplitude by the transformer saturation. Figure 9-54 shows the input and output waveforms that result.

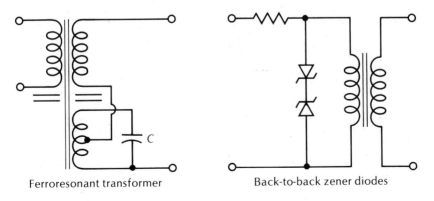

Ferroresonant transformer   Back-to-back zener diodes

Figure 9-53. Transformer preregulators

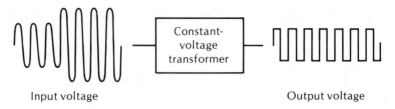

Input voltage   Output voltage

Figure 9-54.   Constant-voltage transformer

Note that changes in input amplitude do not affect the amplitude of the secondary voltage.

Back-to-back zener diodes can be used in the primary as shown in Figure 9-53. Their usage limits the voltage across the primary, and therefore limits the secondary voltage as shown in Figure 9-54.

Fuses and circuit breakers may be used for protection against overvoltage or overcurrent in regulated supplies. However, these devices are too slow for protecting some expensive circuitry. Current limiters serve as adjustable fuses and protect against excessive power output from the supply.

Figure 9-55 shows a special overvoltage protection circuit called a *crowbar*. It consists of an SCR across the output terminals. Although the SCR has a positive voltage on its anode and a negative voltage on its cathode, it is normally in the nonconducting state. A positive gate voltage is required to start the SCR into conduction.

The output of the comparator is negative when the reference voltage is more positive than the sense voltage. However, if the output voltage starts to rise, the sense voltage overrides the reference. A positive gate voltage turns the SCR on and it conducts heavily. The conducting SCR appears as a short

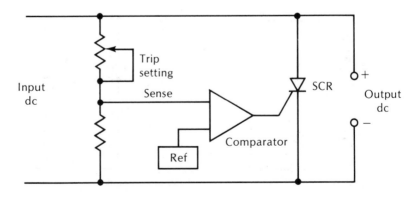

Figure 9-55.   Crowbar protective circuit

circuit across the output, so the output voltage is reduced to about zero volts. This action occurs so rapidly that components across the output terminals receive full protection even though the voltage rise is almost instantaneous.

MAJOR POINTS

1. The so-called power supplies used in industrial electronic equipment are actually power converters. There are four possibilities: ac to ac, ac to dc, dc to ac, and dc to dc.
2. Transformers are used for converting ac from one value to another, for isolation of personnel from the ac power line, and for regulating ac voltage.
3. Rectifier diodes and SCR's are used for converting ac to dc. There are rectifier circuits for single phase and multiphase systems.
4. The output waveform of a rectifier is normally filtered to convert it to a nearly pure dc.
5. Two kinds of filter circuits are being used:

    • *Passive*—with no amplifying components
    • *Active*—with one or more amplifying components

6. The percent ripple, or ripple factor, tells how well a filter circuit is able to convert rectifier pulsations into pure dc. A low value is desirable.
7. The regulation of a power supply is a measure of how well it holds a dc output under a varying load current.
8. Open-loop regulators maintain the output of a power supply constant, but they have no provision for sensing the output.
9. Closed-loop regulators compare the output of the power supply with a known reference. This type of circuit makes a correction if there is an error.
10. Regulator circuits are called *series* type if the regulating component is in series with the power supply load. It is a *shunt* type if the regulating component is in parallel with the power supply load.

PROGRAMMED REVIEW

(Instructions for using this programmed section are given in Chapter 1.)

The important concepts of this chapter are reviewed here. If you have understood the material, you will progress easily through this section. Do not skip this material because some additional theory is presented.

1. Which of the following statements is correct?

   A. Some of the more expensive power supplies can provide a constant voltage and a constant current at the same time. (Go to block 9.)
   B. It is not possible to have a power supply that will provide a constant voltage and a constant current at the same time. (Go to block 17.)

2. The correct answer to the question in block 17 is A. If diodes with different piv ratings are connected in parallel, the piv rating of the combination is equal to the lowest piv rating of the parallel diodes. For example, if two diodes with piv ratings of 500 V and 1000 V are connected in parallel, the piv rating of the parallel circuit is 500 V.

   The forward current rating of parallel-connected diodes is the sum of the diode forward current ratings. A 2 A diode in parallel with a 1 A diode gives a forward current rating of 3 A.

   Here is your next question.
   Which of the following does not require a surge limiting resistor?

   A. A brute force power supply with a simple capacitor filter. (Go to block 16.)
   B. A brute force power supply with a simple inductive filter. (Go to block 6.)

3. The correct answer to the question in block 8 is B. Some filter capacitors are made with a tolerance of −100% to +250%. A larger value can be used without any noticeable effect on the output of a brute force supply. A smaller value should not be used since a greater ripple could result.

   A larger voltage rating would give more protection against breakdown. A lower voltage rating should never be used, since that increases the possibility of circuit failure.

   Here is your next question.
   For the diode shown here, the cathode lead is

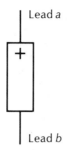

Lead a

Lead b

A. Lead *a*. (Go to block 5.)
B. Lead *b*. (Go to block 10.)

---

4. Your answer to the question in block 12 is wrong. Read the question again, then go to block 26.

---

5. The correct answer to the question in block 3 is A. The + sign on a rectifier diode would indicate the cathode side. This is because the cathode side in rectifier circuits is connected to the positive side of the power supply.

Here is your next question.
To replace the filter circuit in the illustration shown with a capacitor, its capacitance value would have to be

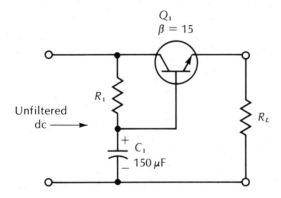

Q_1
$\beta = 15$

R_1

Unfiltered
dc →

R_L

+ C_1
150 μF

A. 150 μF. (Go to block 27.)
B. 2250 μF. (Go to block 13.)

---

6. The correct answer to the question in block 2 is B. A choke input filter will usually limit current surges sufficiently so that a surge limiting resistor is not required.

Here is your next question.
Which is more desirable in a power supply?

A. A low ripple factor. (Go to block 12.)
B. A high ripple factor. (Go to block 19.)

---

7. The correct answer to the question in block 21 is B. A rectifier changes ac to dc and an inverter changes dc to ac.

Here is your next question.
Is the following statement true or false? "It is not possible to receive an electrical shock from equipment that is operated from an isolation transformer."

A. The statement is true. (Go to block 14.)
B. The statement is false. (Go to block 18.)

---

8. The correct answer to the question in block 26 is A. A choke-input filter protects the rectifiers from current surges when the circuit is first energized. These surges can destroy a gas rectifier tube, especially if they occur before the cathode of the rectifier is up to the normal operating temperature.

Here is your next question.
A certain brute force power supply has a defective 1000 μF, 50 V filter capacitor. Which of the following statements is true?

A. Never use a larger capacitance value as a replacement. (Go to block 20.)
B. A higher value of capacitance can be used. (Go to block 3.)
C. A lower voltage rating can be used. (Go to block 24.)

---

9. Your answer to the question in block 1 is wrong. Read the question again, then go to block 17.

---

10. Your answer to the question in block 3 is wrong. Read the question again, then go to block 5.

---

11. Your answer to the question in block 21 is wrong. Read the question again, then go to block 7.

---

12. The correct answer to the question in block 6 is A. In the equation for ripple factor there is a ratio of I_{rms} and I_o. If a dc meter and true rms meter both read the same value, then the ratio of the readings is 1. The value of ripple factor would be equal to zero. This could only occur if there was no ripple, which is the best possible condition.

Here is your next question.

If it is desired that the output voltage of a brute force power supply maintain the same terminal voltage when the load current changes (within limits), then the supply should have

A. A low value of percent regulation. (Go to block 26.)
B. A high value of percent regulation. (Go to block 4.)

---

13. The correct answer to the question in block 5 is B. The electronic filter circuit shown in block 5 is similar to the one in Figure 9-36. The only difference is that the one in Figure 9-36 uses a voltage divider circuit for biasing the transistor. The equivalent capacitance equals the beta value (15) multiplied by the capacitance of C (150 μF):

$$C_{eq} = 15 \times 150 \ \mu F = 2250 \ \mu F$$

Here is your next question.

A 9 V battery with an internal resistance of 15 ohms is delivering 1/10 A to a load. What is the terminal voltage of the battery?

A. 7.5 V. (Go to block 21.)
B. 8.3 V. (Go to block 25.)

---

14. Your answer to the question in block 7 is wrong. Read the question again, then go to block 18.

---

15. Your answer to the question in block 26 is wrong. Read the question again, then go to block 8.

---

16. Your answer to the question in block 2 is wrong. Read the question again, then go to block 6.

---

17. The correct answer to the question in block 1 is B. A constant-voltage supply maintains the output voltage constant regardless of load current changes. A constant current supply maintains the load current constant regardless of load voltage changes. A supply cannot do both things at the same time.

Here is your next question.
When rectifier diodes are connected in parallel,

A. Series equalizing resistors should be used with each diode. (Go to block 2.)
B. They must have identical piv ratings. (Go to block 22.)

18.   The correct answer to the question in block 7 is B. An isolation transformer only protects you from getting a serious electrical shock when you get connected between one ac lead and ground. It does not protect you if you get connected between the two leads of the isolation transformer secondary. Also, a power supply that gets its ac input from an isolation transformer secondary can have a dc output voltage and current capability for delivering a lethal shock.

The most valuable protection you can have when working with any electrical or electronic system is a thorough knowledge of what you are doing.

Here is your next question.
In the power supply circuit shown here, what value of voltage should the voltmeter indicate? (Disregard drops across the diode and surge limiting resistor, leakage in the capacitor, and current drawn by the meter.)

_____ volts. (Go to block 23.)

19.   Your answer to the question in block 6 is wrong. Read the question again, then go to block 12.

20.   Your answer to the question in block 8 is wrong. Read the question again, then go to block 3.

21.   The correct answer to the question in block 13 is A. The circuit is shown in Figure 9-37, where $R_i = 15 \ \Omega$, $V = 9$ V, and $I_L = 0.1$ A. The voltage (V_i) across the internal resistance is equal to $0.1 \times 15 = 1.5$ V. The terminal voltage equals 9 V – 1.5 V = 7.5 V.

Here is your next question.
An inverter changes

A. Ac to dc. (Go to block 11.)
B. Dc to ac. (Go to block 7.)

22. Your answer to the question in block 17 is wrong. Read the question again, then go to block 2.

---

23. Your answer to the question in block 18 should be about 165 V. The rms value of power line voltage is 117 V. The capacitor charges to the peak value of this voltage, or

$$1.414 \times 117 = 165 \text{ volts}$$

Here is your next question.
What is the name of the protective circuit in power supplies that will instantly shut down the output in case of an overvoltage?
(Go to block 28.)

---

24. Your answer to the question in block 8 is wrong. Read the question again, then go to block 3.

---

25. Your answer to the question in block 13 is wrong. Read the question again, then go to block 21.

---

26. The correct answer to the question in block 12 is A. A perfect power supply would have no change in output terminal voltage when the load current varies, and would be rated as having 0% regulation.

Here is your next question.
Which of the following filters must be used with a gaseous rectifier circuit?

A. Choke input. (Go to block 8.)
B. Capacitive input. (Go to block 15.)

---

27. Your answer to the question in block 5 is wrong. Read the question again, then go to block 13.

---

28. The name of the protective circuit is crowbar.

You have now completed the programmed section.

SELF-TEST WITH ANSWERS

(Answers to this test are given at the end of the chapter.)

1. A circuit that converts dc power to ac power is called _____ .

2. An electrostatic shield in a power transformer that prevents electrostatic coupling between windings is called _____ .

3. When the center tap of a transformer secondary winding is grounded, the ends are (a) in phase, or (b) 180° out of phase.

4. The component that protects a technician by isolating him from the power line is called _____ .

5. When a dc voltage or an ac voltage is connected across two capacitors in series, the larger voltage drop will be across the (a) larger capacitance, or (b) smaller capacitance.

6. When rectifier diodes are connected in series, are resistors connected (a) in series, or (b) in parallel with the diodes to equalize the currents through each diode?

7. When diodes are connected in parallel, are resistors connected (a) in series, or (b) in parallel with the diodes to equalize the currents through each diode?

8. Back-to-back SCR's are equivalent to (a) a triac, or (b) a four-layer diode.

9. Rectifier circuits that use gaseous diodes employ (a) choke input, or (b) capacitive input filters.

10. When a half-wave doubler circuit is connected to a 120-volt power line, it should have a dc output voltage of about _____ volts.

11. A half-wave rectifier circuit with a capacitive filter is connected to a 120 V power line. The output voltage should be about _____ volts.

12. Of the single-phase rectifier circuits described in this chapter, which require(s) a transformer for proper operation?

13. How many rectifiers are needed for a full-wave, three-phase circuit?

14. Small capacitance values in parallel with large capacitance electrolytics are used for _____ .

15. Which is usually a simpler circuit: a series regulator or a shunt regulator?

16. Which is usually more efficient: a series regulator or a shunt regulator?

17. A measure of how well a power supply maintains a steady dc output voltage under varying load current conditions is called the power supply _____ .

18. Voltage fluctuations in the dc output voltage of a power supply, and having a frequency that is an exact multiple of the line frequency, are called _____ .

19. A regulator circuit that does not employ a sense circuit is called _____ .

20. Can a regulated power supply act as a constant current supply and a constant voltage supply at the same time? (Yes or No)

Answers to Self-Test

1. An inverter
2. A Faraday shield
3. (b) 180° out of phase
4. An isolation transformer
5. (b) Smaller capacitance
6. (b) In parallel
7. (a) In series
8. (a) A triac
9. (a) Choke input
10. 340
11. 170
12. The two-diode full-wave type
13. 6
14. Grounding parasitics
15. A shunt regulator
16. A series regulator
17. Regulation
18. Ripple
19. An open-loop regulator
20. No

CHAPTER 10

CONTROL CIRCUITS

One distinguishing feature of industrial electronics is the widespread use of control circuitry for operating motors and machines. Early controls used relays extensively, but the modern systems are more likely to be static controls. *Static controls* are simply all-electronic systems which have no moving parts, in contrast to the relay which has a moving armature.

An ON/OFF switch is a control circuit in its simplest form. This is a form of *manual control* where an operator is required to perform the ON/OFF switching and all other functions.

With a *semiautomatic control* the operator decides what changes in operation are necessary and initiates these changes by the use of switches. However, these switches may be located in some remote position with relation to the machine that they control. The switches are not directly connected to the machine, but rather, they initiate some magnetic or electromagnetic system that controls the motor or machine.

With a *fully automatic* system it is no longer necessary for the operator to initiate any changes in operation while the machine is performing its function. The changes are initiated electronically, so the operator usually only controls the starting and, in a few cases, the stopping of the machinery.

The three basic kinds of controls are shown in Figure 10-1. Control systems may also be classified as open-loop or closed-loop types. Two examples are shown in Figure 10-2. In the open-loop control the rheostat is used to set the amount of current flowing through the motor armature or field windings. The amount of current determines the motor speed; the speed is not regulated in any other way. If a mechanical load is connected to the motor, it may tend to turn slower. In order to bring the motor back to speed, the operator would have to manually readjust the rheostat.

In the closed-loop control of Figure 10-2 the speed is continually monitored by a tachometer that is mechanically connected to the motor shaft. The output of the tachometer is a voltage or frequency that varies with the motor speed. In this case the output is amplified, but some tachometers produce enough output to eliminate the need for an amplifier.

A comparator compares the tachometer output with a reference value. If there is any error in speed, the comparator will produce a dc output voltage to control the amount of current to the motor.

With this type of system, if a load is applied to the motor and it tries to slow up, the control system delivers more current to bring the speed back to

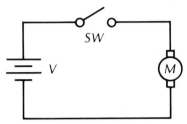

Manual control—The ON/OFF switch directly controls the current to the motor.

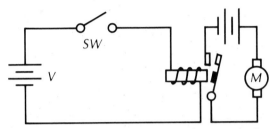

Semiautomatic control—the switch controls the current to the relay. The relay contacts actually control the motor. The switch may be in a remote position.

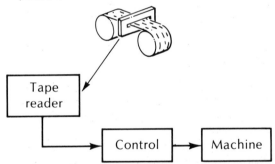

Fully automatic control—holes in the punched tape control the operation of the machine.

Figure 10-1. Three types of control systems

its normal value. Closed-loop systems like the one shown in Figure 10-2 are sometimes referred to as servo systems. When control circuits are used with motors, they regulate speed, the direction of rotation, and torque. When control systems are delivered to machines, a motor control may be one part of a complete system. The subject of this chapter is nonelectronic manual controls.

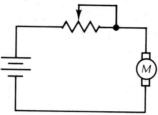

Open-loop control—manual adjustment of the rheostat
sets the motor speed.

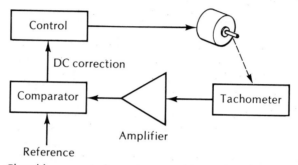

Closed-loop control—motor speed is monitored by the
tachometer, and the electronic system corrects any errors
in speed.

Figure 10-2. Examples of open-loop and closed-loop control systems

THEORY

Manual Starting Controls

The simple ON/OFF switch for the manual control in Figure 10-1 will work
for small dc motors. However, for a large dc motor a special manual starter
that limits current to the armature is normally used.

When a dc motor is first started, the amount of current in the armature is
limited only by the low dc resistance of the wire, the resistance of the
brushes, and the resistance of the lead-in wire. When it is turning at full
speed, the countervoltage limits the armature current. The countervoltage is
induced because the armature consists of a conductor turning through a
magnetic field. This countervoltage is in the direction opposite to the current
in the armature that produces the rotation.

You can think of the countervoltage as being a separate voltage source
that opposes the supply voltage for the armature current. This concept is
illustrated in Figure 10-3. It is physically impossible for the countervoltage
to ever equal or exceed the applied voltage, so current will always flow in the
armature regardless of the amount of countervoltage being induced.

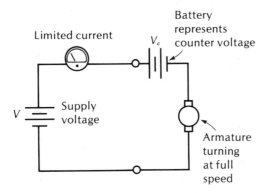

Figure 10-3. Countervoltage

When the motor is first starting, the armature is at rest and no countervolt-age is produced. So, in Figure 10-3 V_c is equal to zero. This means that there would be virtually no limit to the amount of current that can flow in the low-resistance armature, and the armature winding would likely burn out. To prevent burnout, resistors are connected into the armature circuit to limit the current flow until the speed of the armature is sufficiently great to pro-duce a countervoltage for current limiting. The general procedure is to start with a large resistance in the circuit. Then, as the motor speed increases, the series resistance is decreased in steps until the only opposition is the arma-ture countervoltage.

THE TWO-POINT STARTER

A *starter box* is a manual control that is used to insert resistance into the armature circuit, step by step, until the full speed is reached. Figure 10-4 shows a *two-point starter* for a series-wound dc motor. In this example, as in all other cases where a starter box is used, it is assumed that this is a large horsepower dc motor. The starter box in Figure 10-4 is called a *two-point* type because only two leads are connected to the starter box. Closing the main ON/OFF switch in this circuit will not start the motor because the movable arm in the starting box is in the OFF position.

When the arm is moved to position number 2, resistors R_1 through R_5 are connected in series with the field and the armature, and the complete circuit is across the dc source. This setting puts the maximum resistance in the circuit and limits the armature current to a low value. As the speed of the motor increases, the movable arm in the switch is moved to the other connec-tions, one at a time, until the motor reaches full speed (when the arm is at position 7). At that point no resistance is connected in the motor armature and field circuit.

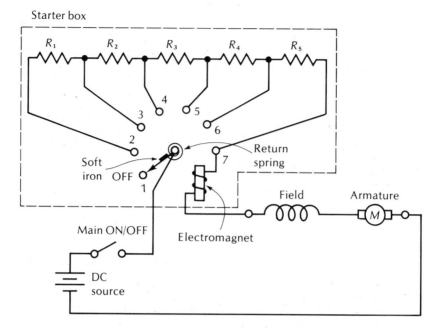

Figure 10-4. Two-point starter for a series-wound motor

Current flowing through the electromagnet produces a magnetic field that attracts the soft-iron slug on the movable arm of the switch, thus causing the switch to be held at position 7. If the main ON/OFF switch is opened, current through the electromagnet will stop flowing and the contact will no longer be held at position 7. The return spring will cause it to automatically rotate to the OFF position. At the same time, the motor will stop because there is no longer a complete circuit for the field and the armature.

The motor circuit is automatically opened in the event that the dc source voltage drops to zero. This feature is important because if the source voltage would drop to 0, and then go back to its maximum value again, the motor would normally try to start if it were still in the circuit. This effort would cause a burnout because there is no countervoltage to limit current flow. So, if the power is lost and then comes back on again, it is necessary to manually restart the motor.

You will remember that a series motor must be operated with a load. Without a load its speed increases until it finally destroys itself. If the load is removed in the series motor of Figure 10-4, the speed increases rapidly, and the countervoltage also increases. The result is equivalent to increasing V_c in Figure 10-3. The armature current decreases to the point where the current in the electromagnet no longer produces enough magnetic field to hold the arm at position 7. Thus, when the motor is operating with no load and the speed begins to increase very rapidly, the starter box automatically shuts the motor off.

THE THREE-POINT STARTER

Figure 10-5 shows a *three-point starter* for a series-wound motor. The difference here is that the holding coil for the movable arm is no longer connected in series with the field and the armature of the motor. The resistor (R_6) in series with the holding coil limits the coil current to a safe value. When the main ON/OFF switch is closed, current flows through the holding coil and the magnetic field is established immediately. To start the motor the arm is moved from position to position clockwise in the same manner as the two-point starter. This movement gradually decreases the resistance in the starter in the motor circuit. Finally, the arm will be in position 7, which connects the motor directly across the dc source. The arm is held in that position by the electromagnet.

This three-point starter will open the motor circuit if the dc source voltage is lost. However, it does not have a provision for removing power to the motor if it is operated without a load.

Figure 10-6 shows a three-point starter for a shunt-wound motor. In a shunt motor a race condition or runaway condition can occur if the shunt

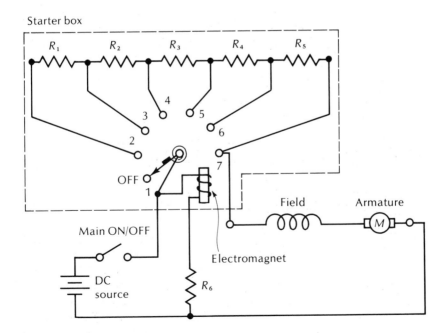

Figure 10-5. Three-point starter for a series-wound motor

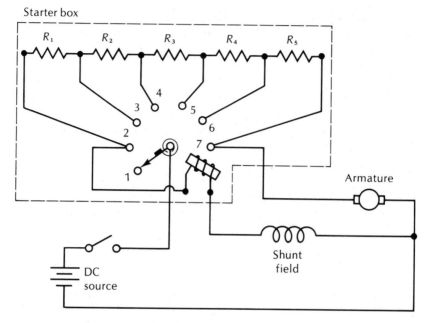

Figure 10-6.   Three-point starter for a shunt-wound motor

field winding is accidentally opened. The three-point starter in the illustration is designed to prevent the runaway condition. As with the other starters, the arm is moved from position to position so that the resistance in the armature circuit is reduced by each step. At the same time, the resistance in the shunt field winding is increased. (The speed of a shunt-wound motor increases as its field current decreases.)

When the arm is in position 7 the holding coil holds it at that position. However, if the shunt field winding should open accidentally, there will be no current in the holding coil and the arm will return to position 1.

THE FOUR-POINT STARTER

Figure 10-7 shows a *four-point starter* for a compound-wound dc motor. The series field and armature are connected to one point. The shunt field is connected to the opposite side of the resistance bank. Current in the holding coil is not flowing until the arm of the starter moves to positions 2 through 7.

In the first position all of the resistors are in series with the series field and armature, but there is no resistance in the shunt field circuit. This produces a large shunt field and a limited series field as the motor begins to turn. As the speed increases, the operator manually turns the switch from position to position until position 7 is reached. Here the holding coil prevents

Starter box

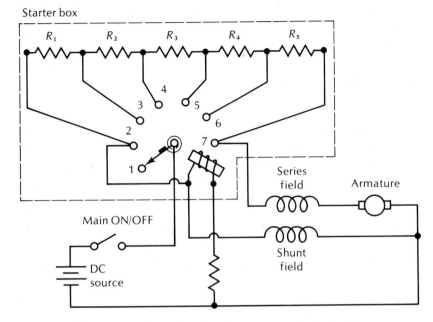

Figure 10-7. Four-point starter for a compound-wound motor

the switch from returning to the zero position with a spring action. At position 7 there is maximum series field current and the shunt field current is reduced due to the resistance of R_1 through R_5.

Only an experienced operator can perform the manual starting with the starter boxes just discussed. If he moves the switch contacts too rapidly, excessive armature current will flow. If he moves them too slowly, a large starting current in the field windings may produce enough heat to burn out the starting resistor elements. The operator turns from position to position while listening to the motor. When it reaches its maximum speed for one position, he goes to the next one.

Semiautomatic controls can be designed for starting large dc motors by using relays or SCR's that are operated in a remote position. It is also possible to design fully automatic starters using electronic timers and other electronic equipment. These electronic starters perform the same functions as the manual starters just described.

STARTING AC MOTORS

Synchronous motors may have zero starting torque unless some provision is made (such as a shading coil) in their design. Large synchronous motors may be started by using a smaller induction motor to turn the armature.

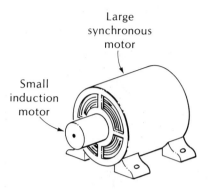

Figure 10-8. Large synchronous motor
with a small induction
motor

Figure 10-8 shows this arrangement. Once the synchronous motor has come up to full speed, the induction motor is automatically switched out. In some cases the induction motor and the synchronous motor are both wound on the same shaft and the same motor shell.

TORQUE AND SPEED CONTROL

There are two general equations for torque and speed control. They are important because they show which factors affect torque and which affect speed. The equations are as follows:

$$\text{Torque} = K_1 \times \text{armature current} \times \text{flux}$$

$$\text{Speed} = \frac{K_2 \times \text{countervoltage}}{\text{flux}}$$

The constants in these equations (K_1 and K_2) depend upon the number of poles, the construction of the motor, and other factors. The term *flux* refers to the magnetic field that is set by the current in the field windings.

The equations show that the torque of a motor can be increased by increasing either the armature current or the field current (flux). The speed equation shows that the speed can be varied by changing either of two factors: the countervoltage or the flux.

Figure 10-9 shows a series-wound motor with two different kinds of speed control. With the field control, a variable resistor is connected directly across the field windings. When the resistance from *a* to *b* is decreased, more current flows through that resistor and less current flows through the field winding. This is the same as decreasing the amount of flux.

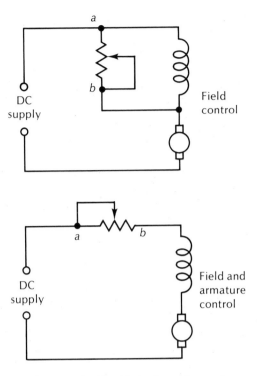

Figure 10-9. Two kinds of speed control

From the equation for speed you can see that if you decrease the flux, you are making the denominator of a fraction smaller. Making the denominator smaller increases the value of the fraction. To summarize, decreasing the flux increases the speed of the motor.

Only a limited amount of speed control can be obtained with field control connections for a series-wound motor. Also shown in Figure 10-9 is a field and armature method of controlling speed. This method has a serious disadvantage in that the total current flows through the resistor. The resistor will dissipate heat, and heat represents a waste of energy in the system. For that reason series motors are not normally controlled in this manner.

Keep in mind that the connections in Figure 10-9 would also work for the so-called universal motor which is a slightly modified series-wound dc motor. These motors are used in applications where it is not possible to remove the load. The connection of the load to the motor is, in itself, one way of regulating the speed. Later in this chapter we will discuss an electronic method of controlling the speed by turning the current on and off rapidly.

Figure 10-10 shows the methods used for controlling the speeds of a shunt-wound motor. In the first illustration, a variable resistor is connected in series with the armature. The equations for torque and speed show that if you increase the armature current, you overcome the countervoltage and increase both the torque and the speed. This is accomplished by reducing the resistance of the variable resistor.

With the field-control method a variable resistor is used to control the amount of current flowing through the field's winding. The equations show

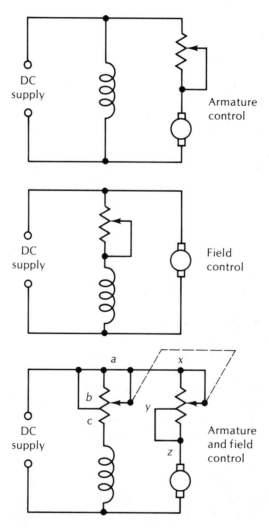

Figure 10-10. Methods of controlling shunt-wound motor speeds

that an increase in flux (corresponding to an increase in field current) will increase the torque and decrease the speed.

In comparing the two methods of controlling speed discussed so far (armature and field control), it is important to understand the limits of their applications. We must first define the term *rated speed*. The rated speed of a motor is that speed which it will achieve when it has the rated field current and the rated armature current flowing in its windings.

If you decrease the field current you will increase the motor speed above its rated speed. Conversely, if you increase the armature current you will decrease the speed below its rated value.

In order to control the speed from some value below its rated value to some value above the rated value, two mechanically connected variable resistors can be used. This is the function of the armature and field control circuits shown in Figure 10-10. For the field winding, one-half of the variable resistor is shorted at point *b*. Moving the arm above point *b* does not change the field current. In the armature control, one-half of the variable resistor is shorted between points *y* and *z*. Moving the arm of the armature control below point *y* will not affect the amount of armature current flowing. The dashed lines in the illustration mean that the two variable resistors are mechanically connected so that one shaft turns both resistors.

Assume that both resistor arms are at the top of their positions. At that point the field current is set by the amount of resistance between *b* and *c*, and the armature current is set by the amount of resistance set by points *x* and *y*. As the two arms are moved downward together, the amount of armature current is increased as more and more of the segments between *x* and *y* are shorted out. However, there is no change at first in the amount of field current.

When the two arms are past the halfway point, moving the arms down will not affect the amount of armature current but field current will start to increase with the reduction of field current resistance. This is because less and less of the variable resistance is being shorted out. In the position where the variable resistors are at points *c* and *z*, there is resistance at both the armature and the field winding.

This type of control makes it possible to vary the speed below the rated speed by increasing the armature current and to vary the speed above the rated speed by decreasing the field current.

Compound motors use variations of armature and field control. Table 10-1 summarizes armature and field control characteristics. The term *constant torque control* is sometimes used for armature speed control because the amount of output torque that a motor can deliver—without overloading—is constant when the armature current is changed from its minimum to its maximum value during a change in speed. If the speed is changed by the field current, then the horsepower rating of the motor is a constant value over the range of speeds controlled by speed control. So, the term *constant horsepower control* is sometimes applied to field current control.

TABLE 10.1. Summary of armature and field control characteristics

Type of Control	Available Torque	Available Power	
Armature	Constant	Increases with armature current	Decreases below rated speed with increase in armature current
Field	Varies inversely with speed	Constant	Increases above rated speed with increase in field current

Control of Ac Motor Speeds

It is more difficult to control the speed of an ac motor since its speed is dependent upon the rate of rotating flux within the motor. That rate, in turn, is dependent upon the frequency of the line voltage.

There is another problem involved in speed control of ac motors. With slow speeds they tend to have a low torque-developing power. This low-torque characteristic is in contrast with a series-wound dc motor which has its maximum torque-developing power at low speeds.

Two factors explain why dc motors are often used in control circuits. The factors are: ease of making speed changes and the ability to develop high torque at low speeds.

This is not to say that ac motor control is not possible. Figure 10-11 shows one method of controlling the speed by controlling the mechanical advantage and the pulley arrangement. A system of this type is sometimes used in drill presses. To increase the speed, the diameter of the spindle pulley is decreased. At the same time, the motor pulley is increased. So, the belt is moved to a lower position to increase speed.

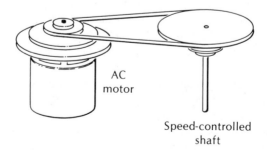

AC
motor

Speed-controlled
shaft

Figure 10-11. Ac motor control

Another way of controlling ac motor speed is to control it by use of a transmission. The function of a transmission is the same as that of a transmission on a car—that is, to change the gear ratios between the drive and the load. As with dc motors, the speed of a universal motor can be controlled by rapidly switching the motor current on and off with a thyristor.

THE WARD-LEONARD SPEED CONTROL

Whenever current flows through a resistance used for controlling speed, power is dissipated in the form of heat, and this power is wasted. A more efficient speed control is shown in Figure 10-12. It is called the *Ward-Leonard speed control.*

In this circuit an ac motor is used to turn a dc generator. The field current for the dc generator is controlled with a variable resistor (R_1). Current from the dc generator flows through the armature of the shunt motor. A small amount of change in the field current of the dc generator will produce a fairly large change in current in the armature.

The field of the shunt motor is powered by a fixed dc source. Since the armature is being controlled, the motor will run at less than its rated speed.

The system in Figure 10-12 is sometimes referred to as an *amplifier* because of the fact that the small change in generator field current causes a large change in motor speed.

Methods of Stopping Motors

An obvious method of stopping a motor is to allow friction to gradually slow it to a stop. However, that method is too slow for some applications. In such cases, special *braking* circuits are used.

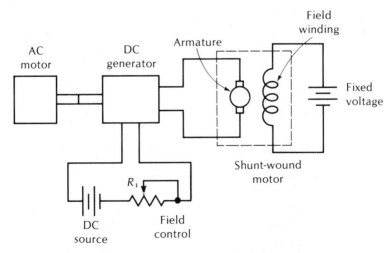

Figure 10-12. Ward-Leonard speed control

A braking system similar to the one used for an automobile can be used. In these applications, a solenoid is used to force two braking surfaces into contact. This method is used for emergency stops, and also as a supplement to other braking systems.

Another method is to use a self-induced voltage to produce an armature current. The armature is still turning after power has been removed from the motor. The rotating armature, turning in a magnetic field, becomes a generator and produces a voltage across the armature terminals. If a resistor or a short circuit is placed across the terminals of the rotating armature, current will flow through the armature winding. The magnetic field around the armature wires opposes the armature motion (in accordance with Lenz's law) and stops the motor. This method of stopping a motor is called *dynamic braking*. A few specific examples will now be given.

DYNAMIC BRAKING FOR A SERIES-WOUND DC MOTOR

Figure 10-13 shows a circuit that provides dynamic braking for a series-wound dc motor. The switching arrangement may be incorporated into a starting switch box. Switches S_1 and S_2 are mechanically connected, or *ganged,* so they move together.

In the first illustration the switch is in the *run* position. The arrows show that the electron current flows through the armature and field winding in series. It is assumed that the motor is already running, so starting circuits need not be shown in this series of braking circuits.

The second illustration shows the switch in the *dynamic braking* position. A series resistor (R_1) has been switched in series with the field winding. This maintains a reduced current through the field, which is necessary because the electromagnetic motor field must act against the armature field to get the braking action.

Variable resistor R_2 is connected directly across the armature, thus providing a current path for the voltage being generated in the armature as it slows. By adjusting R_2, the amount of armature current is controlled, and that controls the rate of braking.

When the motor is nearly stopped the amount of voltage induced in the armature is small. This small voltage is due to Faraday's law, which says that the amount of voltage induced in a conductor depends upon the rate at which the conductor is moving through a magnetic flux. Since the induced voltage is low, the armature current is also low. It follows that the rate of decrease in speed will also be low.

In the third illustration the armature and field winding are both disconnected, and the *friction brake* circuit is in operation. As mentioned before, the friction brake may be a brake lining and metal surface pressed in contact in an arrangement similar to the one used in automobile brakes. The friction brake is engaged with a solenoid in the circuit.

In the last illustration of Figure 10-13 the motor and its brake circuitry are disconnected from the dc power line.

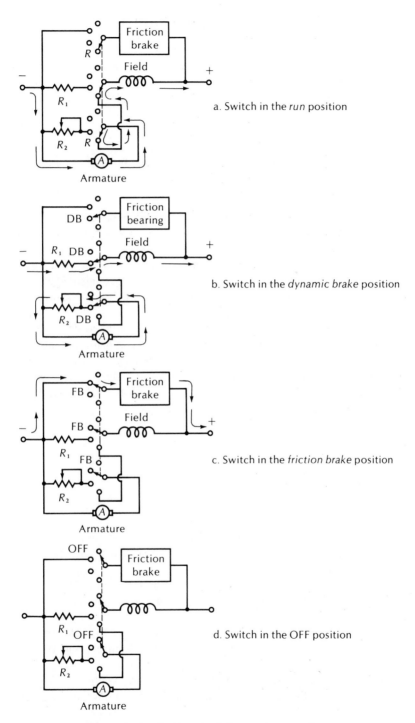

a. Switch in the *run* position

b. Switch in the *dynamic brake* position

c. Switch in the *friction brake* position

d. Switch in the OFF position

Figure 10-13. Braking circuit for a series-wound motor

DYNAMIC BRAKING FOR A SHUNT-WOUND MOTOR

Figure 10-14 shows a dynamic braking circuit for a shunt-wound motor. The *friction braking* and *off* positions are not shown because they are identical to the ones shown in Figure 10-13.

In the *run* position the armature and field are in parallel. Arrows show the electron current paths.

In the *dynamic braking* position the current paths are the same as for the series-wound motor.

A separate field winding may be used for dynamic braking of dc motors.

DYNAMIC BRAKING FOR AC MOTORS

The dynamic braking methods just described are not suitable for ac motors. Instead, the motor is stopped by a method called *plugging*. This method involves reversing, or attempting to reverse, the motor.

In a two-phase motor the power connection to one of the windings is reversed to bring the motor to a stop. This technique also applies to capacitor-type motors in which the capacitor is used to get the split phase. However, it is not used with capacitor-start motors.

Squirrel-cage types can be stopped by changing from ac to dc current in one of the windings. This changeover creates a magnetic field, and the moving conductors have a voltage induced in them as they move through that field. Since the conductors are shorted at the ends, an induced current flows. The magnetic field of the induced current reacts with the field from the dc current to brake the motor.

As with dc motors, braking can be supplemented with a friction brake.

MAJOR POINTS

1. Manual controls require the operator to perform each function with switches. Semiautomatic controls use relays for remote switching. Fully automatic controls operate without the need for someone to perform switching procedures.

2. Open-loop controls do not have any provision for altering their output. Closed-loop controls can alter their output to compensate for inaccuracies.

3. Large dc machines must be started with resistance in their armature circuit. The resistance limits armature current until a countervoltage develops.

4. When a countervoltage develops in the armature winding, it opposes the power supply voltage and limits the armature current to a safe value.

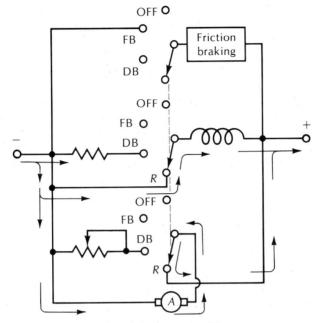

a. Switch in the *run* position

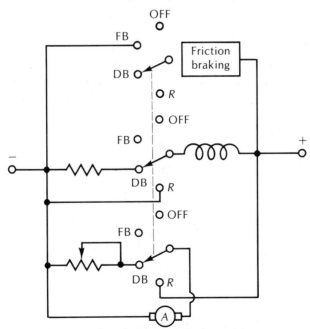

b. Switch in the *dynamic brake* position

Figure 10-14. Braking circuit for a shunt-wound motor

5. A two-point starter box is used with a series-wound motor. It has a provision for preventing a runaway in the event that the load is removed from the motor.
6. Three-point starters are used for series-wound and shunt-wound dc motors.
7. Four-point starters are used with compound-wound motors.
8. The general equation for dc motor torque is

$$\text{Torque} = K_1 \times \text{armature current} \times \text{flux}$$

This equation indicates that the torque can be increased by increasing either the armature current or the field current.
9. The general equation for dc motor speed is

$$\text{Speed} = \frac{K_2 \times \text{countervoltage}}{\text{flux}}$$

This equation indicates that the speed of a dc motor can be increased by increasing the armature current or reducing the field current.
10. Variable resistors can be used for torque and speed control, but they waste power in the form of heat.
11. Rated speed is the speed that a motor achieves when it is operated with its rated armature current and rated field current.
12. Armature current control is sometimes called constant-torque control. Varying the armature current causes the speed to vary below the rated speed.
13. Field current control is used for controlling speed above the rated speed.
14. A Ward-Leonard speed control is an efficient way to regulate the speed of a dc motor by supplying the motor field current from a dc generator.
15. Dynamic braking of commutating motors consists of connecting a resistor or short circuit across the armature. Voltage induced in the armature causes a current to flow. The field of the induced current opposes the armature motion in accordance with Lenz's law.
16. Ac motors may be stopped by plugging—that is, by attempting to reverse their direction.
17. Friction brakes are used as an aid in stopping both dc and ac motors.

PROGRAMMED REVIEW

(Instructions for using this programmed section are given in Chapter 1.)

The important concepts of this chapter are reviewed here. If you have understood the material, you will progress easily through this section. Do not skip this material because some additional theory is presented.

1. If a dc motor is operated from a remote position, as shown in the upper illustration, the voltage drop in the line will cause a loss of electrical power to the motor. A better system would use a low-power relay that is operated from a remote position as shown in the lower illustration. This is an example of

   A. A manual control. (Go to block 9.)
   B. A semiautomatic control. (Go to block 17.)

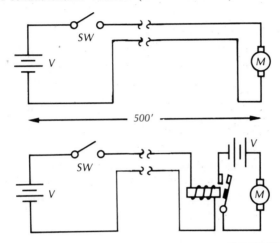

2. The correct answer to the question in block 12 is B. Changing the field current automatically changes the amount of flux in the motor.

   Here is your next question.
   Which of the following statements is correct?

   A. When the armature countervoltage is greater than the motor power supply voltage the motor speed is decreased. (Go to block 8.)
   B. It is not possible for the armature countervoltage to exceed the motor power supply voltage. (Go to block 16.)

3. Your answer to the question in block 14 is wrong. Read the question again, then go to block 10.

4. Your answer to the question in block 17 is wrong. Read the question again, then go to block 11.

5. Your answer to the question in block 24 is wrong. Read the question again, then go to block 12.

6. The correct answer to the question in block 16 is B. The number of points actually refers to the number of connections to the starter box, and is not necessarily related to the number of things controlled.

Here is your next question.
The solid-state equivalent of a thyratron is

A. an SCR. (Go to block 14.)
B. a triac. (Go to block 21.)

7. The correct answer to the question in block 10 is B. See Figure 10-12.

Here is your next question.
An induced current in the armature has a magnetic field that slows the motor. This is in accordance with

A. Lenz's law. (Go to block 13.)
B. The law of inertia. (Go to block 22.)

8. Your answer to the question in block 2 is wrong. Read the question again, then go to block 16.

9. Your answer to the question in block 1 is wrong. Read the question again, then go to block 17.

10. The correct answer to the question in block 14 is B. The starter does not *limit* resistance. Instead, it increases resistance when the motor is first started.

Here is your next question.
The Ward-Leonard system is used for controlling the speed of

A. An ac motor. (Go to block 19.)
B. A dc motor. (Go to block 7.)

11. The correct answer to the question in block 17 is B. A transducer senses the output of the closed-loop system and feeds back voltage, current, resistance, inductance, or capacitance in an amount that is related to the output.

Here is your next question.
When a dc motor is running at its rated speed, decreasing the field current will

A. Increase its speed. (Go to block 24.)
B. Not affect its speed. (Go to block 15.)
C. Decrease its speed. (Go to block 23.)

12. The correct answer to the question in block 24 is B. Numerical control is fully automatic because the operator needs only to start the system.

Here is your next question.
There are two ways to control the speed of a dc motor when it is running at its rated speed. They are: change the field current or

A. Change the amount of flux in the motor. (Go to block 18.)
B. Change the armature voltage. (Go to block 2.)

13. The correct answer to the question in block 7 is A. According to Lenz's law, the magnetic field of an induced current always opposes the motion that produced it.

Here is your next question.
Which of the following will reverse the direction of rotation of a dc motor?

A. Reverse the direction of current in the armature *and* field winding. (Go to block 25.)
B. Reverse the direction of current in the armature *or* field winding. (Go to block 26.)

14. . The correct answer to the question in block 6 is A. In noncommutating motors, plugging is one method used to apply a stopping force.

Here is your next question.
The purpose of the two-point starter for dc motors is to

A. Limit circuit resistance to a low value until the motor reaches its rated speed. (Go to block 3.)
B. Limit current until the motor reaches its rated speed. (Go to block 10.)

15. Your answer to the question in block 11 is wrong. Read the question again, then go to block 24.

16. The correct answer to the question in block 2 is B. An inductive countervoltage in an armature can never be larger than the initial voltage. If this were possible, the motor would stop or reverse.

Here is your next question.
Is this statement correct? A three-point starter is better than a two-point starter because it controls more things.

A. The statement *is* correct. (Go to block 20.)
B. The statement *is not* correct. (Go to block 6.)

---

17. The correct answer to the question in block 1 is B. With a semiautomatic motor control a switch energizes a relay in a remote position. The relay contacts complete the motor circuit.

   Here is your next question.
   Which of the following *requires* a transducer?

   A. Open-loop control. (Go to block 4.)
   B. Closed-loop control. (Go to block 11.)

---

18. Your answer to the question in block 12 is wrong. Read the question again, then go to block 2.

---

19. Your answer to the question in block 10 is wrong. Read the question again, then go to block 7.

---

20. Your answer to the question in block 16 is wrong. Read the question again, then go to block 6.

---

21. Your answer to the question in block 6 is wrong. Read the question again, then go to block 14.

---

22. Your answer to the question in block 7 is wrong. Read the question again, then go to block 13.

---

23. Your answer to the question in block 11 is wrong. Read the question again, then go to block 24.

---

24. The correct answer to the question in block 11 is A. Decreasing the field current reduces the flux, so less countervoltage is developed in the armature. The armature current increases and so does the speed.

   Here is your next question.
   Numerical control is an example of

   A. Semiautomatic control. (Go to block 5.)
   B. Fully automatic control. (Go to block 12.)

25. Your answer to the question in block 13 is wrong. Read the question again, then go to block 26.

26. The correct answer to the question in block 13 is B. This information was given in an earlier chapter on motors.

You have now completed the programmed section.

SELF-TEST WITH ANSWERS

(Answers to this test are given at the end of the chapter.)

1. Which type of control is used to vary the speed of a dc motor when it is operating below its rated speed?

2. Which type of dc motor is used (with some modifications) on ac power?

3. A four-point starter is used with a _____ dc motor.

4. Name two things that the amount of torque in a dc motor depends upon.

5. Name two things that the speed of a dc motor depends upon.

6. Attempting to reverse the direction of rotation of an ac motor as a method of stopping it is called _____ .

7. Stopping a dc motor by placing a resistor or short circuit across the armature is called _____ .

8. The braking effect of placing a short circuit across the armature will be greatest when the speed of the armature is (a) high or (b) low.

9. A starting circuit is needed for large motors to limit armature starting current. What limits the armature current when the motor is running at full speed?

10. In the Ward-Leonard speed control, an ac motor is used to turn a (a) dc motor or (b) dc generator.

Answers to Self-Test

1. Armature
2. Series-wound
3. Compound-wound
4. Armature current and flux
5. Armature current and flux
6. Plugging
7. Dynamic braking
8. (a) high
9. The countervoltage
10. (b) Dc generator

CHAPTER 11

ELECTRONIC CONTROLS

The use of electromechanical controls was discussed in the previous chapter. They are important for a complete understanding of electronic systems.

Electronics has not eliminated the use of electromechanical controls. Industrial electronics technicians will likely encounter such equipment in their work. They should understand their purpose and theory of operation in order to better understand complete control systems.

This chapter deals with the use of electronics in industrial applications, and specifically in control systems. Although it wouldn't be possible to cover all of the uses, the examples given here are typical.

The ultimate purpose of industrial electronic control is to regulate the action of machinery or of a process. This objective is always accomplished by controlling some parameter such as voltage, current, or frequency. The approach used here is to show how the parameter is controlled, then to show how the controlled parameter is used for some industrial electronic purpose.

Keep in mind that some basic electronic circuits are essentially the same in all systems. For example, oscillators, low-frequency amplifiers, high-frequency amplifiers, and logic circuits are the same in industrial electronic systems as they are in any other system. The major difference between one system and another is not in their basic circuitry, but rather, in their applications.

The block diagram material will help you to analyze the newer systems that use modular construction and integrated circuits. There are also circuit discussions in this chapter.

Control of Current

There are four basic methods of controlling current (and therefore controlling power) in industrial electronic systems:

1. Control the circuit resistance.
2. Control the voltage.
3. Control the ON vs OFF times.
4. Permit the current to flow for a given amount of time.

The methods used for controlling voltage will be discussed later in this chapter.

CONTROL OF CIRCUIT RESISTANCE

A rheostat is a variable resistor connected so that it directly controls circuit current. The rheostat in Figure 11-1 is an example. Moving the arm toward point *a* reduces the circuit resistance and increases the current through the load.

The two-step control circuit of Figure 11-1 can be set for either of two load current values. This type of current control is less expensive than using a rheostat.

Figure 11-2 shows a motor speed control that uses two different values of resistance to determine the current through the armature of a PM (permanent-magnet field) motor.

In the first position, R_1 is in series with the motor and power source. In position 2, resistor R_2 is switched into the circuit and R_1 is switched out. This reduces the resistance (1 kΩ instead of 2 kΩ).

In position 3 the two resistors are in parallel, thus reducing the circuit resistance even further. (The resistance of two resistors connected in parallel is always less than the smaller of the two resistance values.) In position 3 the motor current is maximum, and therefore its speed is maximum.

The resistance values corresponding to each position are:

Switch Position	Resistance in the Circuit
1	2 kΩ
2	1 kΩ
3	0.66 kΩ

The disadvantage of resistance controls is that the resistors cause a waste of power. Current flowing through a limiting resistor causes power to be

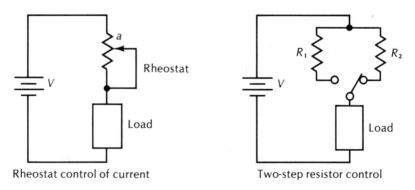

Rheostat control of current   Two-step resistor control

Figure 11-1.   Control of load current by varying resistance

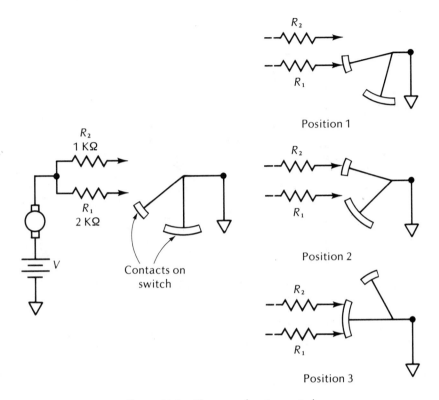

Figure 11-2. Three-speed motor control

dissipated in that resistor in an amount equal to I^2R. For controlling large currents the resistors are not only wasteful, but also expensive and bulky.

The control circuit of Figure 11-3 is an alternative to direct resistance control. Variable resistor R_1 in the base circuit of the darlington amplifier controls a small current. The low value of base current in turn controls a high value of collector current. The high emitter current is the sum of the base and collector currents. The most important advantage is that the variable resistor can have a low power rating even though it is controlling a high current value.

The circuit of Figure 11-4 is a modification of the one just discussed. This is an example of a closed-loop speed control. A tachometer generator is placed in series with the rheostat. The polarity of the generator is such that it subtracts from the power supply voltage.

Assume that the motor is turning at the desired speed, and the base voltage of the darlington amplifier is 10 volts positive with respect to ground. If

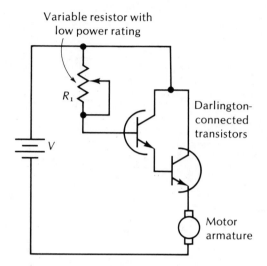

Figure 11-3. Transistor motor control

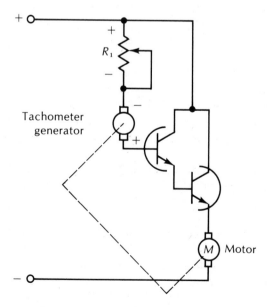

Figure 11-4. An automatic speed control

the motor speed increases for any reason, the tachometer voltage will also increase. Since the generator voltage subtracts from the supply voltage, it follows that an increase in generator voltage will result in a decrease in the base voltage. The decrease in base voltage will lower the collector and emitter currents, and reduce the motor speed to its regulated value.

Conversely, a decrease in motor speed will decrease the tachometer generator voltage, resulting in a more positive base voltage. This result, in turn, increases the collector and emitter currents, and increases the motor speed back to its regulated value.

When a mechanical load is placed on the motor, it will try to slow down, but the feedback circuit just described will bring the motor speed back to its regulated value. On the other hand, when a mechanical load is removed from the motor it will try to go faster. But, the feedback circuit will slow it down to regulated speed.

Adjusting R_1 sets the regulated speed. When the circuit is first energized, no voltage is produced by the generator, so the base current is limited only by R_1. The result is a high collector and emitter current, and a high starting current for the motor.

Power amplifiers, like the darlington transistor, are usually connected in series with the load when it is necessary to control a large amount of power. Remember that high currents are associated with high output power. The power amplifier may be thought of as being a variable resistor.

CONTROL OF CURRENT ON AND OFF TIMES

The average value and the rms value of a current can be controlled by turning it on and off. Figure 11-5 shows two ways of controlling current.

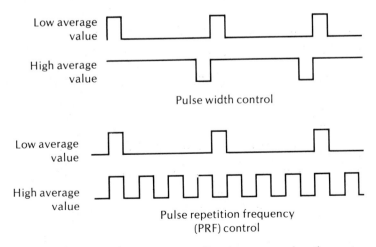

Figure 11-5. Two methods of controlling the average value of current

Pulse width control is sometimes called *pulse width modulation* (PWM). The greater the amount of ON time, the higher the average (and rms) value of the current. This type of control is obtained with two basic circuits: a *pulse generator* that converts the dc power to a pulsating dc, and an *ON/OFF control* that determines the ON time for each cycle. Figure 11-6 shows the basic plan.

In this circuit, as with others in the chapter, a power amplifier may be used for current control. It is not shown in the block diagram of Figure 11-6.

Figure 11-7 shows the basic layout for frequency control of pulses. The difference between this circuit and the one in Figure 11-6 is in the method of control. This is called the PRF (pulse repetition frequency) method of control.

Figure 11-8 shows how a multivibrator can be used to obtain pulse width control. This *NPN* transistor free-running multivibrator is also called an *astable* multivibrator. It has two amplifiers: Q_1 and Q_2. When Q_1 conducts to saturation, it cuts Q_2 off. Conversely, when Q_2 is saturated, Q_1 is cut off. The OFF time for Q_1 is determined by the time constant of C_2 and R_3; and the OFF time of Q_2 is set by the time constant of C_1 and R_2.

Assume that Q_1 has just started to conduct to saturation. The voltage drop across R_1 makes the collector voltage low. This low voltage is coupled to the base of Q_2 through C_1, and causes Q_2 to be cut off.

Eventually, C_1 will charge through R_2 sufficiently to put a positive voltage on the base of Q_2. Q_2 conducts and shuts off Q_1 because of the low voltage coupled through C_2.

This action continues, with each transistor cut off whenever the opposite one is conducting.

The ON/OFF time is determined by the time constants, which in turn are determined by the settings of R_2 and R_3. These resistors are mechanically

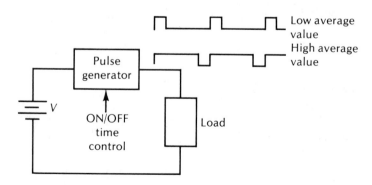

Figure 11-6. Control of average current value by varying pulse width

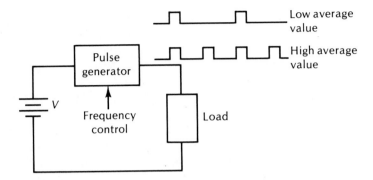

Figure 11-7. Control of average current value by varying frequency

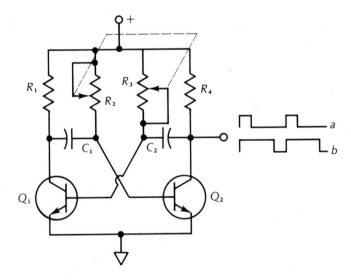

a. When R_3 is adjusted so that the time constant of R_3 and C_2 is long, Q_1 is OFF for a long period of time, and Q_2 is ON for a long period of time. When Q_2 is ON, the drop across R_4 makes the output low.

b. When R_3 is adjusted so that the time constant of R_3 and C_2 is short, Q_1 is OFF for a short period of time, and Q_2 is ON for a short period of time. When Q_2 is ON, the drop across R_4 makes the output low.

Figure 11-8. The ON/OFF time of Q_2 determines the average value of output voltage

coupled so that an increase in the resistance of one results in a decrease in the resistance of the other. That way, the total ON plus OFF time for either transistor is unchanged when the setting is changed. Since the ON plus OFF time is equal to the time required for one cycle, which is a constant, the frequency is not affected by adjusting R_2 and R_3.

The output waveforms shown in Figure 11-8 are ideal. In practice they are not exactly square waves because of the time required in switching from one transistor to the other. A Schmitt trigger can be used to square the waveforms. A Schmitt trigger is a circuit that has only two stable states: ON and OFF. The switching time from one state to the other is very short— being in the order of nanoseconds for some integrated circuit versions.

Figure 11-9 shows a unijunction transistor (UJT) frequency control circuit. The frequency of this oscillator is set by the time constant of R_1, R_2, and C_1. The purpose of R_2 is to prevent the arm of R_1 from being set to the positive power supply voltage. That would likely destroy the UJT.

The output pulses from the UJT oscillator have a very short duration. In fact, they look like spikes on an oscilloscope display. A one-shot multivibrator has only one stable state—OFF. A pulse causes it to cycle on for a set period of time, then it turns off again. The overall result is that the spikes

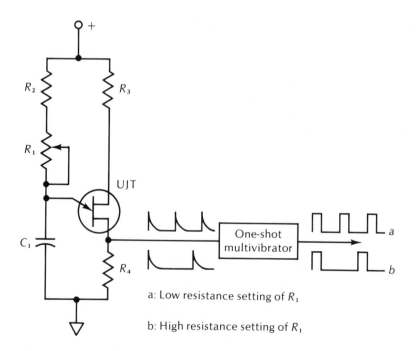

a: Low resistance setting of R_1

b: High resistance setting of R_1

Figure 11-9.   A method of obtaining a variable PRF

from the UJT are converted to positive pulses. The pulse frequency is determined by the frequency of oscillation of the UJT circuit.

TIMER CONTROL OF CURRENT ON TIME

Figure 11-10 shows another method of controlling current in order to control
power to a load. In this case a timer is used to gate the current ON and then
OFF one time. At the end of the ON time, the timer must be reset for the
next application.

An example of a system that uses a timer control for current ON time is
shown in Figure 11-11. This is a spot welder. In the current cycle an enormous current (30,000 A or greater) is caused to flow through the work pieces
in a narrow beam. The high current melts the metal in the pieces being
welded, causing them to be joined together. The complete welding cycle is
shown in Figure 11-11.

Control of Voltage

Voltage control in industrial electronic systems usually involves the use of a
fast switching device such as a thyratron or solid-state thyristor.

DIODE CONTROL

A solid-state diode provides an efficient way to get two-speed control. Figure
11-12 shows the circuit. In position *a* the motor runs at its maximum speed
when the ON/OFF switch is in the ON position. When the speed control
switch is turned to position *b*, the semiconductor diode is placed in series
with the motor. As shown by the waveforms, current can only flow for one-
half of each complete cycle. This current limiting has the effect of reducing

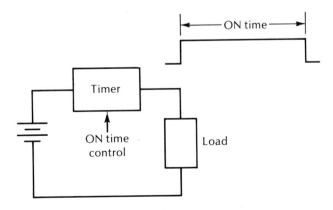

Figure 11-10. Timer control of load current

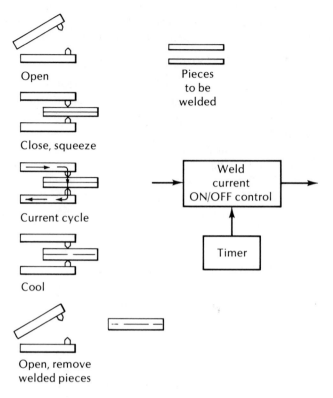

Open

Close, squeeze

Current cycle

Cool

Open, remove
welded pieces

Pieces
to be
welded

Weld
current
ON/OFF control

Timer

Figure 11-11.   Use of a timer in spot welding

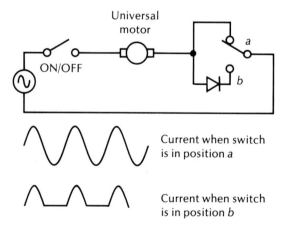

Universal
motor

ON/OFF

a

b

Current when switch
is in position a

Current when switch
is in position b

Figure 11-12.   A simple two-speed solid-state control

the average current delivered to the motor, and hence reduces its motor speed.

The advantage here is that in position *b* the only resistance in the circuit is the dc resistance of the diode when it is forward biased. This resistance is so small it can be neglected when compared to that in a resistance control circuit. To summarize, the effect is to produce a two-speed motor control without the waste of power associated with resistive controls.

SPEED CONTROL WITH AN SCR

Before discussing the SCR motor control circuit, it is necessary to review the basic *RC* circuit shown in Figure 11-13. An ac generator is supplying sine wave voltage *(V)* to the circuit. In the resistive branch comprised of R_2 and R_3 the voltage at the junction (point *b*) is in phase with the supply voltage. In the reactive circuit comprised of R_1 and *C,* the voltage at point *a* is lagging behind the supply voltage. The amount of lag depends upon the amount of resistance of the circuit.

The waveforms of this circuit are shown in Figure 11-13. You can see that the voltage at point *a* is lagging behind the voltage at point *b*.

Now consider the circuit of Figure 11-14. In this circuit there is an SCR and a phase control comprised of R_1, R_2, and *C*. The voltage at point *x* will lag behind the generator voltage. On the half cycle when the anode of the SCR is made positive, it is held off from conduction because the gate is not conducting. The gate is connected to variable resistor R_2 through a neon lamp.

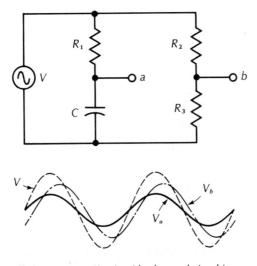

Figure 11-13. Circuit with phase relationships

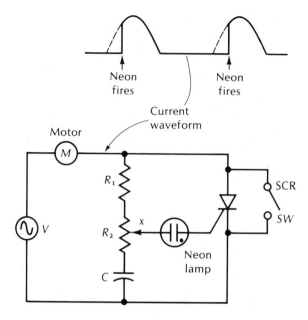

Figure 11-14. SCR motor speed control

You will remember that a neon lamp is a form of breakover device. It will not conduct until the voltage across it has reached a certain minimum value. The voltage at point x starts to go positive sometime later than the voltage at the anode, but the gate still cannot conduct because the neon lamp has not fired. When the firing potential of the neon lamp is reached, it conducts and the gate becomes positive. The positive gate causes the SCR to switch on. The current waveform of the SCR and motor circuit is shown in Figure 11-14.

The circuit of Figure 11-14 offers a limited control, and has the disadvantage that the current is flowing through the motor only during the positive half cycle. The switch across the SCR makes it possible to obtain the maximum speed from the motor.

SPEED CONTROL WITH A TRIAC

A similar circuit is shown in Figure 11-15. Here a triac is used in place of the SCR. You will remember that a triac can conduct in both directions. In other words, it is bilateral. The disadvantage of half-wave rectification by the SCR in the previous circuit is eliminated by using a triac.

Instead of a neon lamp, a diac is used. A diac is the solid-state equivalent of the neon lamp, but has a breakover voltage that is much lower in value. When the voltage at anode 1 (A_1) is positive, the gate of the triac is held off

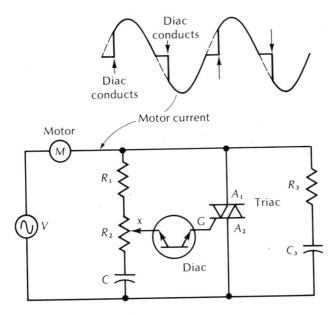

Figure 11-15. Trial motor control

from conduction until the voltage at point x equals the diac breakover voltage. Once the diac begins to conduct, gate current (G) causes the triac to conduct.

On the next half cycle, anode 2 (A_2) is made positive with respect to the gate. When the breakover voltage of the diac is again reached, current again flows through the motor, but this time the direction of current is reversed. The waveform of the motor current is shown in the illustration.

The RC circuit (R_3 and C_3) connected across the triac is called a *snubber*. Its purpose is to prevent the triac from conducting because of a countervoltage surge from the motor winding. You can think of a snubber as a circuit that provides a path for the inductive kickback voltage around the triac.

CLOSED-LOOP VOLTAGE CONTROL OF SPEED

In both the triac and the SCR motor speed controls, the speed is dependent upon the setting of the variable resistor (R_2). In other words the setting of R_2 determines the time at which the gate of the triac or SCR is made positive enough to conduct. By controlling the precise time of firing, it is possible to make a closed-loop motor speed control, as shown in Figure 11-16.

The tachometer generator provides an ac voltage with a frequency that is proportional to speed. The output frequency of a tachometer is compared with a reference frequency in the comparator. If the two frequencies are the same, then the output of the thyristor firing circuit is not changed.

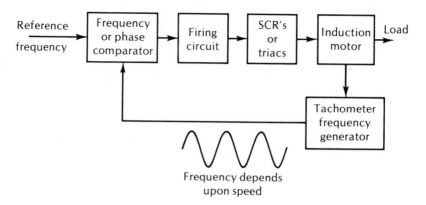

Figure 11-16. Closed-loop SCR speed control

If the motor speed is low, the tachometer frequency will be below the reference frequency. The frequency difference produces a change in the firing circuit that causes the SCR to fire sooner. There is a resulting increase in motor speed which brings the tachometer frequency back up to the value equal to the reference signal. Conversely, if the tachometer output frequency is above the reference, then the firing circuit will be modified to allow the SCR or triac to conduct for a shorter period of time. The decrease in conduction reduces the induction motor speed.

Frequency and Phase Control

In the chapter on motors it was explained that the speed of a synchronous motor is dependent upon the frequency of the input power. So, a simple speed control can be made with a sine wave oscillator that has a variable frequency and a power amplifier. Figure 11-17 shows such a speed control in block diagram form.

The disadvantage of the system in Figure 11-17 is that the speed is not regulated. Any drift in the output frequency of the variable frequency oscillator (VFO) will result in a change of motor speed.

A phase-locked loop is ideal for regulating the motor speed in the system of Figure 11-17. Figure 11-18 shows the basic parts of the phase-locked loop. It can be made with discrete components, and it is also available in a number of different integrated circuit families. The one shown in Figure 11-18 is an integrated circuit version.

The phase comparator has two input signals. One is from an external frequency source such as an oscillator; the other is from the voltage-controlled oscillator (VCO). When these signals are in phase, there is a zero volt output from the phase comparator.

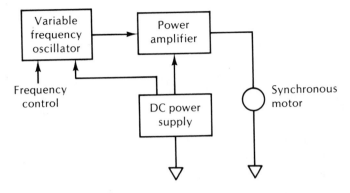

Figure 11-17.   Simple frequency speed control

If the signals are out of phase, there will be a dc voltage out of the comparator. The amount of dc voltage depends upon the amount of phase difference.

The lowpass filter assures that only the dc error voltage from the phase comparator is passed to the error correction amplifier. This device is a dc amplifier that takes the relatively low value of dc out of the comparator and changes it to a larger voltage value for correcting the VCO frequency.

The voltage-controlled oscillator has a free-running frequency that is set by an *RC* network. This frequency is normally set to be approximately equal to the input frequency when the loop is used as shown in Figure 11-18.

The dc correction voltage from the error amplifier adjusts the VCO frequency until it is exactly equal to the input frequency.

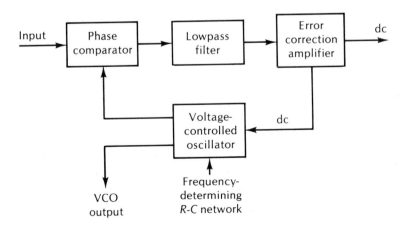

Figure 11-18.   Parts of a phase-locked loop

Figure 11-19 shows a phase-locked motor speed control. The VCO in the block diagram of Figure 11-18 has been replaced by a tachometer. Its output is a frequency that is directly related to motor speed.

The tachometer frequency is compared to a reference frequency from a stable oscillator. If the signals are the same in frequency and phase, there is no correction voltage. If they are not the same, a dc voltage proportional to the difference is delivered to the error correction amplifier by way of the lowpass filter. The error correction amplifier delivers a correction voltage to adjust the bias of the power amplifier in order to adjust the motor speed.

When the motor speed is correct, there is no error voltage, and no correction bias is delivered to the power amplifier.

A counting circuit can be inserted between the tachometer output and the phase comparator as illustrated in Figure 11-20. The purpose of the counter is to divide the tachometer frequency.

As an example, suppose the counter divides the frequency by 4. This means the motor will have to increase its speed by exactly four times its value in Figure 11-19. In other words, the motor has to run four times as fast so that when the counter divides the tachometer output by 4 there will still be a phase lock at the phase comparator.

The VCO output in the circuit of Figure 11-18 can be used to operate a *digital motor* (also called a *stepping motor*). Many of these devices are not actually forms of motors in the conventional meaning of the term. They are more nearly like stepping switches. Their shaft turns a fixed number of degrees for each pulse input.

Since the VCO output is a square wave (in most circuits), it can be used to accurately turn a stepping motor at a very accurate speed.

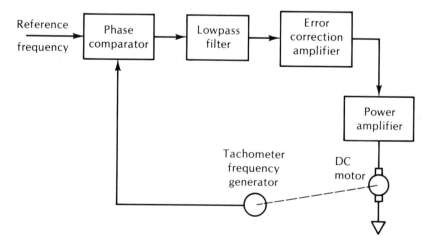

Figure 11-19. Phase-locked motor speed control

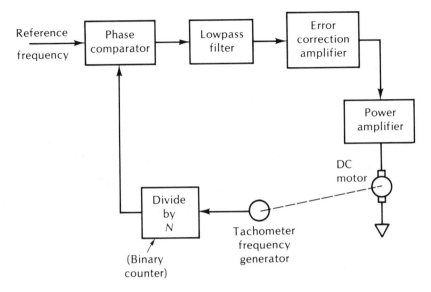

Figure 11-20. Variable frequency phase-locked speed control

NUMERICAL CONTROL

Numerical control is a system that employs punched tapes or magnetic tapes to automatically operate machines. An example of a punched tape code is shown in Appendix I. Each hole represents a binary 1, and a binary 0 is represented by places where there is no hole.

Instead of the letters shown in the illustration, the hole positions could combine to represent some manufacturing process such as drilling, reversing a motor, moving a drill press spindle, or moving a piece of work on an assembly line.

Punched tape requires a light-to-electricity transducer. The principle is illustrated in Figure 11-21. As the tape is moved past the reader, light shines through the holes onto the transducers which generate a voltage or change in resistance proportional to the light.

Both the magnetic and punched tape systems are digital in that only two conditions—ON and OFF—are recorded. Figure 11-22 shows a block diagram of a system for controlling machines with a tape.

The first step is to have a blueprint that describes the finished product. From this blueprint a programmer operates a tape puncher that puts the proper holes in the tape for operating the machine. The tape is placed on a

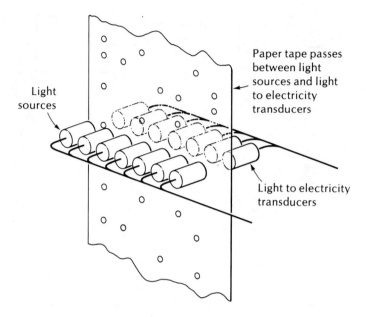

Figure 11-21. Optical tape reader

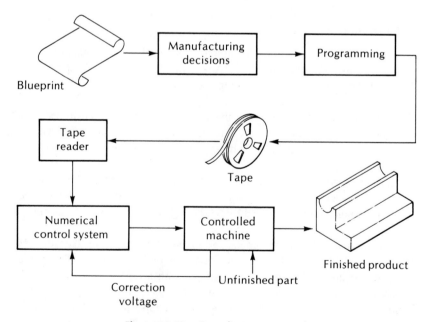

Figure 11-22. Open loop tape control

tape reader that converts the holes into binary digits that correspond to various codes for operating the machine. The numerical control system takes these digits and delivers instructional signals to the controlled machine.

The system represented in Figure 11-22 is an open-loop numerical control. Note that there is no feedback for correcting the manufacturing process. A closed-loop system is shown in Figure 11-23. It also employs a magnetic or punched tape. The tape reader delivers its signal to a comparator. The comparator has two inputs. One is a feedback signal that tells the equipment about the actual position of the work piece. The input from the tape tells the desired position. There is an error signal from the comparator which is proportional to the difference in the two positions. This error signal is amplified and delivered to a drive motor which operates the machine table. The table repositions the work and continues to move the work until the actual position and the desired position input signals are identical. At that point there is no output signal from the comparator and the work piece is no longer being moved. Additional hole combinations then operate the machine.

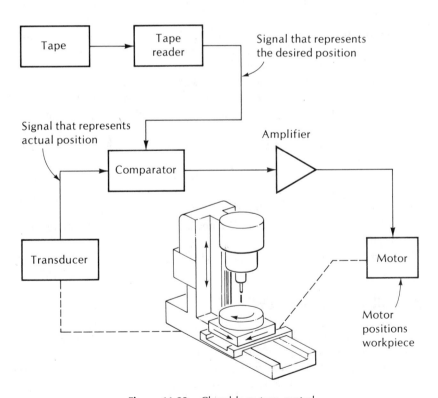

Figure 11-23. Closed loop tape control

MAJOR POINTS

1. When electronics is used for control in industrial electronics, it accomplishes its task by controlling voltage, current, frequency, or some other parameter.
2. Controlling power to a load is accomplished by controlling load current.
3. A number of ways are used to control current. Any of the following may be used:

    *   Control the circuit resistance.
    *   Control the circuit voltage.
    *   Control the ON vs OFF times of pulsed circuit current.
    *   Use a timer to set current duration.

4. Rheostat and fixed-value resistor controls are easy to design and are inexpensive, but they waste power.
5. Power amplifiers are more efficient, and rheostats with a low power rating can be used to control load current.
6. Power amplifiers are connected in series with the load in order to control current.
7. There are two ways to use pulsed current to control power to a load. They are PWM and PRF (pulse width modulation and pulse repetition frequency control).
8. Timed current flow is used in spot welders.
9. Thyratron and thyristor controls do not waste power in the form of heat. They control load current by switching the voltage on and off with precise timing.
10. SCR's are unilateral devices. In other words, they conduct only in one direction.
11. Triacs are like back-to-backs SCR's. They are bilateral (two-way) switching devices.
12. Phase-locked loops can be used for precise speed control.
13. Digital, or stepping, motors advance a fixed amount of rotation for each input pulse.
14. Numerical control uses a digital code to control an industrial function.

PROGRAMMED REVIEW

(Instructions for using this programmed section are given in Chapter 1.)

The important concepts of this chapter are reviewed here. If you have understood the material, you will progress easily through this section. Do not skip this material, because some additional theory is presented.

1. A disadvantage of using variable resistors for controlling motor speed is that

    A. Power is lost in the form of heat when current flows through the resistor. (Go to block 9.)
    B. It is much more expensive than other methods of control. (Go to block 17.)

2. The correct answer to the question in block 20 is B. A diac operates at a lower voltage than a neon, but they have similar characteristics. Both are used for gating SCR's and triacs.

    Here is your next question.

    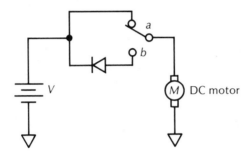

    When the switch is changed from position *a* to position *b* in this circuit, the motor will

    A. Turn slower. (Go to block 10.)
    B. Stop. (Go to block 15.)
    C. Reverse its direction of rotation. (Go to block 5.)

3. The correct answer to the question in block 22 is A. The toggled flip-flop divides the frequency by 2, so the motor will have to run at twice the speed in order for a phase lock to occur.

    Here is your next question.
    A motor that turns a given number of degrees is called a

    A. One-shot. (Go to block 8.)
    B. Stepping motor. (Go to block 13.)

4. Your answer to the question in block 13 is wrong. Read the question again, then go to block 20.

5. Your answer to the question in block 2 is wrong. Read the question again, then go to block 15.

6. Your answer to the question in block 12 is wrong. Read the question again, then go to block 23.

7. Your answer to the question in block 9 is wrong. Read the question again, then go to block 22.

8. Your answer to the question in block 3 is wrong. Read the question again, then go to block 13.

9. The correct answer to the question in block 1 is A. Rheostat controls are less expensive than digital controls, but not as efficient.

   Here is your next question.

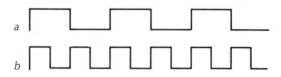

   For the current waveforms shown in this block, the ON times and OFF times are exactly equal. Which is the following statements is true?

   A. The low-frequency waveform of *a* will produce greater heat in a resistor. (Go to block 7.)
   B. The high-frequency waveform of *b* will produce greater heat in a resistor. (Go to block 11.)
   C. They will produce equal amounts of heat in a resistor. (Go to block 22.)

10. Your answer to the question in block 2 is wrong. Read the question again, then go to block 15.

11. Your answer to the question in block 9 is wrong. Read the question again, then go to block 22.

12. The correct answer to the question in block 15 is A. The thyratron, SCR, and triac are all rapid-switching devices. However, the triac is bilateral, while the thyratron and SCR are both unilateral.

Here is your next question.

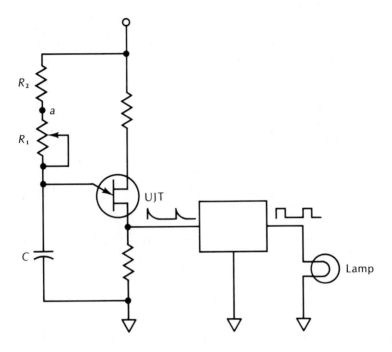

Moving the arm of R_1 toward point a in this circuit will

A. Cause the lamp to glow brighter. (Go to block 23.)
B. Not affect the lamp brightness. (Go to block 6.)
C. Make the lamp glow dimmer. (Go to block 14.)

---

13. The correct answer to the question in block 3 is B. Stepping motors are also called *digital motors* and *step motors*.

Here is your next question.

What type of circuit can be used to convert the input spikes to output pulses as shown in this illustration?

A. UJT. (Go to block 4.)
B. One-shot multivibrator. (Go to block 20.)

---

14. Your answer to the question in block 12 is wrong. Read the question again, then go to block 23.

15. The correct answer to the question in block 2 is B. When the diode is in the circuit, the motor will stop because the diode is reverse biased and no current can flow.

Here is your next question.
The solid-state equivalent of a thyratron is

A. An SCR. (Go to block 12.)
B. A triac. (Go to block 18.)

16. Your answer to the question in block 22 is wrong. Read the question again, then go to block 3.

17. Your answer to the question in block 1 is wrong. Read the question again, then go to block 9.

18. Your answer to the question in block 15 is wrong. Read the question again, then go to block 12.

19. Your answer to the question in block 22 is wrong. Read the question again, then go to block 3.

20. The correct answer to the question in block 13 is B. The complete circuit is shown in Figure 11-9.

Here is your next question.
The solid-state equivalent of a neon lamp is

A. A triac. (Go to block 24.)
B. A diac. (Go to block 2.)

21. Your answer to the question in block 23 is wrong. Read the question again, then go to block 25.

22. The correct answer to the question in block 9 is C. The waveforms both have equal ON and OFF times, so they will produce the same amount of heat.

Here is your next question.

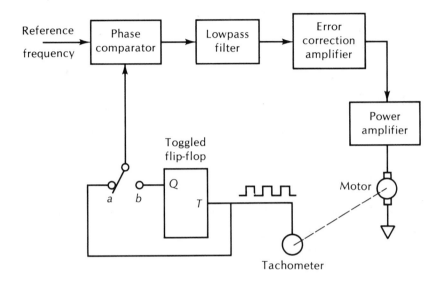

In the block diagram shown here the switch connects the output of the tachometer. When the switch is moved from position *a* to position *b*, the motor speed will

A. Double. (Go to block 3.)
B. Become one-half its original value. (Go to block 16.)
C. Not change. (Go to block 19.)

---

23. The correct answer to the question in block 12 is A. Moving the arm toward point *a* reduces the resistance and also reduces the time for one cycle. This raises the frequency and raises the average value of the output.

Here is your next question.
Which of the following is an oscillator that is designed to change frequency when there is a change in the dc input correction voltage?

A. VFO. (Go to block 21.)
B. VCO. (Go to block 25.)

---

24. Your answer to the question in block 20 is wrong. Read the question again, then go to block 2.

---

25. The correct answer to the question in block 23 is B. VCO is the abbreviation for voltage-controlled oscillator. A change in dc control voltage delivered to the VCO causes its frequency to change.

You have now completed the programmed section.

SELF-TEST WITH ANSWERS

(Answers to this test are given at the end of the chapter.)

1. The circuit that prevents the countervoltage of a motor armature from triggering an SCR or triac too soon is called a _____ .

2. Which of the following diodes is the solid-state equivalent of a neon lamp?
    a. A diac.
    b. A four-layer diode.

3. A variable resistor in series with the load connected, so that it controls load current, is called a _____ .

4. An advantage of thyristor controls over resistor controls is that they are more _____ .

5. Two types of ON/OFF current control are _____ and _____ .

6. A system that uses a timer to control current is _____ .

7. A component that behaves like back-to-back SCR's is the _____ .

8. What is the name for the type of motor that rotates a fixed number of degrees with each input pulse?

9. In a phase-locked loop what are two input signals to the comparator?

10. A punched tape is used to operate machinery in a _____ system.

Answers to Self-Test

1. Snubber
2. (a) A diac
3. Rheostat
4. Efficient
5. PWM and PRF
6. Spot welder
7. Triac
8. A digital, or stepping, motor
9. Reference and VCO
10. Numerical control

CHAPTER **12**

TROUBLESHOOTING: THEORY AND PRACTICE

It is not always possible to draw a fine line between good and bad troubleshooting procedures. Some examples will be given in this chapter to show that an acceptable method of locating trouble in one system can result in the destruction of a device in another system.

Troubleshooting is a dynamic subject in electronics. It changes from decade to decade as new devices and new circuits are introduced. The so-called *shotgun approach* demonstrates this fact. Basically, this approach involves replacing all of the parts in the circuit rather than taking the time to determine which of the individual parts is at fault. In the earlier days of tube circuitry this was considered an absolutely terrible practice. Any technician that had to resort to shotgunning had, at the very least, overlooked the value of logically determining the fault. However, in many modern systems that use integrated circuits and packaged subassemblies (referred to as *modules*), it is not even possible to locate the faulty component and replace it. Troubleshooting in these systems stops at the time when the faulty circuit is located. Then, that circuit is replaced as one package. So, what was once a bad practice has become, in many cases, an acceptable practice.

As a general rule technicians are very cautious against making any modifications in circuitry. Company policy often prohibits such modifications, and in any event, it is not normally recommended. The use of rubber bands, hairpins, sealing wax, and baling wire has no place in the standard approach to keeping expensive electronics equipment in operation. However, their use may sometimes be called *clever* or *innovative*—particularly when used to save a day of production.

This chapter discusses troubleshooting practices from a number of different viewpoints. The quick-check methods are useful for quickly locating a trouble, but they can result in unnecessary loss of equipment and time when their limitations are not understood.

Troubleshooting in systems, units, and circuits is discussed, with the traditional signal injection and signal tracing techniques emphasized. These methods are useful in both linear and digital systems.

Feedback systems require a special consideration, as do oscilloscope practices. Both are covered in the chapter.

THEORY

Quick-Check Troubleshooting Methods

The quick-check methods described here can be useful only if their limitations are known and the proper precautions are taken to protect the equipment. These methods are not intended to replace the more detailed tests described later is this chapter.

QUICK-CHECK METHODS THAT CAN DESTROY DEVICES

An example of how a technique changes with the introduction of new devices is illustrated in Figure 12-1. When a tube is conducting an excessive amount of current, it is considered a good practice to measure the grid voltage. If the coupling capacitor (C) is leaky (that is, conducting dc current), it will permit a positive dc voltage to appear on the grid. This effect would account for the excessive tube current conduction.

In the MOSFET circuit, measuring the gate voltage as shown in Figure 12-1 can destroy the device. The reason is that the electrostatic charge on the probes of a voltmeter can be sufficiently high to destructively break down the insulating region between the gate and channel. This is especially true of the early types of MOSFET's and older voltmeters. Thus, a technique that worked for tubes destroys the MOSFET's. It should be mentioned here that newer instruments are purposely designed to eliminate high-voltage static charges on the probes. Furthermore, many MOSFET's have built-in zener diodes between the gate and the channel leads. These zener diodes conduct the high voltages from the gate lead to the source or drain leads to prevent

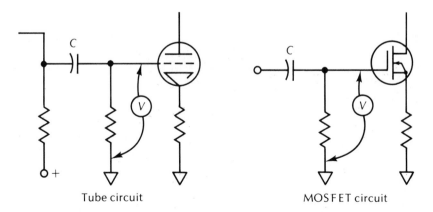

Tube circuit   MOSFET circuit

Figure 12-1. One of these measurements is acceptable; the other is not recommended

them from destroying the insulating gate. However, as a general practice, technicians are still wary of measuring MOSFET gate voltages.

It used to be a common practice (but not necessarily a good practice) to move a tube in and out of its socket to produce rapid on and off transient voltages. These transient voltages could be traced with an oscilloscope through coupling circuits to determine if there were open circuits between two points. If the same practice is used with bipolar transistors, it can very quickly destroy the device. The countervoltages across circuit inductances destroy the *PN* junctions within the transistor.

Soldering guns of the type shown in Figure 12-2 provide quick heat energy for soldering and have the advantage that no power is dissipated when they are not in use. The loop on the gun is actually the secondary of a transformer and consists of a single turn. When the trigger is pulled on the gun, the primary of the transformer located inside the gun induces a secondary voltage. Because of the enormous stepdown ratio, the loop current is very high and produces heat.

Guns like these are fine for soldering electrical equipment and appliances. They were used extensively for repair work in early tube circuits. However, they should never be used for bipolar transistor or FET circuits. The electromagnet field around the gun's secondary turn can induce voltages in coils in the circuitry. The resulting countervoltages can destroy transistors. Also, the isolated secondary winding can have a very high electrostatic charge that is destructive to MOSFET's.

One troubleshooting method that is sometimes used is to start by checking and replacing all of the tubes, transistors, and FET's. But because that is a time-consuming procedure, many technicians prefer to check voltages to determine if the amplifying device is working properly. This method is not recommended in applications where a component has obviously burned up and the circuit is not operating. Figure 12-3 shows an example. The tube is grid-leak biased by capacitor C and resistor R_1.

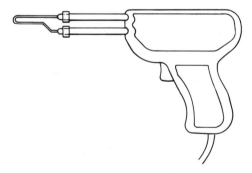

Figure 12-2. This type of soldering gun should not be used on solid-state circuits

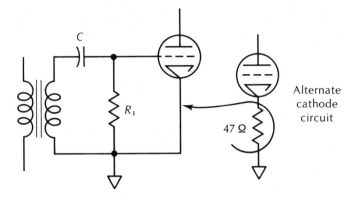

Figure 12-3. Replacing all bad tubes as a first step to troubleshooting is not always a good practice

The bias voltage is supplied by the input signal. If the signal is lost, the dc bias voltage is lost, and the tube will be destroyed. Most tubes are quickly destroyed if the grid is operated at zero volts with respect to the cathode, because the negative grid voltage limits the amount of plate current. When the grid voltage is reduced to zero, the plate current increases to a destructive value. When the tube is in a glass envelope, the effect of the excessive current can be seen by the tube plate turning red-hot before it is destroyed.

Suppose the technician has started the procedure by replacing all bad tubes. If the signal is lost at the input, a new tube will also burn out because the source of the trouble has not been fixed. The technician may have spent a considerable amount of time checking the tubes, but he has not even approached the solution.

Some designers, realizing that this can happen, place a small resistor in the cathode circuit. This is shown in the inset of Figure 12-3. This cathode resistor has a low power rating. It can safely pass the cathode current when a signal is being delivered to a tube. However, if the signal is removed, the plate current (also, the cathode current) increases rapidly and burns out the resistor. This result is normal and desirable because a resistor is much cheaper to replace than a vacuum tube. It is now apparent why technicians are cautioned never to make changes in circuit values. By visual inspection the technician may spot the burned out resistor and replace it with another 47-ohm resistor. That will also burn out immediately. If he decides to use a resistor with a higher power rating, it may outlast the more expensive tube.

Figure 12-4 shows a very popular quick check for bipolar transistor circuits. The procedure is to cause a temporary short between the emitter and the base while measuring the collector voltage. An emitter-base short should shut the transistor off because the only time a transistor can produce a collector current is when current flows through the emitter-base junction. In the

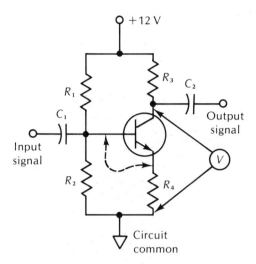

Figure 12-4. The emitter-base short circuit test

circuit of Figure 12-4 the short circuit, shown with a dashed line, should cause the collector voltage to rise to the $B+$ voltage, or 12 V.

A momentary short between the base and collector, rather than between the base and emitter, can destroy the transistor. This is one of the reasons that the emitter-base short circuit test should be used with caution. Another reason is that in some circuits the quick check will not work. In fact, the test may destroy a transistor.

Two examples of such circuits are shown in Figure 12-5. In the class B amplifier the base voltage is maintained at zero volts in the absence of a signal. When there is an input signal, the base goes positive on one half cycle and negative on the next half cycle. Each half cycle that makes the base positive causes the transistor to conduct, and current flows through the emitter resistor to produce an output signal. The effect is to produce only positive half cycles in the output.

An emitter-to-base short in this circuit will not produce any change in the voltmeter reading at the collector. The voltage delivered to the collector is the same dc value regardless of whether the transistor is conducting or not. If the technician is not aware of this fact, he may spend a lot of time replacing the transistor and find out later that this step did not help him in his troubleshooting procedure.

In the direct-coupled amplifier circuit of Figure 12-5, an emitter-base short at transistor Q_1 will quickly destroy transistor Q_2. Transistor Q_1 con-

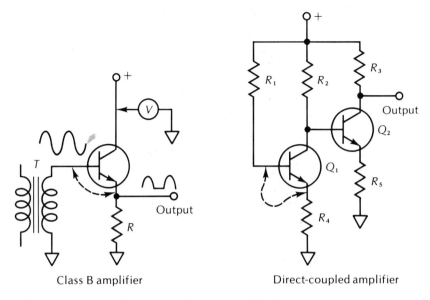

Class B amplifier   Direct-coupled amplifier

Figure 12-5.   Destructive emitter-base test

ducts its collector current through resistor R_2. Note that the base voltage of transistor Q_2 is primarily due to the applied $B+$ voltage minus the voltage drop across resistor R_2 that is due to the collector voltage. (The base current of Q_2 is also flowing through R_2, but this current is so small that it can be disregarded.) If transistor Q_1 is shorted from emitter to base during a quick check, it is shut off. The collector current of Q_1 stops flowing, and the base voltage of Q_2 rises immediately to the $B+$ value. This action destroys transistor Q_2. If there was no trouble at the circuit before the emitter-to-base short circuit test was used, there certainly will be afterward.

A DANGEROUS QUICK CHECK

Figure 12-6 shows a quick check that can be very dangerous. Technicians sometimes develop a bad habit of placing their finger on the grid or base or gate of the amplifying device. The disturbance caused by the presence of their finger at the input should cause an output change and this output change is observed to determine if the stage is operating. This test is based upon two incorrect assumptions:

1. The grid, base, or gate voltage is always too small to cause any personal injury.
2. Solid-state circuits always employ low voltage supplies.

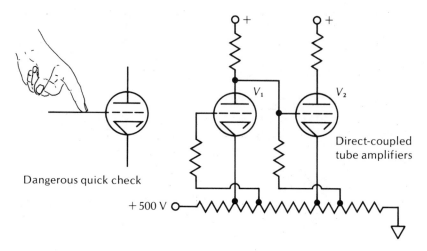

Dangerous quick check

+ 500 V

Direct-coupled tube amplifiers

Figure 12-6. An improper test

Both assumptions are absolutely false! To understand why, look at the direct-coupled amplifier circuit also shown in Figure 12-6. In order to directly couple the tubes, the grid of V_2 must be at the plate voltage of V_1. In other words, the grid of V_2 must be highly positive. The grid of V_1 must be even more positive.

The cathodes and grids are located on a large voltage divider from a 500 V power supply. The grid of the first tube may be almost 500 V with respect to ground. Touching this grid with your finger can cause a severe shock.

It is also possible for this condition to exist in transistor and FET circuits. Furthermore, the collector or gate of solid-state devices may be operated with voltage values well over 100 V. There is the possibility of touching the wrong terminal accidentally, and a severe shock will result.

This quick-check method should be discouraged because it can lead to personal injury. In place of using a finger, technicians sometimes use a screwdriver blade. It has approximately the same effect. The presence of the screwdriver will cause a change in the output signal and this signal can be readily observed with an oscilloscope. A disadvantage of the screwdriver method is that it can accidentally produce a short circuit. In a bipolar transistor an accidental momentary short between the base and collector will destroy that transistor.

QUICK CHECKS USING VOLTAGE, CURRENT, AND RESISTANCE CHECKS

Ohmmeter measurements in a transistor circuit must be made with care. The ohmmeter works on the principle that it supplies a voltage to a resistor and measures the resulting current flow. This current flow is indexed on the meter scale to indicate the number of ohms of resistance.

When the ohmmeter delivers a voltage to a transistor or solid-state diode circuit, it can turn the device on; that is, it can forward bias it into conduction. This conduction will produce a faulty resistance reading. An example is shown in Figure 12-7. The circuit is deenergized as required for making ohmmeter measurements in any circuit. That is why the supply voltage is marked "Zero volts, supply off."

The ohmmeter is being used to measure the resistance of R_2, but the positive terminal of the ohmmeter supply is connected to the base side of this resistor. This positive voltage forward biases the emitter-base junction of the transistor, and current flows through this junction as shown by the dashed line. Therefore, resistor R_4 and the emitter-to-base resistance of the transistor are in parallel with resistor R_2 as seen by the ohmmeter. The resistance reading for R_2 will be too low and a technician who is not aware of this problem can spend valuable time replacing R_2.

By reversing the ohmmeter leads the technician can eliminate the faulty reading if the emitter-base junction is not faulty. Even at best, the technician will have to interpret the readings in relationship to the circuit. In some circuits it is not possible to eliminate the problem by reversing the ohmmeter leads.

Instrument manufacturers have solved the difficulty by making ohmmeters with an ohms switch that has two positions: *high ohms* and *low ohms*. (Different manufacturers may use different terms for these two positions.) In the low-ohms position the ohmmeter supplies such a low amount of voltage to the circuit that it cannot accidentally forward bias a silicon or germanium *PN* junction. With the high-ohms scale it is possible to supply power to determine if a component is breaking down with current flow.

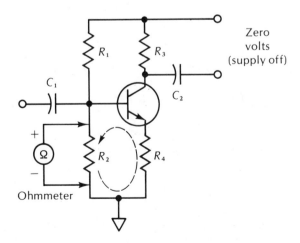

Figure 12-7. Misleading ohmmeter measurement

If a silicon transistor is operating properly, the emitter-to-base voltage should be about 0.7 V and about 0.2 V for the germanium transistor. One method of troubleshooting in bipolar transistor circuits, then, is to measure the emitter-to-base voltage. Two methods for measuring this voltage are illustrated in Figure 12-8.

With the direct method, the voltage across the emitter and base junction is measured as shown. Some voltmeters cannot accurately measure this low voltage, so the indirect method must be used. This method involves measuring both the base voltage and the emitter voltage and then subtracting the two values. In either case the voltmeter method will determine if there is a forward emitter-to-base junction voltage.

Having measured this voltage and determined that the transistor is properly biased for conduction, the next step is to measure the collector voltage. If the transistor is operating properly, there should be a voltage drop across resistor R_3 so the collector voltage should be less than the applied voltage. If this is a class A amplifier, almost half of the supply voltage will be dropped across resistor R_3.

Suppose the voltmeter shows there is a proper emitter-to-base voltage but that there is no collector current as evidenced by the absence of a voltage drop across R_3. This means that the collector circuit of the transistor is open.

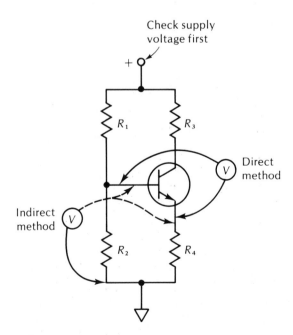

Figure 12-8. Voltmeter check of a bipolar transistor amplifier

Before replacing the transistor, give a quick visual check of the connection between the collector and resistor R_3 and make sure that there is a direct connection. Do not overlook the possibility in checks such as these that transistors in sockets (as well as tubes and FET's in sockets) can be inserted into a pin that is not making good connection.

Measuring the voltages around the transistor without understanding the circuit can give misleading information. Figure 12-9 shows two different ways to obtain dc operating voltages for a bipolar *NPN* transistor. In the

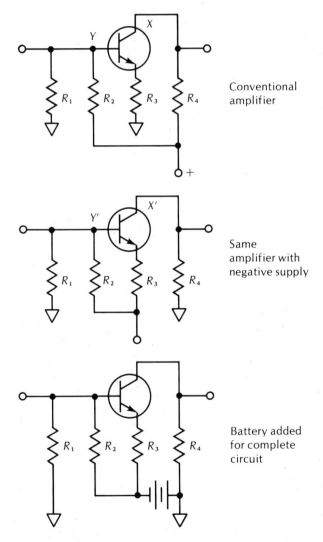

Conventional amplifier

Same amplifier with negative supply

Battery added for complete circuit

Figure 12-9. The power supply connection affects polarities

conventional method the emitter is grounded through an emitter stabilization resistor R_3, and the base voltage is obtained through the voltage divider action of R_1 and R_2. The collector goes to the positive source through the load resistor R_4. A dc voltage measurement at points X and Y in this conventional circuit should show a voltage drop across the collector resistor. Point X should be a voltage positive with respect to common, and point Y should also be positive with respect to common, but point X should be more positive than point Y.

Now, look at the circuit that uses the negative emitter supply in Figure 12-9. The collector lead is grounded through load resistor R_4. If measurements are taken with respect to ground, the collector (point X') and the base (point Y') will both be negative with respect to common.

These voltage polarities would seem to be wrong for *NPN* transistor circuits. However, the transistor is properly biased for operation in this case. To understand why, the emitter supply circuit is redrawn with a battery. Note that the positive side of the battery is connected to common. Therefore, the base and collector *are both positive with respect to the emitter.*

This type of circuit is sometimes employed where both *NPN* and *PNP* transistors are used with the same negative power supply. The *PNP* transistor can be connected in a conventional manner to the power supply and the *NPN* transistor is connected in the unconventional supply setup. If a positive supply is used, then the *NPN* transistor would be connected in the conventional manner and the *PNP* would use an emitter supply.

QUICK CHECKS IN LOGIC CIRCUITS

It is possible to troubleshoot a logic circuit using a voltmeter, but this is strongly discouraged by experienced technicians. One reason is that the difference between a 1 and a 0 in a logic circuit may be difficult to distinguish with a voltmeter reading. Suppose, for example, that the circuit is made with TTL logic gates. Logic 1 is supposed to be +5 V, but what would a reading of 4.6 V signify? Should it be called logic 1, or logic 0, or a defect in the circuit? A logic 0 may be represented by 3 V, 3.2 V, 3.4 V, etc. What does a reading of 2.5 V mean? The technician spends too much time trying to decide whether the voltage measurement represents 1 or 0, and not enough time in troubleshooting.

Another problem is that the technician must look up to the voltmeter reading each time he touches the voltmeter probe to a point on the circuit. To test a 16-pin DIP integrated circuit he must stop and look up 16 times. This is an annoying waste of time. A better procedure is to use a logic probe like the one shown in Figure 12-10. It simply has light-emitting diodes that represent HI, LO, and PULSE.

When the probe is touched to a contact where a logic 1 is present, the HI LED is ON. If there is a logic 0 at that point, the LO LED is ON. If there is a series of pulses or square waves, and the PULSE/MEM switch is in the PULSE position, the PULSE LED is ON, and the light blinks at a rate of

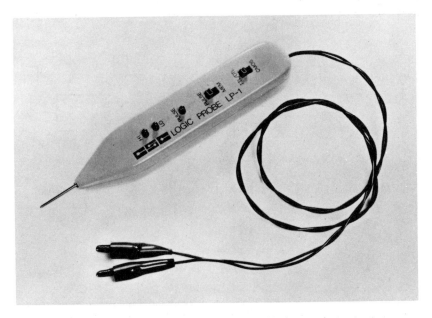

Figure 12-10. A low-cost logic probe for troubleshooting logic circuits

about 3 Hz. This rate is not affected by the pulse frequency, so a pulse frequency up to 10 MHz can be detected. When the switch is in the MEMORY position the probe can detect and store a one-time function, such as the output of a one-shot multivibrator.

Figure 12-11 shows how a probe is used to check an AND gate. When the probe shows that the inputs are at logic 1, the output is also at logic 1, then the gate is good. When the gate is defective the logic probe will show that the inputs or the output is not correct.

A logic pulser, like the one shown in Figure 12-12, may be used in conjunction with the logic probe to produce the desired input logic levels. This logic pulser and logic probe usually come in a single kit. The pulser injects a signal 1 when the pushbutton is operated.

A variation of the logic probe is the logic clip shown in Figure 12-13. It permits all of the input and output signals to be used simultaneously in logic circuits which are in a DIP package. Lights appear at all of the eight terminals in which a logic 1 is present.

Troubleshooting in Closed-Loop Circuits

None of the quick-check methods described so far can be used for closed-loop circuitry. There are usually a number of circuits of this type in industrial electronic systems, so they warrant a separate discussion in a troubleshooting chapter.

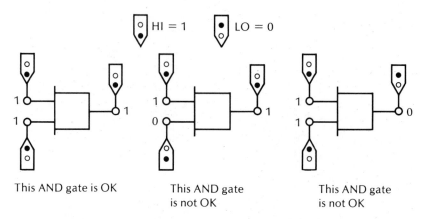

HI = 1   LO = 0

This AND gate is OK   This AND gate is not OK   This AND gate is not OK

Figure 12-11. Use of a logic probe

Figure 12-14 shows a block diagram of a regulated power supply that uses closed-loop circuitry. It is assumed in this discussion that the output voltage is not correct. The problem could be a defective series-pass amplifier, a defective sense circuit, or a defective voltage amplifier. The difficulty in troubleshooting this (or any) closed-loop circuit is that each section depends upon the others for its operation. The series amplifier gets its input voltage from the voltage amplifier. The voltage amplifier receives its voltage from

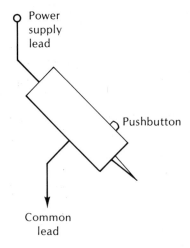

Power supply lead

Pushbutton

Common lead

Figure 12-12. A logic pulser used to inject a logic signal

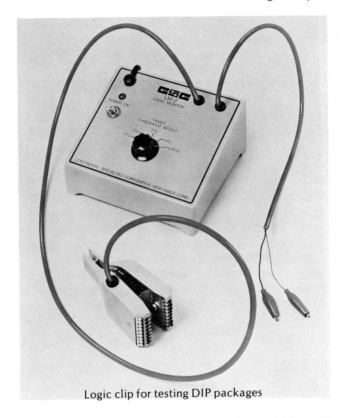

Logic clip for testing DIP packages

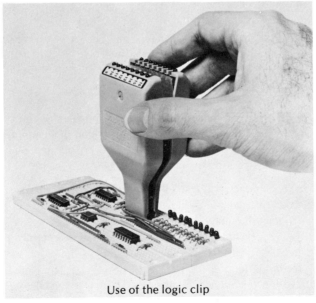

Use of the logic clip

Figure 12-13. A logic clip and its use for testing a DIP package

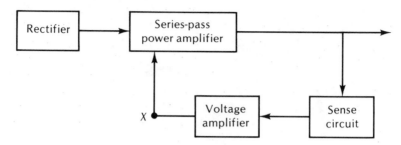

Figure 12-14. A closed-loop power supply regulator

the sense circuit, which in turn receives its voltage from the power amplifier.

If any one of these stages is defective, then all of the stages in the closed loop will receive an improper voltage, and the output voltage will be the wrong value. Voltage measurements around the loop will not isolate the defective stage.

There are several ways to determine the circuit at fault:

1. Replace each unit, one at a time, until the output voltage is correct. This procedure works well in modular systems where the circuits can be replaced with very little soldering.

2. Clamp the feedback voltage at some convenient point. An example of such a point is shown in Figure 12-14. Point X should be a dc voltage that controls the series-pass power amplifier. A dc power supply connected between this point and ground will set the gain of the amplifier and defeat the feedback loop. Then, measurements can be taken in each section to determine the cause of trouble.

3. Open the feedback loop. This is the most popular way to troubleshoot closed loops. A logical place to open the loop of Figure 12-14 would be point X. A dc power supply is used to provide the control voltage for the series-pass power amplifier.

In some closed-loop systems, all (or part of) the feedback signal is ac rather than dc. An example is shown in Figure 12-15. This is a system designed to maintain the motor speed at a constant speed. A voltage-controlled oscillator (VCO) develops the basic motor signal. It is amplified by a power amplifier and delivered to the motor.

In the feedback loop the VCO frequency (f_1) is compared in a discriminator with a countdown frequency (f_2) from a crystal oscillator. The output of the discriminator is a dc voltage which is amplified and used to set the VCO frequency. A gain control at the input of the amplifier makes it possible to set the VCO frequency to produce a desired motor speed.

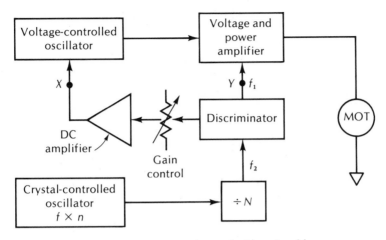

Figure 12-15. Motor speed control with a closed loop

This loop can be opened at point X and a power supply dc substituted to control f_1. Another approach is to open the loop at point Y and use a signal generator to inject frequency f_1 into the discriminator. Both methods of troubleshooting are effective. After the loop is opened, the circuit is tested for incorrect operation and location of fault.

The Oscilloscope as a Troubleshooting Instrument

The use of voltmeters, ammeters, and ohmmeters is normally covered in basic dc and ac courses. The oscilloscope, because of its importance in troubleshooting, is discussed here.

The most obvious use of the oscilloscope is for observing waveforms on a time axis. However, the oscilloscope is also capable of showing waveforms on a frequency axis. To understand the difference, refer to Figure 12-16. Two different waveforms are shown, and they have two different frequencies. Amplitudes are plotted vertically on the three-dimensional axis.

Viewing the waveforms from point A will result in a *time domain display,* also shown in Figure 12-16. In this display, time gets later as you move from left to right and the amplitude increases as you go in the vertical direction. Frequency f_2 is twice frequency f_1 as indicated by the fact that f_2 goes through two complete cycles in the time that it takes f_1 to complete one cycle.

With a triggered sweep oscilloscope the control for the time axis on a time domain display is normally marked in seconds, milliseconds, or microseconds. The frequency can be determined accurately from this equation:

$$T = \frac{1}{f}$$

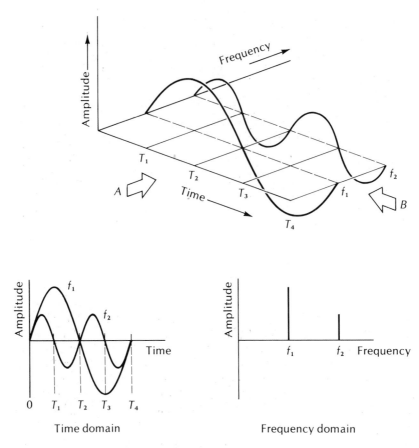

Figure 12-16.   Oscilloscope displays

where f is the frequency in hertz and T is the period, or time, for one complete cycle.

For a recurrent sweep oscilloscope, the sweep control is usually marked with a very rough indication of frequency, but is not accurate enough for making measurements.

If the waveforms in Figure 12-16 are viewed from the standpoint of B, then a *frequency domain display* is obtained. This is also shown in Figure 12-16. In this case the frequency increases as you move from left to right along the horizontal axis, and the amplitude increases as you go in a vertical direction. Since you are viewing the waveforms from the edge, they appear as straight lines. Theoretically they would be straight lines above and below the axis but the type of probe often used for this kind of display eliminates the lower half.

To obtain the frequency domain display, the beam of the oscilloscope is moved from left to right and back along the frequency axis of the three-dimensional illustration. To obtain the time domain display, the beam of the oscilloscope is moved from left to right along the time axis and back. Both types of displays are used extensively but the time domain display is more popular.

Figure 12-17 shows how oscilloscopes are used for time domain and frequency domain displays. For the time domain display the internal sweep of the oscilloscope is used. This sweep, which has a sawtooth waveform, is synchronized with the waveform to be observed. The display is a voltage waveform. For the frequency domain display the beam of the oscilloscope must sweep in step with the frequency of the generator. This means that as the generator frequency increases from the lowest to highest value, the beam moves from left to right on the oscilloscope display.

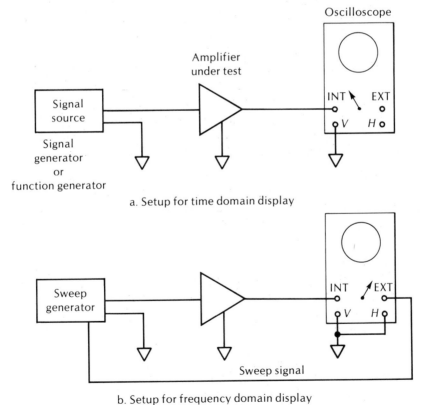

Figure 12-17. Test setups for time and frequency domain displays

This scope sweep is obtained from the sweep generator (or from a function generator which has a sweep capability). The sweep moves the beam of the oscilloscope back and forth as the frequency goes up and down. The amplifier passes a signal amplitude in accordance with its frequency response. The plot on the oscilloscope will be a typical Bode plot for an amplifier. (Bode plots were discussed in the chapter on operational amplifiers.)

VOLTAGE AND FREQUENCY MEASUREMENTS

The oscilloscope can be used to measure frequency and voltage. It is preferred over meters for making these measurements when the waveforms are nonsinusoidal. The reason is that most voltmeters are calibrated to read only pure sine wave voltages.

Figure 12-18 shows a sine wave voltage display. The oscilloscope is calibrated in the horizontal direction to read so many microseconds or milliseconds per inch or centimeter. Suppose, for example, that the display shown in Figure 12-18 is obtained and the horizontal sweep is calibrated to read 10 μs per division. Since it takes four complete divisions to complete the cycle, it is obvious that the cycle requires 40 μs (0.000040 second) for its period. The frequency is calculated as follows:

$$T = \frac{1}{f} = \frac{1}{0.00004}$$

$$= 25,000 \text{ Hz}$$

$$T = 25 \text{ kHz}$$

The vertical axis of the oscilloscope is normally calibrated to read so many volts or millivolts per inch or centimeter. Suppose, for example, that

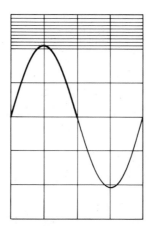

Figure 12-18. Time domain display for measuring voltage and frequency

in the display of Figure 12-18 the oscilloscope has been calibrated to read 500 mV/inch. The divisions are inches as shown in the illustration. The peak voltage is slightly over 2 inches, or 1000 mV. The graticule divisions show that the peak is one-tenth of a division over 1000 mV. If a complete division is 500 mV, then one-tenth of a division would be 50 mV. This is more conveniently written as 1.05 V.

Any spikes or interference on the waveform will be immediately notice-able and measurable on an oscilloscope display, but a voltmeter will measure only the rms value of the complete waveform. This is an advantage of using the oscilloscope for voltage (and current) measurements.

MEASUREMENT OF CURRENT

Oscilloscopes are designed to read voltage waveforms. In order to read a current waveform, it is necessary to convert the current to a voltage. The method is shown in Figure 12-19. Here it is desired to observe the current waveform of the charge and discharge current for capacitor C. To obtain this waveform a very small resistance value is placed in series with the current line. In this case 10 ohms is used, but 1 ohm is also a possible value. The oscilloscope is actually displaying the voltage waveform across this 10-ohm resistor. Since the voltage and current are in phase in the resistor, the voltage waveform will correspond to the current waveform.

The oscilloscope can be calibrated directly to read current magnitude. By Ohm's law, the current through a resistor is equal to the voltage across it divided by the resistance of the resistor. Returning to the display of Figure 12-18, if this 500 mV/inch display is obtained by probing across a 10-ohm resistor, then the 500 mV/inch will actually correspond to 500 mV divided by 10, or 50 mA/inch.

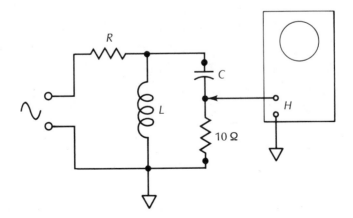

Figure 12-19. Setup for observing capacitor current waveform

MEASUREMENT OF FREQUENCY AND PHASE

It has been noted that the frequency of a waveform can be measured accurately with a triggered sweep scope by determining the time for one complete cycle. The frequency can also be determined by the Lissajous manner illustrated in Figure 12-20. In this setup the unknown frequency (f_1) is applied to the vertical input terminals and a known frequency (f_2) from a signal generator or a function generator applied to the horizontal terminals. The external sweep provision is used on the oscilloscope. When the known frequency is adjusted to obtain either a circle or a straight line at 45°, then the two frequencies are exactly equal.

To obtain the proper display for this type of Lissajous pattern, it is necessary to adjust the vertical gain of the oscilloscope so that f_1 produces a given sweep, say 2 inches, and then adjust the horizontal sweep of the oscilloscope so that f_2 produces exactly the same amount of sweep in the horizontal direction.

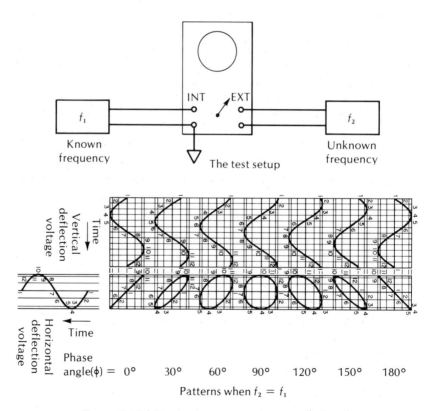

Figure 12-20. Lissajous test setup and patterns for $f_1 = f_2$

If f_1 and f_2 have the same frequency and a phase difference of 0° or 180°, the display will be a straight line; and if they differ by 90° or 270°, the display will be a perfect circle. The various patterns in Figure 12-20 show examples of phase angles. If it is desired to obtain the exact phase angle by the Lissajous method, the calculation shown in Figure 12-21 can be used. A reasonably accurate estimate can be made by comparing the pattern with those shown in Figure 12-20. The Lissajous setup can also be used to determine the ratio of two frequencies from the pattern like the one shown in Figure 12-22. The method of making these measurements is given in the illustration.

One additional note should be given with reference to the Lissajous pattern method. If f_2 in the setup of Figure 12-20 is known to be a perfect sine wave, when f_1 is at 90° difference from f_2 it should produce a perfect circle. If there is distortion in f_1, the circle will not be perfect. The presence of distortion by this method is more accurate than by observing the waveform on a time domain display. Distortion of 10% in a sine wave is not readily apparent on a time domain display, but a distortion of 10% in one of the waveforms of a Lissajous pattern will be easily recognized.

LOGIC DOMAIN DISPLAYS

Oscilloscopes can be used to display logic signals in a setup called the *logic domain display*. With this type of display there is a separate sweep for each terminal under observation. Some oscilloscopes have as many as 16 sweeps that can be observed simultaneously.

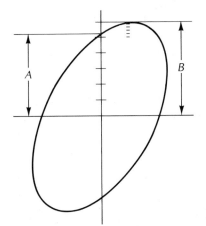

ϕ is the phase angle between f_1 and f_2

$$\text{Sin } \phi = \frac{A}{B}$$

$$\phi = \arcsin \frac{A}{B} \text{ (also written } \sin^{-1} \frac{A}{B}\text{)}$$

Example: $A = 5.2$, $B = 6.0$

$$\phi = \arcsin \frac{5.2}{6.0} = \arcsin 0.866 = 60°$$

Figure 12-21. Calculation of phase angle between f_1 and f_2 using test setup of Figure 12-20

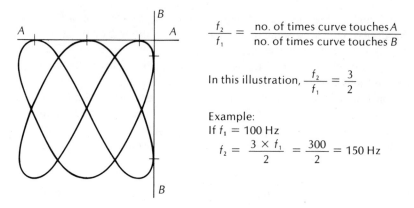

$$\frac{f_2}{f_1} = \frac{\text{no. of times curve touches } A}{\text{no. of times curve touches } B}$$

In this illustration, $\dfrac{f_2}{f_1} = \dfrac{3}{2}$

Example:
If $f_1 = 100$ Hz

$$f_2 = \frac{3 \times f_1}{2} = \frac{300}{2} = 150 \text{ Hz}$$

Figure 12-22. Calculation of frequency using test setup of Figure 12-20

Figure 12-23 shows the logic domain display for a simple three-input OR gate. Waveforms *A, B,* and *C* represent the inputs and *D* represents the output. Note that the output is at logic 1 at all times except during the period of time when the inputs are all at zero.

Logic domain displays have an advantage over the use of logic probes in that they can readily identify glitches. This capability is shown in Figure 12-24. The inputs and outputs of a NOR circuit are being observed. The truth table for the NOR circuit is given in the illustration, along with the two different symbols. The input signals *A* and *B* are supposed to be 180° out of phase. If they were exactly 180° out of phase, there would be no output at any time. However, waveform *B* is slightly behind *A* so there is a brief period

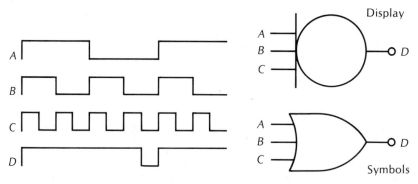

Figure 12-23. Logic domain display

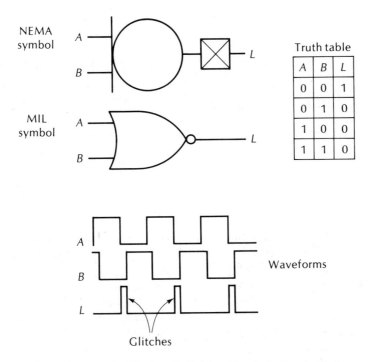

NEMA
symbol

MIL
symbol

Truth table

A	B	L
0	0	1
0	1	0
1	0	0
1	1	0

Waveforms

Glitches

Figure 12-24.   Identification of glitches with a logic domain display

when both inputs go to zero. During this brief period of time the output produces an undesired pulse which is referred to as a *glitch*. Glitches may be only a few microseconds in duration. This would not be sufficient time to produce a display on a logic probe or logic clip, but they show up well on the oscilloscope.

General Troubleshooting Procedure

When troubleshooting an electronic system, the usual procedure is to go from general facts about the system's behavior to the specific device at fault. This procedure is illustrated in Figure 12-25.

The first step is to define the problem. The operator of the system can often supply useful information about its behavior. Troubleshooting manuals for the system usually describe the starting point and the important check points.

If the faulty unit cannot be identified from the symptoms, two kinds of tests can be used. Both are illustrated in Figure 12-26.

With the signal injection method, the output of the system is monitored while signals are injected to the input of each unit. If there is no output when the signal is injected at point *A*, then the next step is to inject the

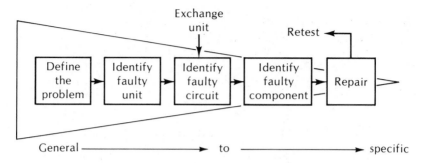

Figure 12-25. General troubleshooting procedure

signal at points B and C until no output is observed. For example, if there is an output when the signal is injected at point B, but no output when it is injected at point C, then unit 1 is at fault.

With the signal tracing method an input signal is delivered to the first stage at all times. An oscilloscope (or other suitable piece of test equipment) is used to follow the signal from point A to point B to point C, etc., until the signal is lost. For example, if the signal is present at point A but not at point B, the trouble is in unit 2.

Once the faulty unit is located, the next step is to find the faulty circuit within that unit. The techniques of signal injection and signal tracing can be used to locate faulty circuits. Each numbered section in Figure 12-26 represents a circuit rather than a unit.

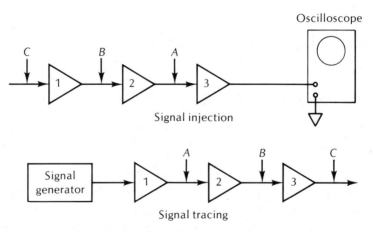

Figure 12-26. Signal injection and signal tracing

Once the faulty circuit is located, the defective device is located by one of the techniques described earlier in this chapter.

After the circuit has been repaired, the last step is to retest it. Retesting is shown in the block diagram of Figure 12-25.

MAJOR POINTS

1. Quick-check methods are useful for troubleshooting when their limitations are fully understood.
2. An emitter-base short circuit shuts a transistor off. The resulting change in collector voltage is an indication that the transistor is doing its job. However, the test can be misleading or destructive. Always analyze the circuit before making such a test.
3. Voltage, current, and resistance checks are more reliable, and safer than quick checks. However, such tests require more time for interpretation.
4. Never touch anything in an energized circuit—tube or solid state. This is a dangerous troubleshooting procedure.
5. Logic circuit inputs and outputs are at 1 or 0. A logic probe senses these conditions. A logic pulser is used to produce these signals at gate inputs.
6. Logic clips are used to check all pins of a DIP package at one time.
7. Troubles in closed-loop circuits can be determined by substituting each section, one at a time. A dc supply can be used to clamp a dc voltage in the feedback loop. A favorite technique is to open the loop and substitute a voltage or frequency to check the operation.
8. Oscilloscopes are used to measure time, frequency, voltage, current, and phase angle.
9. Logic domain displays on an oscilloscope show glitches that may escape detection with a logic probe or logic clip. However, the probes and clips are much faster and require less interpretation.
10. For troubleshooting systems the accepted technique is to go from a general analysis of the overall system, to the defective unit, to the defective circuit, and finally to the defective device.

PROGRAMMED REVIEW

(Instructions for using this programmed section are given in Chapter 1.)

The important concepts of this chapter are reviewed here. If you have understood the material, you will progress easily through this section. Do not skip this material, because some additional theory is presented.

1. The figure below shows a truth table for

A	B	L
0	0	0
0	1	1
1	0	1
1	1	0

    A. An inclusive OR gate. (Go to block 9.)
    B. An exclusive OR gate. (Go to block 17.)

2. The correct answer to the question in block 17 is B. If you missed this question you should review the material in Chapter 3.

    Here is your next question.
    The figure below shows a speed control for a small synchronous motor. The power supply provides dc voltages for the oscillator, amplifiers, and motor. Adjusting the oscillator frequency changes the motor speed.
    The complaint is that the motor speed is not constant. Which of the following is a logical possibility?

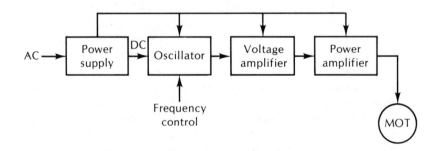

    A. The oscillator frequency is not stable. (Go to block 6.)
    B. The power supply voltage is too high. (Go to block 27.)

3. The correct answer to the question in block 6 is A. Refer to the illustration in this block which shows the transistor amplifier with the battery for the power supply.

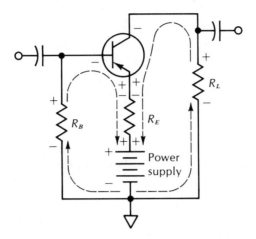

The arrows in this illustration show the path of electron flow. Notice that the electrons are flowing away from ground through load resistor R_L, making the collector side of R_L positive with respect to ground. The only way to obtain a voltage drop across a resistor is to have current flow through it. If the voltage at the collector is zero volts, it means that there is no current flowing through R_L.

Here is your next question.
Assuming the oscillator frequency is not stable in the circuit of block 2, which of the following is a better test?

A. Observe the oscillator waveform on a triggered sweep oscilloscope. (Go to block 34.)
B. Use a Lissajous pattern to check the oscillator frequency against an audio oscillator. (Go to block 7.)

---

4. Your answer to the question in block 11 is wrong. Read the question again, then go to block 8.

---

5. The correct answer to the question in block 7 is B. The input and output signals in this conventional amplifier should be 180° out of phase. The Lissajous pattern for a 180° out-of-phase signal is a straight line.

Here is your next question.
In the circuit for block 6 the power supply voltage should be

A. Positive with respect to ground. (Go to block 10.)
B. Negative with respect to ground. (Go to block 23.)

---

6. The correct answer to the question in block 2 is A. As the oscillator frequency drifts up and down, the motor speed which is dependent upon frequency will also increase and decrease. While it is possible that an incorrect power supply voltage could result in the oscillator being off frequency, it would not cause the oscillator frequency to drift up and down. Therefore, Answer B is incorrect.

Here is your next question.
In the amplifier circuit shown below, a dc voltage measurement shows the collector voltage to be 0 V. Which of the following is correct?

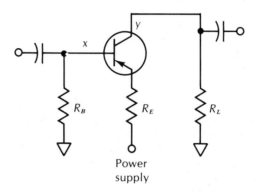

Power
supply

A. The voltage reading shows that the transistor is not conducting. The collector voltage of this *PNP* transistor should be positive with respect to ground. (Go to block 3.)
B. This is a normal voltage reading when the collector is connected to ground through a load resistor. (Go to block 22.)
C. The circuit cannot possibly operate this way. (Go to block 26.)

7. The correct answer to the question in block 3 is B. The Lissajous pattern is a good way to check the oscillator frequency. By comparing the oscillator frequency with the frequency of a stable audio oscillator, it can be easily seen if the oscillator is drifting off frequency. It is not possible to tell this by observing the oscillator waveform on an oscilloscope. The reason is that the oscilloscope —especially a triggered sweep scope—will lock onto the oscillator frequency on either a positive or a negative half cycle. If the frequency drifts, the sweep circuit will readjust so as to maintain the display.

Here is your next question.
A Lissajous pattern is used to compare the input and output signals

(points x and y) in the circuit of block 6. Assuming the amplifier is working properly, the Lissajous pattern should be a

A. Circle. (Go to block 29.)
B. Straight line. (Go to block 5.)

8. The correct answer to the question in block 11 is B. With an oscilloscope it should be possible to view the trigger pulses to the SCR. A number of things can go wrong with the triggered circuit, and the manual control circuit is only one of them. If you start by checking the manual control and you find that it is good, you still do not know if the trigger circuit is operating properly. That is why choice A is not as good as choice B.

   Here is your next question.
   In the circuit in block 10 the complaint is that the motor is not running, and the SCR is suspected. Which of the following is *not* correct?

   A. A short across the anode-cathode of the SCR might burn up the bridge rectifier and prove nothing. (Go to block 21.)
   B. Replace the motor and see if the SCR can make the new motor operate correctly. (Go to block 35.)

9. Your answer to the question in block 1 is wrong. Read the question again, then go to block 17.

10. The correct answer to the question in block 5 is A. The power supply voltage should be positive with respect to ground. Observe that the transistor is a *PNP* type. Therefore, its collector must be negative with respect to its emitter. Another way of saying this is that the emitter must be positive with respect to the collector. If the collector is at or near ground potential, then the emitter must be made more positive than the collector. This case is easy to visualize in the circuit in block 3 in which the battery has been inserted at the power supply terminal. Note that the collector and base leads are both connected to the negative side of this power supply.

Here is your next question.

To observe the current waveform of the motor in the circuit below:

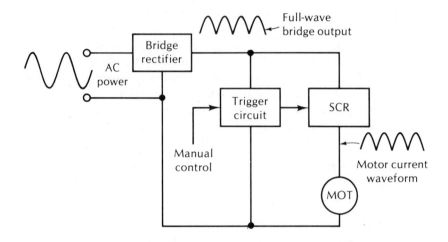

A. Connect the oscilloscope across the SCR. (Go to block 16.)
B. Place a small resistor in series with the motor and SCR, and observe the voltage waveform across this resistor. (Go to block 14.)

---

11.   The correct answer to the question in block 14 is B. The answer is correct for two reasons. First, the smaller the resistor connected into a circuit for observing current waveform the better the situation, and the less disturbance the resistor will make on the actual true current that flows. That is why it is desirable to have ammeters and milliammeters with low resistances. The second reason the answer is correct is because a 10-ohm resistor makes it easy to calibrate the oscilloscope for reading peak values of current. It is not impossible to do this with a 15-ohm resistor, it is only that it is not as convenient.

Here is your next question.

The complaint for the circuit in block 10 is that the motor does not run. Which of the following is a better test of the trigger circuit?

A. Make a resistance and voltage test of the manual control circuit. (Go to block 4.)
B. Use an oscilloscope to determine if a signal is being delivered to the SCR. (Go to block 8.)

12. Your answer to the question in block 18 is wrong. Read the question again, then go to block 24.

13. Your answer to the question in block 21 is wrong. Read the question again, then go to block 18.

14. The correct answer to the question in block 10 is B. The voltage drop across the small resistor will be directly related to the current through it and this is the standard method of looking at a current waveform. Connecting the oscilloscope across an SCR will not necessarily show an accurate view of the waveform. The reason is that the SCR is not a linear device, which means that the current through it does not vary directly as the voltage across it.

    Here is your next question.
    To measure current with an oscilloscope, it would be better to

    A. Connect a 15-ohm resistor in series with the current being measured. (Go to block 30.)
    B. Connect a 10-ohm resistor in series with the current being measured. (Go to block 11.)

15. Your answer to the question in block 18 is wrong. Read the question again, then go to block 24.

16. Your answer to the question in block 10 is wrong. Read the question again, then go to block 14.

17. The correct answer to the question in block 1 is B. With an exclusive OR gate there is an output only when the input terminals are in an opposite condition.

    Here is your next question.
    An advantage of dc motors over ac motors in control systems is that they

    A. Require practically no maintenance. (Go to block 33.)
    B. Have a high torque at low speeds and are easily reversed. (Go to block 2.)

18. The correct answer to the question in block 21 is B. The output at L should be a combination of A and B and C or D. $(A \times B) \times (C +$

D) = L. The left side of the equation is sometimes simplified or rewritten as shown here:

Left side $$\frac{C + D \times AB}{ABC + ABD}$$
of equation:

Therefore, $ABC + ABD = L$

Note that ordinary mathematics is used for combining the *ABCD* inputs.

Here is your next question.
A logic probe shows the output of gate *C* in block 21 to be 0 at all times, regardless of which signals or combinations of signals are delivered to gates *A, B, C*, and *D*. Which of the following is the *least likely* to be correct?

A. The gate power supply voltage is not correct. (Go to block 12.)
B. Both input gates *(A* and *B)* are defective. (Go to block 15.)
C. Gate *A* or gate *B* is faulty. (Go to block 24.)
D. Gate *C* is defective. (Go to block 25.)

---

19.   Your answer to the question in block 20 is wrong. Read the question again, then go to block 31.

---

20.   The correct answer to the question in block 24 is A. In order to get an output from gate *A* it is necessary to have inputs at both *A* and *B*. Likewise, it is necessary to have inputs of 1's at both terminals at gate *C*. Since the input is delivered only to *A*, the output of gate *A* is zero. Therefore, one input to gate *C* is zero and the output should also be zero.

Here is your next question.
If the gates in block 21 are TTL types, their power supply voltage should be

A. 3.6 V. (Go to block 19.)
B. 5 V. (Go to block 31.)
C. 5–15 V. (Go to block 32.)

---

21.   The correct answer to the question in block 8 is A. Placing a short circuit across the anode and cathode could burn up the bridge rectifier if the motor terminals are for some reason or another shorted. This would mean that the bridge rectifier is delivering current to a load of zero ohms because there is no resistance by the

SCR and there is certainly no resistance to the motor, and therefore the bridge rectifier would be destroyed. There is always a danger in making a short circuit test of any kind. Each short circuit test must be considered very carefully before proceeding. Usually this is more time-consuming than simply replacing the component. If the new motor is not available, it is possible to replace the old motor with a large resistance value to simulate it and then proceed with the check-by-check procedure. It is not possible to destroy a new motor by placing it in the circuit, because even if the SCR is completely shorted out, the motor will not be receiving greater than normal peak voltages and currents. Remember that the trigger circuit can adjust the current waveform so that it flows for the full half cycle.

Here is your next question.
In the circuit below, the output signal at L should be

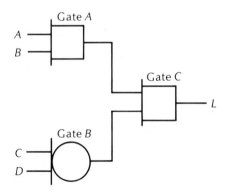

A. $(A + B) \times (C \times D)$. (Go to block 13.)
B. $(A \times B) \times (C + D)$. (Go to block 18.)

---

22. Your answer to the question in block 6 is wrong. Read the question again, then go to block 3.

---

23. Your answer to the question in block 5 is wrong. Read the question again, then go to block 10.

---

24. The correct answer to the question in block 18 is B. The least likely answer is choice B. In the first place it is possible for a faulty gate to have a 1 output at all times. Therefore, if gate A is faulty and the output is 1 at all times, there will be a 1 at L when the output of either A or B is delivered to gate C. The worst thing about choice B is the fact that it requires both gates to go bad at the same time. These are two different kinds of gates—AND and OR—and they are unlikely to be in the same integrated circuit. The likelihood of two different integrated circuits going bad at the same time is remote compared to the likelihood of one (gate C) going bad. Therefore, D is a better choice than B. Finally, the power supply voltage is a likely possibility, and when troubleshooting logic circuits this is the first place to begin.

Here is your next question.
In the circuit of block 21 there is a 1 at input terminals A, C, and D. A logic probe at the output (L) should show that

A. There is a 0. (Go to block 20.)
B. There is a 1. (Go to block 28.)

---

25. Your answer to the question in block 18 is wrong. Read the question again, then go to block 24.

---

26. Your answer to the question in block 6 is wrong. Read the question again, then go to block 3.

---

27. Your answer to the question in block 2 is wrong. Read the question again, then go to block 6.

---

28. Your answer to the question in block 24 is wrong. Read the question again, then go to block 20.

---

29. Your answer to the question in block 7 is wrong. Read the question again, then go to block 5.

---

30. Your answer to the question in block 14 is wrong. Read the question again, then go to block 11.

---

31. The correct answer to the question in block 20 is B. The normal input voltage to TTL circuits is 5 V, and this should be a regulated supply voltage.

Here is your next question.
To obtain the oscilloscope waveform shown below, the horizontal sweep is calibrated for 1 ms per division. The frequency of the waveform is

A. 100 Hz. (Go to block 37.)
B. 500 Hz. (Go to block 38.)

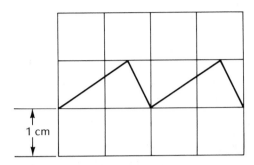

1 cm

---

32. Your answer to the question in block 20 is wrong. Read the question again, then go to block 31.

---

33. Your answer to the question in block 17 is wrong. Read the question again, then go to block 2.

---

34. Your answer to the question in block 3 is wrong. Read the question again, then go to block 7.

---

35. Your answer to the question in block 8 is wrong. Read the question again, then go to block 21.

---

36. Your answer to the question in block 38 is wrong. Read the question again, then go to block 40.

---

37. Your answer to the question in block 31 is wrong. Read the question again, then go to block 38.

---

38. The correct answer to the question in block 31 is B. Note that it takes two divisions to complete one cycle, therefore, the period *(P)* of the waveform is 2 ms (0.002 second). The frequency of a waveform is one divided by a period *(f = 1/T)*. The frequency is calculated as follows:

$$f = \frac{1}{T} = \frac{1}{0.002}$$

$$= 500 \text{ Hz}$$

Here is your next question.
The illustration in block 31 shows a waveform on an oscilloscope screen. The scope's vertical input is calibrated for 10 V/cm. A VOM used to measure this voltage would indicate

A. 30 V. (Go to block 36.)
B. 21.2 V. (Go to block 39.)
C. Neither of these answers is correct. (Go to block 40.)

---

39. Your answer to the question in block 38 is wrong. Read the question again, then go to block 40.

---

40. The correct answer to the question in block 38 is C. The volt-ohm-milliammeter is calibrated to read the voltage for pure sinusoidal waveforms only. The waveform in the display in block 31 is a sawtooth, and it is not possible to get an accurate reading on a VOM by direct methods. If you use choice B (21.2 V) by multiplying 30 times 0.707 to get the rms value, your answer is also incorrect. That only works for pure sine waves. The rms value of a sawtooth waveform is not seven-tenths of the peak voltage. To avoid faults like this, it is best to make voltage measurements on an oscilloscope when nonsinusoidal waveforms are involved.

You have now completed the programmed section.

SELF-TEST WITH ANSWERS

(Answers to this test are given at the end of the chapter.)

1. Is it possible for a bipolar transistor to have voltages high enough to cause a shock? (Yes or No)

2. The emitter-base voltage of a silicon transistor used as an amplifier should be approximately _____ V.

3. An emitter-base short circuit can destroy a transistor in a _____ circuit.

4. If two tubes are *RC* coupled, an indication of a leaky coupling capacitor is _____ .

5. In a *PNP* transistor circuit there is a positive voltage on the collector and on the base. Does this mean that the circuit must be defective? (Yes or No)

6. The grid voltage of a vacuum tube amplifier is measured to be +300 volts. Does this mean that the circuit must be defective? (Yes or No)

7. An ohmmeter has a *high* and *low* power switch. What is the purpose of having a low power position?

8. A logic probe shows that one input lead to a two-input OR gate is at logic 1 and the other lead is at logic 0. The output should be at logic

    _____ .

9. Name a disadvantage of logic probes and logic clips compared to oscilloscope readings in the logic domain.

10. Name an advantage of logic probes and clips over oscilloscopes with logic domain displays.

11. What is the most commonly accepted procedure for troubleshooting a closed-loop system?

12. In a certain troubleshooting procedure the course of an input signal is followed through a system with an oscilloscope. This procedure is called _____ .

13. In a certain troubleshooting procedure the output of a defective unit is monitored while a signal is introduced at various points. This procedure is called _____ .

14. A sine wave is displayed on an oscilloscope. This display is in the _____ domain.

15. Two different sine waves are represented by vertical lines on an oscilloscope display. This display is in the _____ domain.

16. For the display of Figure 12-27 the oscilloscope is calibrated to read 5 V/inch. A volt-ohm-milliammeter would measure this voltage to be about _____ V.

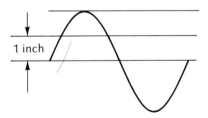

Figure 12-27.

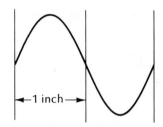

Figure 12-28.

17. For the display of Figure 12-28 the oscilloscope sweep is calibrated for 15 ms/inch. What is the frequency of the wave?

18. A Bode plot on an oscilloscope is obtained with a _____ display.

19. Explain how a current waveform is obtained on an oscilloscope.

20. A circular pattern on an oscilloscope display is an example of a _____ pattern.

Answers to Self-Test

1. Yes
2. 0.7
3. Direct-coupled
4. A positive voltage on the grid of the second tube
5. No
6. No
7. To permit resistance measurements without forward biasing *PN* junctions
8. 1
9. They do not show the presence of glitches.
10. They are simpler to use.
11. Open the loop and substitute a voltage (or frequency) for operating the loop.
12. Signal tracing
13. Signal injection
14. Time
15. Frequency
16. 7
17. 33.3 kHz
18. Frequency domain
19. A small resistor is placed in the circuit and the voltage waveform across the resistor is displayed. This waveform is the same as the current waveform.
20. Lissajous

CHAPTER 13

SOME PRACTICAL APPLICATIONS

By far the most important asset of an industrial electronics technician is a knowledge of electronics. If he knows how the components work, how the circuits work, how to locate troubles by measurements, he can work with a complete system.

It would be a waste of time to describe the operation of large, complex systems in a textbook. Rapid advances in electronics could easily make such a system obsolete before the book comes off the press. Furthermore, there is little chance of the reader ever seeing the system described.

Each company has closely guarded proprietary circuits that cannot be published. However, when a technician works on a system, he can avail himself with the system theory by reading the company technical publications. His knowledge of components, circuits, and basic systems makes it possible for him to understand the operation and repair of complex equipment.

The purpose of this chapter is to show how components and circuits can be combined into some basic systems. Some technicians like to keep a notebook of applications. The systems described in this chapter can be the start of such a notebook. There is no program or self-test for this section since it is only the beginning of your notebook.

ELECTRONICALLY CONTROLLED RELAYS

Earlier in the book it was shown that a thyristor, such as an SCR or triac, can be used to simulate relay operation. By way of review, Figure 13-1 shows a simple relay circuit for controlling a lamp and the equivalent SCR circuit. In both cases the switch is connected into a low power circuit and controls a large amount of power to a load.

In one sense the circuits can be considered to be amplifiers. A very low amount of power at the switch controls a large amount of power to the load. An advantage is that an economical switch with a low power rotary can be used to control power.

There is another advantage to the circuits in Figure 13-1: the switches may be located at some remote point from the power being controlled.

In the early 1960s the demise of the relay was forecast because it was thought that the electronic circuits were superior in a number of ways. First, they did not rely upon motion and mechanical operation, and therefore their

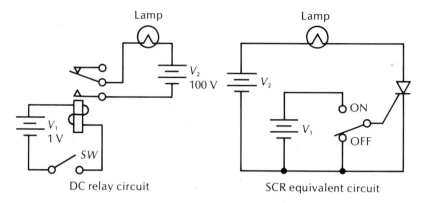

Figure 13-1. Static switches

reliability was considered to be greater. Furthermore, they are not as affected by shock and vibration, whereas large amounts of shock and vibration can cause relay contacts to bounce. In addition, electronic engineers tend to make systems completely electronic, even if (in some cases) a mechanical or electromechanical device might be superior. For those reasons, and others, it was felt that the relay had a limited future.

But in the late 1970s the relay was still a strong contender in control circuits. Some of its values could not be overlooked. For one thing, it is easy to troubleshoot relay circuits. A second reason is that with one single relay it is possible to control a very large number of output circuits, completely independent of one another. This feature is called the *fanout* of the relay, and it is difficult to match electronically.

A third reason for relays still being a contender in control systems is the fact that changes in their basic design over the period of the 1960s and 1970s have made them extremely compact. In fact, a relay can be purchased in the same, physical-sized package as an integrated circuit. They can be operated with much lower input powers than the early, heavy-duty models associated with the 1940s and '50s. Also, their reliability has been increased enormously.

An interesting alternative to the choice between relays and electronic controls is to combine them into a single circuit. Such combinations take advantage of the best features of each system. Examples will be discussed in this section.

Pulse-Operated Relay

It is sometimes desirable to operate an electromechanical relay with a very low amplitude voltage pulse. Assume that a relay is needed in some particular application because of its high fanout capability. However, most relays with a high contact power rating cannot be operated from low amplitude pulses.

To get around this problem, the circuit of Figure 13-2 can be used. Here, the input pulse is delivered to the base of a darlington-connected amplifier through limiting resistor *R*. The darlington amplifier is made with two transistors connected to produce a very large output power and a very high gain. The small positive voltage causes the darlington to conduct through the relay coil. The contacts are close, and the external circuitry is energized.

The diode across the relay coil is used to prevent the inductive counter-voltage from destroying the darlington transistors when the circuit is turned off. (Remember that stopping the current in any inductive circuit causes an inductive kickback that can destroy the bipolar transistance.) In the circuit of Figure 13-2, the inductive kickback voltage forward biases the diode and produces a short circuit across the coil, thus eliminating the large reverse voltage on the darlingtons.

An advantage of the circuit in Figure 13-2 over an SCR circuit is that the pulse can turn the relay on and can also turn it off. In SCR and triac circuits, an incoming pulse latches the SCR on (in a dc circuit) and additional circuitry must be used to turn it off. The relay is not latched on by the positive-going pulses.

Figure 13-3 shows an electronic relay circuit for operation on an ac line. Again, low voltage turns the circuit on and off. A bipolar transistor is used to amplify the control voltage and to energize the relay coil.

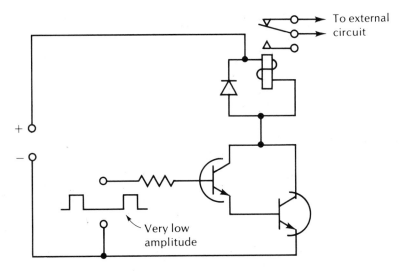

Figure 13-2. Bipolar on-off switch

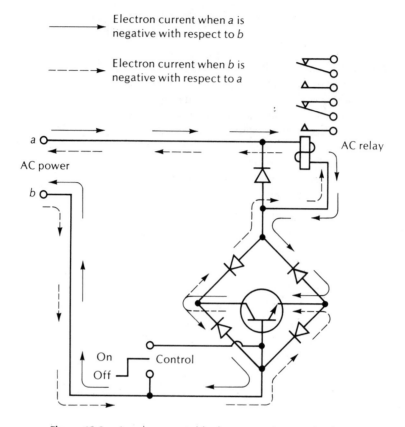

Figure 13-3. Ac relay operated by low-power dc control voltage

When the transistor is off—that is, when its base voltage is 0 V—there is no complete circuit through the bridge, and the relay is not energized. A positive pulse causes the transistor to conduct and produces a forward and reverse path for ac operation. The conduction paths are shown for both half cycles, so when the base of the transistor is positive, the ac relay is energized and the external circuitry is turned on.

Light-Operated Relay

Figure 13-4 shows how electronics can be combined with an electromechanical relay to produce light-operated relay control. The light-activated transistor Q_1 has a very high impedance in the absence of light. Therefore, when there is no light the base of Q_2 is virtually open circuited, and no conduction can take place between its emitter and collector. In other words, the transistor Q_2 is like an open switch when no light falls on the base of Q_1.

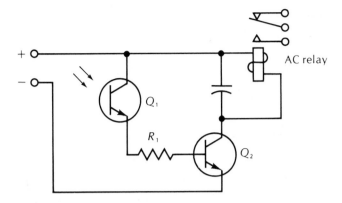

Figure 13-4. Light-operated relay circuit

When light shines on the base of Q_1, its impedance becomes very low. This low impedance produces a forward current in the base of Q_2 and causes that transistor to conduct. The coil is then energized and the relay contacts closed.

The capacitor across the coil in Figure 13-4 absorbs the kickback voltage that occurs when the system is deenergized, thus preventing the voltage from destroying transistor Q_2.

Semiconductor Switch

Figure 13-5 shows a semiconductor circuit capable of delivering 5 amperes to a load. Although the current through the load is delivered through an SCR, and the applied voltage is dc, it is possible to turn this circuit *on* and *off* with an input pulse.

Assume first that the input pulse is its positive-most value, or *on*. This turns transistor Q_1 on to saturation and grounds the gate of transistor Q_3. It also provides a discharge path for capacitor C_2 through diode D_1. The overall effect is to prevent SCR Q_3 from conducting.

The positive voltage is also applied through resistor R_5, and capacitor C_3 begins to charge in a positive direction. When the voltage reaches the breakover voltage value of D_2, SCR Q_4 is fired. The output of Q_4 is delivered to the gate of Q_2 and turns that SCR on. The current is now passing through the 5 ampere load.

When an SCR is *on*, with a positive voltage on its anode, it is normally necessary to turn the power supply off to delatch the SCR. However, in the circuit of Figure 13-5 a second SCR is used to turn the first one off. The procedure is called *commutating*. The commutating SCR is Q_3. When the input pulse goes back to the low, or *off* position, transistor Q_1 is cut off, and

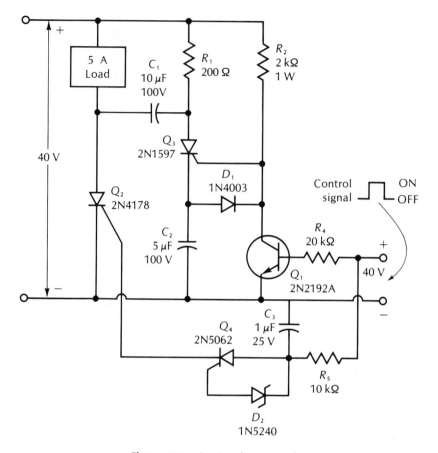

Figure 13-5. Semiconductor switch

the gate of Q_3 is now positive through resistor R_2. This causes Q_3 to conduct heavily with a negative-going voltage at the bottom side of R_1. This negative-going voltage is applied to the anode of Q_2. When the anode of Q_2 is driven highly negative, it is shut off. The circuit has the same function as a dc relay which can be triggered on and off with an input pulse delivered to an amplifier.

MOTOR CONTROLS

Electronic circuits are used in a wide variety of applications for control of motors, speed, torque, and direction. Figure 13-6 shows how two opera-

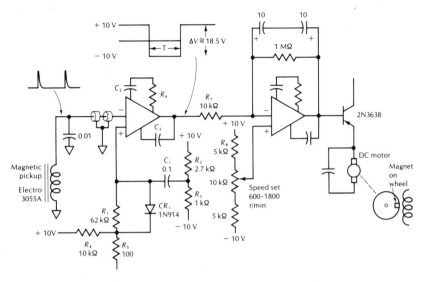

Figure 13-6. Op amp motor speed control

tional amplifiers can be used to control the speed of the motor in a closed-loop feedback system.

Operational amplifier number 1 is used as a *one-shot multivibrator*. This circuit is also known as a *pulse stretcher*. The input signal is a series of very narrow pulses. Regardless of the pulse width of the input signal, the output always has the same (longer) pulse width. The input and output waveforms are shown in the illustration.

The time duration for the output pulse is determined by capacitor C_1 and resistors R_1, R_2, and R_3. The pulse frequency is determined by the motor speed. This speed, in turn, determines how many times per second the magnet moves by the pickup coil.

The average difference between the output of op amp number 1 and a reference voltage set by resistor R_6 determines the average value of voltage delivered to the dc motor.

If the motor begins to run too fast, the pulse repetition frequency out of op amp number 1 will be too high. The high pulse frequency causes the average value of voltage input to op amp number 2 to be too high. Since op amp number 2 is connected as an inverting amplifier, the output current to the transistor will be reduced. The overall effect is a lower value of dc voltage delivered to the dc motor, which reduces the speed back to its current regulated value.

In the speed control system of Figure 13-6, it is assumed that the dc motor is a fractional horsepower type. The circuit is not capable of delivering a high dc current for the operation of larger motors.

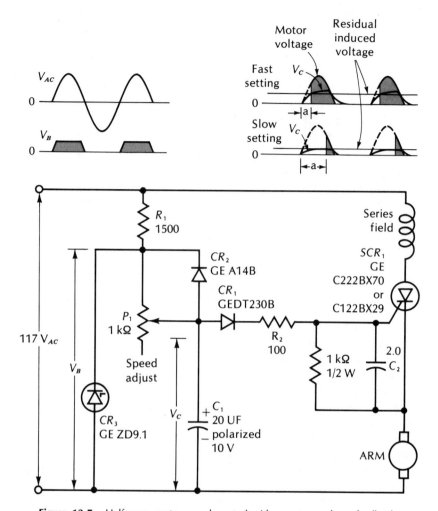

Figure 13-7. Half-wave motor speed control with armature voltage feedback

Figure 13-7 shows a motor speed control that utilizes the countervoltage of the armature to regulate speed. Resistors R_1 and R_2 and the combination of zener diode CR_3 and electrolytic capacitor C_1 form a network that determines at what point the SCR fires on the input ac cycle. Note that a separate series and field and armature windings are necessary for operating with this speed control.

The countervoltage from the armature is delivered to the gate of the SCR by the 1 kΩ resistor and capacitor C_2 in parallel. Note that this countervoltage decreases the forward voltage drop across R_1, causing the SCR to fire sooner. If the motor speed increases, the increased countervoltage tends to hold the SCR off, making it fire later. The effect is to delay the average value of voltage delivered to the armature in series field, and therefore, to slow the motor to its required speed.

This type of circuit regulates against variations in speed due to load changes. If the load increases, the speed of the motor tends to decrease but reduces the countervoltage and causes the SCR to fire sooner. When the SCR fires sooner, the average value of current delivered to the armature and field increases and that in turn increases the speed back to its normal operating value.

LIGHT-OPERATED ALARM SYSTEM

Many alarms are designed to be dependent upon a constant light source. The one in Figure 13-8 is an example. In this case the light source is focused on the base of the light-dependent, darlington-connected transistors. As long as the light source is present, the gate of the SCR is grounded through the

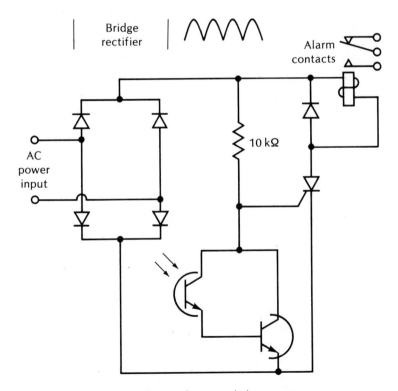

Figure 13-8. Light-operated alarm system

saturated darlingtons. If the light is broken, the gate of the SCR is no longer grounded. Instead, it receives a positive voltage through the 10 kΩ resistor.

Once the positive voltage is applied to the SCR, conduction takes place through the relay terminal. The relay can be used for turning on an alarm, or for some other control system.

The diode across the relay coil is to prevent the countervoltage from the relay coil from destroying the SCR.

TIMING CIRCUITS

In many applications a control system is necessary to introduce time delays. Motor-starting circuits are one example. If power is delivered to a large dc motor before a countervoltage can be built up, it is possible to burn out the windings. Therefore, a time delay is introduced to permit the motor to start out with low dc power and then gradually shift in higher and higher amounts of power. The required time delay can be accomplished by the operator of the starter circuit, but it can also be accomplished electromechanically or electronically.

Another example of the use of a time delay would be in a controlled machine where it is desired to allow the spindle to come up to full speed before drilling starts. Figure 13-9 shows how time delays can be introduced mechanically and electronically. In the thermal time delay circuit the bimetal contact is exposed to a heated coil. When the system is first energized, it takes a period of time for the heater to bring the temperature of the bimetal strip to a high enough value so that the contacts mate.

A second method of accomplishing time delay is to use a motor-driven contactor, also shown in simplified form in Figure 13-9. When the switch is closed, the motor starts turning the contact *a*. When it has reached a turn, it mates with contact *b* and produces an output. The amount of time delay depends on the speed of the motor, which can be controlled in a number of ways as described earlier in this book.

An electronic time delay is also shown in Figure 13-9. In this case, a unijunction transistor (UJT) is used to turn the SCR *on*. When the switch is first closed, capacitor *C* begins to charge through resistor *R*. When the capacitor reaches a sufficiently high value (called the *breakover point of the UJT*), it will fire and produce a positive-going pulse as shown in the illustration. The positive-going pulse fires the SCR and causes the lamp to light. The lamp will stay on as long as the switch is closed because, once gated, the SCR cannot be turned off by its gate voltage.

The amount of delay time in the circuit of Figure 13-9 is dependent upon the time consonant of the resistance capacitor circuit. In general, the larger the resistance or capacitance is made, the longer the time delay.

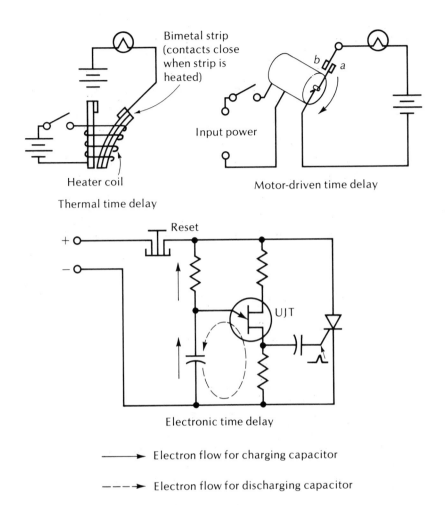

Thermal time delay

Motor-driven time delay

Electronic time delay

⟶ Electron flow for charging capacitor

----▸ Electron flow for discharging capacitor

Figure 13-9. Time delays

LONG TIME DELAY CIRCUIT

Consider the problem of obtaining a long time delay with the simple UJT circuit of Figure 13-10. When the switch is closed, capacitor C begins to charge through resistor R. The charging curve is exponential as shown in the illustration. If the UJT is adjusted to fire when the capacitor voltage reaches point a on the curve, the operation is unstable. The reason is that there is very little difference between the amplitude at point a and the amplitude at point b. Small changes in temperature, component values, and power supply

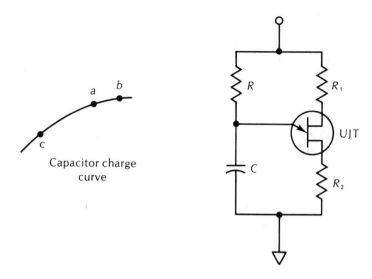

Figure 13-10. Circuit unsuitable for long time delays

voltages will have a large effect on the amount of time delay obtained when operation is in the region of a and b.

It is much better to design the UJT to fire at point c in Figure 13-10. At point c the voltage change per unit of time is quite large and the firing can be made more stable. Of course, if resistor R is many megohms in resistance value and capacitor C is many microfarads in value, it would be possible to obtain such a long delay that operation at point c is easily obtained.

The problem with using a large capacitor is that they usually have a problem of high leakage current. Leakage currents can alter the operation because they heat the capacitor and change its value.

Usually, when long time delays are desired, it is a common practice to use a constant-current generator in series with the charging capacitor in the UJT current to provide a linear charging current. An example of such a circuit is shown in Figure 13-11. Here the capacitor used to charge and fire the UJT is C_E. This capacitor charges transistor Q_1. Note that it is an *NPN* transistor, so a positive emitter voltage is required for operation.

The base voltage of Q_1 is adjusted by variable resistor R_1. The base voltage determines the collector current of the transistor, so it also controls the rate of charge of C_E. Because of the very large value of emitter resistance, the transistor behaves very much like a constant-current generator in series with the capacitor.

The voltage across C_E is presumed to be insufficiently large to supply the peak point required to trigger the UJT. Therefore, an amplifier (Q_2) has been

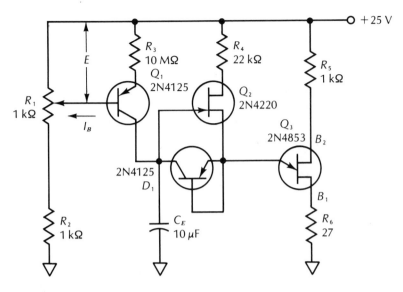

Figure 13-11.   Circuit for long time delays

added to increase the desired peak point current. This is a JFET amplifier, and it cannot handle the large discharge current through its gate-to-source terminal. For that reason, a diode has been placed in parallel to gate-to-source connections to provide a low opposition path of the discharge current to a unijunction Q_3.

When the circuit of Figure 13-11 is properly designed, and if the ambient temperature is not excessive, it is possible to obtain accurate time delays up to 10 hours.

DIGITAL INSTRUMENTS

Most digital instruments employ a frequency counter that is operated by accurately determined clock pulses, and the accuracy of the frequency counter goes through a decoder and display circuit.

The object of the digital instrument is to allow the frequency counter to count up to a point where it is shut off by the parameter being measured. To understand how this works consider first the basic comparator of Figure 13-12. Any operational amplifier will work as a comparator, but the integrated circuits designed specifically as comparators have very fast crossover points, very short crossover times.

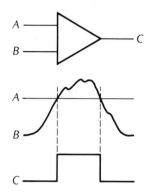

Figure 13-12. Voltage comparator

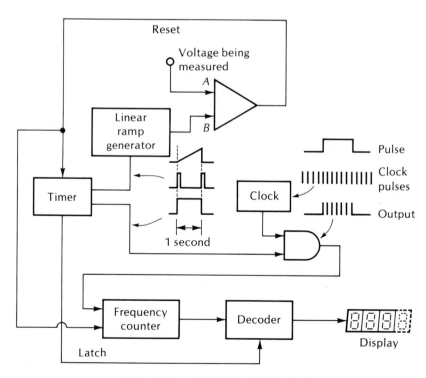

Figure 13-13. Model of a digital voltmeter

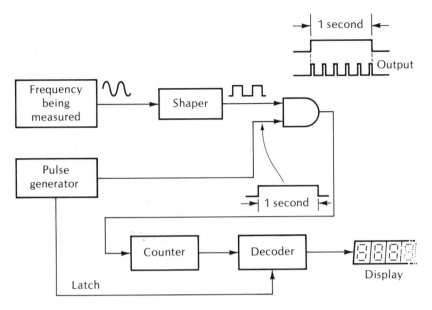

Figure 13-14. Model of a frequency counter

In the comparator of Figure 13-12, the input voltage to point *A* is a dc value. The input to *B* is an analog voltage with continually changing value. Whenever the voltage at *B* is equal to or above the voltage at *A*, the output at point *C* goes positive.

The three waveforms show how the voltages at *A*, *B*, and *C* are related. Note that when waveform *B* crosses over the dc value of *A*, the output at *C* goes positive and remains positive until the voltage at *B* goes below the value voltage at *A*. This simple voltage comparator, along with a frequency counter, can be used to make a digital voltmeter.

Figure 13-13 shows a digital voltmeter simplified block diagram. A clock pulse is delivered to a frequency counter which enables the counter to count to a number well beyond the highest value of voltage to be measured. The pulse generator produces a very accurately timed pulse for a given delay. For convenience, let's say this delay is 1 second. The pulse generator also delivers a pulse to start the linear ramp generator. The output of this ramp generator is a constantly increasing voltage with a fixed slope. The ramp must be very linear in order to obtain accurate measurements.

When the pulse generator starts the linear ramp, the voltage will increase until it equals the dc voltage being measured. At that instant the output goes positive and delivers a latch voltage to the decoder. At the same time the positive output of the comparator delivers a reset voltage to the counter and ramp. Suppose, for example, the ramp voltage has a maximum of 10 V. As the voltage progresses from 0 to 2 V, the frequency counter will go from 0 to 2. When it goes from 2 to 4 V, the frequency counter at that same period will count from 2 to 4. The result is that when the counter is stopped by a positive voltage from the comparator, the display will show a numerical value equal to the measured voltage.

The latch holds the display during the next count.

Figure 13-14 shows the basic principle of a frequency counter. The input frequency can be a sine wave or any other wave shape. Such wave shapes are not suitable for operating a counting circuit. Therefore, a waveshaping network is used which converts the input signal to a series of pulses with the pulse repetition rate equal to the frequency being measured.

The pulses are delivered to the AND, and at the same time, a positive pulse is delivered to the AND. The AND output will be a series of pulses delivered to the counter. The counter will count for a specific period (in this case, 1 second). At the end of 1 second, the counter is shut off, but a latch voltage is delivered to the display to hold the display while the next count is being made.

Since the input pulse is exactly 1 second long, it follows that the count and display will show a number of pulses per second. This, of course, is equal to the number of cycles per second (or hertz) of the input signal.

NUMBERING SYSTEM AND CODES

Binary and octal numbers are used in a wide variety of logic and control systems. When you count in the decimal system, you use ten symbols: 0 through 9. In the binary system, only two symbols are used: 0 and 1. The octal system uses eight: 0 through 7.

The following table shows how equivalent numbers are represented by each system. The basic idea is the same in each system. Starting with zero, you count through the numbers. When all the symbols are used up, you begin a new column on the left, starting with 1. Then you repeat the process. Keep the pattern in mind while you look at the table.

Decimal	Binary	Octal
0	0	0
1	1	1
2	10	2
3	11	3
4	100	4
5	101	5
6	110	6
7	111	7
8	1000	10
9	1001	11
10	1010	12
11	1011	13
12	1100	14
13	1101	15
14	1110	16
15	1111	17
16	10000	20
17	10001	21
18	10010	22
19	10011	23
20	10100	24

When the binary system is used with a machine, there is a disadvantage. Machines represent 0 with a lack of current. A machine can't tell, however, whether a lack of current means 0, or whether a breakdown (like a power failure) is producing the 0. A special code has been made to avert this problem. It is called the excess 3 code because 0 is represented by binary 3, 1 by binary 4, and so on. The table below shows how the code is set up.

Decimal	Excess-3 Code
0	0011
1	0100
2	0101
3	0110
4	0111
5	1000
6	1001
7	1010
8	1011
9	1100

Another disadvantage of the binary system is that it requires a lot of digits to represent larger numbers. For example, it takes five digits (10000) to represent the decimal number 16. The binary number

$$1\ 1\ 0\ 1\ 0\ 1\ 1\ 0\ 0\ 1\ 0\ 1\ 1\ 0\ 1\ 0\ 1\ 1\ 1\ 0$$

represents decimal number

$$1755996$$

You can see that large binary numbers are hard to read and convert. An easier way of coding is the *binary coded decimal system* (BCD), in which each decimal number is represented by four binary numbers.

Decimal	Binary-Coded Decimal
0	0000
1	0001
2	0010
3	0011
4	0100
5	0101
6	0110
7	0111
8	1000
9	1001

Using this system, the decimal number 237 is represented by

$$
\begin{array}{ccc}
0\ 0\ 1\ 0 & 0\ 0\ 1\ 1 & 0\ 1\ 1\ 1 \\
\downarrow & \downarrow & \downarrow \\
2 & 3 & 7
\end{array}
$$

This actually requires more binary numbers, but is much easier to convert to decimal values.

Letters of the alphabet, and instructions, can also be represented with binary numbers. There are many different codes used for this purpose. The EIA code (Figure I-1) is typical. The black circles represent 1's and spaces represent 0's. This code is used on punched tape.

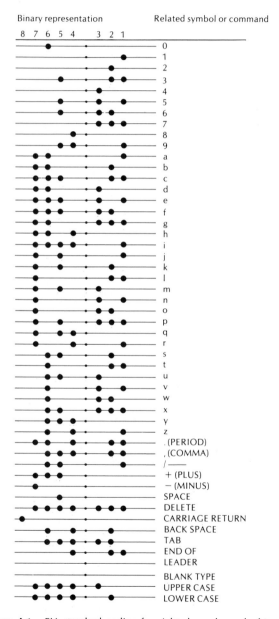

Figure I-1. EIA standard coding for eight-channel punched tape

INDEX